Luis M. Botana (Ed.)
Environmental Toxicology

Also of interest

Chemistry for Environmental Scientists
Möller, 2022
ISBN 978-3-11-073514-7, e-ISBN 978-3-11-073517-8

Environmental and Biochemical Toxicology.
Concepts, Case Studies and Challenges
Gailer, Turner (Eds.), 2012
ISBN 978-3-11-062624-7, e-ISBN 978-3-11-062628-5

Food Safety and Toxicology.
Present and Future Perspectives
Ijabadeniyi, Folake Olagunju (Eds.), 2023
ISBN 978-3-11-074833-8, e-ISBN 978-3-11-074834-5

Hazardous Substances
Risks and Regulations
Schupp, 2020
ISBN 978-3-11-061805-1, e-ISBN 978-3-11-061895-2

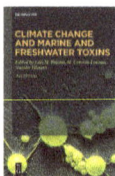

Climate Change and Marine and Freshwater Toxins
Botana, Louzao, Vilarino (Eds.), 2020
ISBN 978-3-11-062292-8, e-ISBN 978-3-11-062573-8

Environmental Toxicology

Non-bacterial Toxins

2nd Edition

Edited by
Luis M. Botana

DE GRUYTER

Editor
Prof. Dr. Luis M. Botana
Departamento de Farmacología
Facultad de Veterinaria
Universidad de Santiago de Compostela
27002 Lugo
Spain
luis.botana@usc.es

ISBN 978-3-11-101431-9
e-ISBN (PDF) 978-3-11-101444-9
e-ISBN (EPUB) 978-3-11-101476-0

Library of Congress Control Number: 2023948420

Bibliographic information published by the Deutsche Nationalbibliothek
The Deutsche Nationalbibliothek lists this publication in the Deutsche Nationalbibliografie;
detailed bibliographic data are available on the Internet at http://dnb.dnb.de.

© 2024 Walter de Gruyter GmbH, Berlin/Boston
Cover image: StefanSorean/iStock/Getty Images Plus
Typesetting: Integra Software Services Pvt. Ltd.
Printing and binding: CPI books GmbH, Leck

www.degruyter.com

Contents

List of contributors

Olga Aguín
Estación Fitopatolóxica de Areeiro
Diputación de Pontevedra
Subida a la Robleda s/n
36153 Pontevedra, Spain
olga.aguin@depo.es

Amparo Alfonso
Departamento de Farmacología
Facultad de Veterinaria
Universidad de Santiago de Compostela
Campus s/n
27002 Lugo
Spain
amparo.alfonso@usc.es

Carmen Alfonso
Laboratorio CIFGA S.A.
Avenida Benigno Rivera, 56
27003 Lugo
Spain
carmen.alfonso@cifga.com

Rebeca Alvariño
Departamento de Fisiología
Facultad de Veterinaria
Universidad de Santiago de Compostela
Campus s/n
27002 Lugo
Spain
rebeca.alvarino@usc.es

Ana M. Botana
Dpto. Química Analítica, Nutrición y
Bromatología
Fac. Ciencias
Campus de Lugo
Universidad de Santiago de Compostela
27002 Lugo
anamaria.botana@usc.es

Luis M. Botana
Departamento de Farmacología
Facultad de Veterinaria
Universidad de Santiago de Compostela
Campus s/n
27002 Lugo
Spain
luis.botana@usc.es

Alejandro Cao
Departamento de Farmacología
Facultad de Veterinaria
Universidad de Santiago de Compostela
Campus s/n
27002 Lugo
Spain
alejandro.cao.cancelas@usc.es

Celia Costas
Departamento de Farmacología
Facultad de Veterinaria
Universidad de Santiago de Compostela
Campus s/n
27002 Lugo
Spain
celia.costas.sanchez@usc.es

Eva Cagide
Laboratorio CIFGA S.A.
Avenida Benigno Rivera 56
27003 Lugo
Spain
eva.cagide@cifga.com

Vanesa Ferreiroa
Departamento de Producción Vegetal y
Proyectos de Ingeniería
Facultad de Veterinaria
Universidade de Santiago de Compostela
Campus s/n
27002 Lugo
Spain

https://doi.org/10.1515/9783111014449-203

Jesús M. González Jartín
Departamento de Farmacología
Facultade de Farmacia
Universidade de Santiago de Compostela
15782 Santiago de Compostela
Spain
jesus.gonzalez@usc.es

M. Carmen Louzao
Departamento de Farmacología
Facultad de Veterinaria
Universidad de Santiago de Compostela
Campus s/n
27002 Lugo
Spain
mcarmen.louzao@usc.es

J. Pedro Mansilla
Estación Fitopatolóxica de Areeiro
Diputación de Pontevedra
Subida a la Robleda s/n
36153 Pontevedra
Spain
pedro.mansilla@depo.es

Nadia Pérez-Fuentes
Departamento de Farmacología
Facultad de Veterinaria
Universidad de Santiago de Compostela
Campus s/n
27002 Lugo
Spain
nadiagabriela.perez.fuentes@usc.es

Sandra Raposo-García
Departamento de Farmacología
Facultad de Veterinaria
Universidad de Santiago de Compostela
Campus s/n
27002 Lugo
Spain
sandra.raposo.garcia@usc.es

María J. Sainz
Departamento de Producción Vegetal y
Proyectos de Ingeniería
Facultad de Veterinaria
Universidade de Santiago de Compostela
Campus s/n
27002 Lugo
Spain
mj.sainz@usc.es

Carmen Vale
Departamento de Farmacología
Facultad de Veterinaria
Universidad de Santiago de Compostela
Campus s/n
27002 Lugo
Spain
mcarmen.vale@usc.es

Natalia Vilariño
Departamento de Farmacología
Facultad de Veterinaria
Universidad de Santiago de Compostela
Campus s/n
27002 Lugo
Spain
natalia.vilarino@usc.es

Natalia Vilariño and Alejandro Cao

1 High-throughput detection methods

1.1 Introduction

High-throughput detection methods are experimental techniques that involve auto-mated tools and allow rapid acquisition of data related to the presence or absence of a certain molecule or group of molecules. They are often referred to as high-throughput technologies (HTTs) or high-throughput screening (HTS) in the literature. These meth-ods provide a remarkable increase of productivity when compared with traditional de-tection. Currently, high-throughput terminology is applied to processing of hundreds of samples in a short period of time.

These methods often comprise several approaches to achieve high workflow rates, such as laboratory automation, miniaturization and parallelization of detection technologies, effective experiment design, continuous processing, and computerized data interpretation. Although many of them have been developed for pharmaceutical screening of high numbers of products, they are also suitable for the detection of en-vironmental contaminants.

This chapter provides an overview of laboratory automation, detection technolo-gies, and experimental design compatible with high-throughput detection using non-analytical methods. Analytical methods will be presented in Chapter 2.

1.2 Laboratory automation

Laboratory robots are essential tools for handling high numbers of samples. Simulta-neous assay of hundreds of samples is often performed in 96-, 384-, or 1536-well plates (Figure 1.1). Manual handling of 384- or 1536-well plates or several 96-well plates in parallel is impractical if not impossible (Figure 1.1) [1, 2]. Therefore, automated liquid handling is necessary for actual high throughput. Adequate programming will allow to perform an experimental protocol simultaneously in a high number of samples. Currently, automated liquid handlers, dispensers, and workstations can effectively dispense liquids with high precision and good reproducibility in 96-, 384-, or 1,536-well plates (for a review of liquid handling technologies and principles, see [2]). Auto-mated workstations also perform sample transfer, tip replacement, incubations, or mixing and shaking, depending on the device design and programming (Figure 1.2A). These instruments have pipetting arms with many-channel heads that allow parallel transfer of samples or reagents (Figure 1.2B and C). They are often endowed also with gripper arms to perform different tasks such as moving labware (tubes or plates) to different positions within the station or removing plate lids. The pipetting and gripper

https://doi.org/10.1515/9783111014449-001

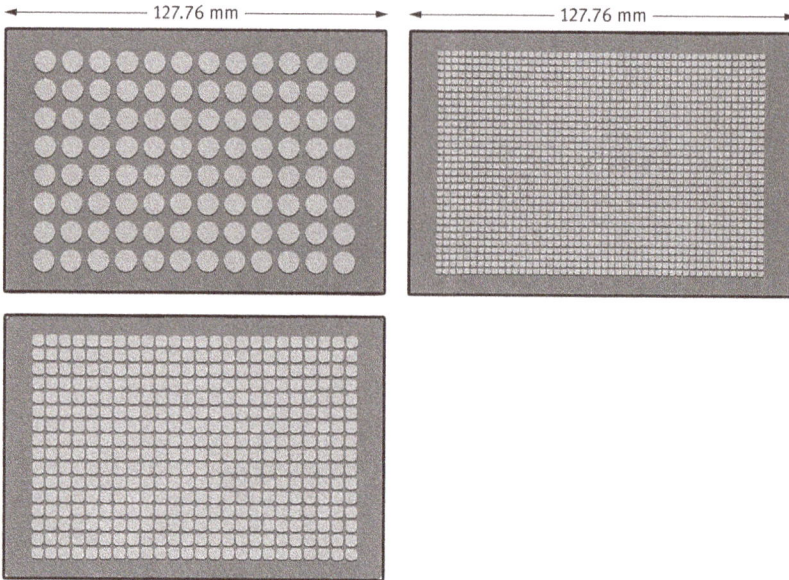

Figure 1.1: Distribution of wells in 96-, 384-, and 1586-well plates.

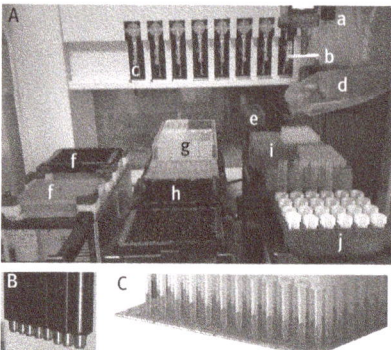

Figure 1.2: (A) Automated liquid handling workstation: moving pipetting head (a), moving gripper hand (b), syringe pumps and valves (c), disposal station (d), tip washing station (e), shakers and temperature control blocks (f) and static positions for liquid containers (g), plates (h), tips (i), and tube racks (j); (B) 8-channel pipetting head; and (C) 96-channel pipetting head.

arms can move along X-, Y-, and Z-axes across a stationary deck. In some instruments, volume measurement systems have been incorporated as feedback quality controls for precision and accuracy of pipetting or dispensing [2].

Automated laboratory workstations have been used mainly to execute detection assays in high numbers of samples before taking them to a reader. However, sample preparation previous to assay or analysis is often a bottleneck in most laboratories to ensure efficient workflow [3]. Sample preparation automation is critical to ensure high-throughput turnover of detection methods, including quantitative analytical techniques. Automated solid-phase extraction (SPE) procedures for sample cleanup have been developed for online and offline processing. Robots may be used for exchange of disposable online SPE columns. Alternatively, versatile liquid handling workstations allow simultaneous offline sample processing through 96- or 384-well SPE plates (Figure 1.3) using robot-operated vacuum manifolds [1]. Both online and offline SPE provide similar limits of detection, ranges, accuracy, and reproducibility; however, online methods allow lower amounts of sample [3].

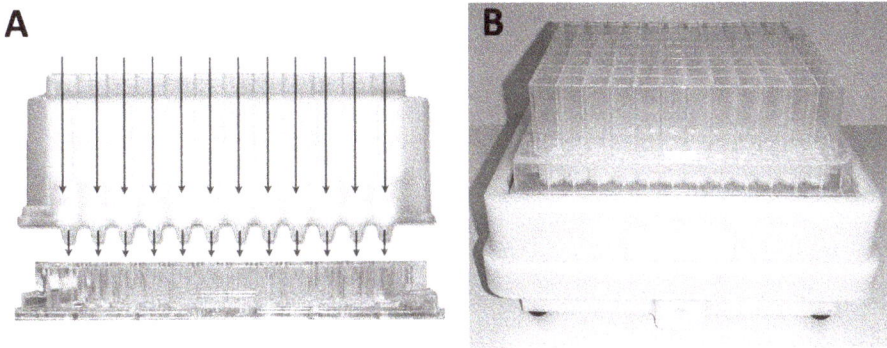

Figure 1.3: Sample preparation by SPE cleaning: (A) schematic representation of simultaneous SPE cleaning in 96-well SPE plates and (B) vacuum manifold for SPE cleaning with 96-well plates.

Many workstations include bar code readers which are critical for automated sample identification and tracking. The generation of log files related to protocols and sample tracking is also useful for quality control and auditing.

Laboratory robots and workstations are required to warrant high productivity rates in high-throughput detection methods and improve reproducibility, but they also reduce contamination, increase safety, and eliminate human error [1, 4]. Moreover, these techniques help to optimize laboratory personnel hands-on time and save costs by scaling down volumes. However, programming of complex experimental procedures is not trivial. In spite of user-friendly platforms, propagation of errors can alter thousands of results, and miniaturization of plate wells makes these experiments highly sensitive to dust, air bubbles, and evaporation [2, 4]. In addition, liquid handling conditions must be adjusted to avoid dripping and cross-contamination, and optimized sometimes for application-specific solutions, such as aqueous buffers containing different detergent concentrations, high protein content, or viscous material [1, 2]. In addition, liquid handling is sensitive to some variables that must be considered for optimal reproducibility such as temperature, air pressure, vibrations, humidity, and disposable pipette tips, which should be the tips recommended by the manufacturer of the liquid handling system to ensure adequate fit and standard parameters of material, shape, and wettability that can affect the accuracy and repeatability of the pipetting process [5].

1.3 Miniaturization and parallelization of detection technologies

Miniaturized assay formats for high throughput require readers capable of simultaneous detection of miniaturized well plates. There are many instruments in the market that provide data for 96-, 384-, and 1536-well plates in less than a minute. Many plate readers are designed to perform different detection technologies. Multimode plate readers may include several of the following: absorbance, fluorescence, luminescence, fluorescence polarization, time-resolved fluorescence (TRF), fluorescence resonance energy transfer (FRET), time-resolved fluorescence resonance energy transfer (TR-FRET), and image-based cytometry.

Spectrophotometric absorbance measures how much light a chemical substance absorbs in solution (Figure 1.4A). The absorption spectrum (wavelength range of the light absorbed) is characteristic of each molecule. This type of optical readout has a higher probability of interferences because colored molecules and pigments are frequent and abundant in environmental samples.

Fluorescence occurs as a result of absorption of the energy of an excitation photon by a fluorophore (molecule with fluorescence properties), which creates an excited electronic state S_1' (Figure 1.4B and C). Some energy is dissipated coming to an S_1 energy state, and after a few nanoseconds, the fluorophore returns to its ground state S_0 by emitting a photon that has lower energy, and therefore higher wavelength, than the excitation one (Figure 1.4B and C). Fluorescence intensity assays are also known as FLINT assays. When fluorescence intensity is measured, excitation occurs simultaneously to emission detection. TRF consists of introducing a delay between excitation and emission measurements (usually 50–150 µs). TRF reduces the background signal significantly because it avoids interference of short-lived fluorescence emission by components of complex samples. However, TRF requires specific fluorophores with prolonged emission properties such as the lanthanides europium and terbium, which must be in the form of chelates or cryptates (Figure 1.5). These complexes act as an antenna that collects and transfers the energy to the rare earth ions, which are not fluorescent by themselves. The instrumentation and reagents needed for TRF are more expensive.

Fluorescence polarization and anisotropy are based on the fact that if a fluorophore is excited by polarized light, the emitted light will be depolarized if the molecule is rapidly rotating (Figure 1.4D). Small molecules rotate rapidly, while large molecules rotate slowly. Therefore, the interaction of small fluorophore-labeled molecules with large molecules can be detected by a reduction of emitted light depolarization (Figure 1.4D and E). In these techniques, fluorescence emission is measured with

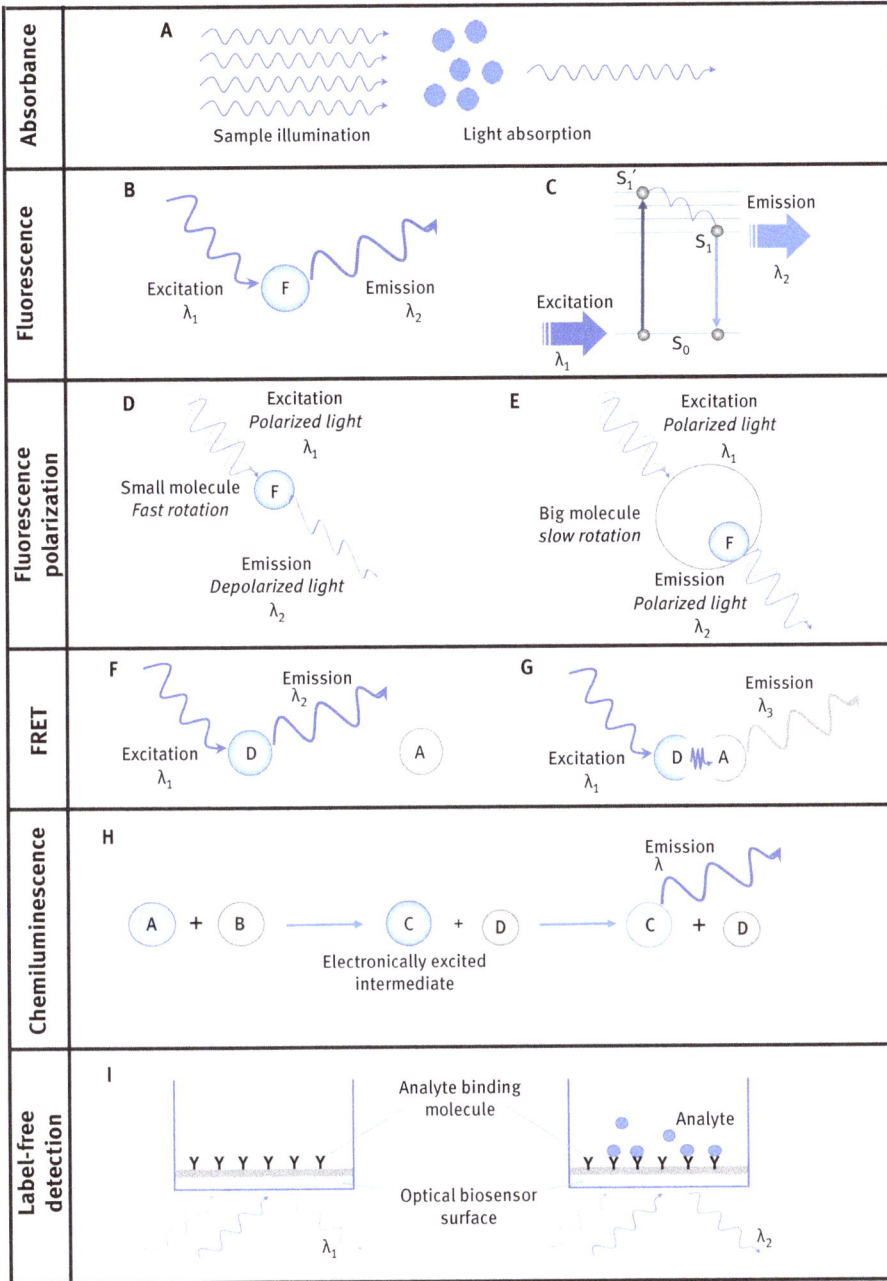

Figure 1.4: Schematic representation of (A) absorption, (B, C) fluorescence, (D, E) fluorescence polarization, (F, G) fluorescence resonance energy transfer, (H) chemiluminescence, and (I) label-free optical detection. See main text for detailed description.

polarizers parallel and perpendicular to the polarized excitation light, and their values are calculated using the following equations: fluorescence polarization = $(I_{parallel} - I_{perp})/(I_{parallel} + I_{perp})$; fluorescence anisotropy = $(I_{parallel} - I_{perp})/(I_{parallel} + 2I_{perp})$.

FRET (fluorescence resonance energy transfer, also known as Förster's resonance energy transfer) consists in energy transfer from an excited fluorophore (donor) to another molecule (acceptor) without the emission of a photon when they are in close proximity. The transference of energy occurs at small distances comparable to macromolecule size (Figure 1.4F and G), and when the acceptor fluorophore returns to its ground energy state, it emits a photon of different wavelength than the donor. Therefore, the interaction of one molecule labeled with the donor fluorophore with another molecule labeled with the acceptor fluorophore will result in quenching of donor fluorescent signal and appearance of acceptor fluorescence (Figure 1.4G). The wavelength of the light emitted by the donor should overlap with the acceptor excitation spectrum. The same principle used for TRF applies to TR-FRET, which uses lanthanides as donors. TR-FRET provides additional information about conformation, flexibility, and equilibrium populations of interacting molecules [6].

Luminescence is the emission of light by a molecule. It really comprises any emission of light from an excited molecule when it returns to its ground energy state independently of the process that leads to the high energy state. Frequently, the emission is triggered by the generation of a high-energy intermediate as a result of a chemical reaction. This process is called chemiluminescence (Figure 1.4H). The chemical reaction may be catalyzed by an enzyme. The detector is configured to measure the intensity of the emitted light. From a physicochemical point of view, fluorescence is a type of luminescence; however, it is usually considered separately for the purpose of readers' capabilities.

Some readers also incorporate label-free detection. This technology is based on the changes in light refraction caused by mass modifications in close proximity (150 nm) to the refractive surface. The changes in the refraction index are detected by a change of reflected light wavelength (Figure 1.4I) [7]. The advantage of these techniques is the capability of avoiding labeling with fluorescent dyes or other molecules, which may alter the results in some instances. However, label-free technology requires the use of special plates with an optical biosensor surface integrated in the clear bottom of black microplates. Electric impedance biosensing can also be used for HTS in cell-based assays using special microplates to evaluate cell viability and growth [8].

Another technology available in multimode readers is imaging of the plate bottom, including fluorescence intensity, bright field, and digital phase contrast imaging. Three color images can be obtained for a 384-well plate in a few minutes. Imaging is mainly used for cell-based applications. This kind of data can also be collected using high-content imagers which are specialized instruments for this purpose [9].

These readers can be incorporated into robots or automated workstations, or be upgraded with plate stackers to provide a continuous workflow suitable for high-throughput requirements.

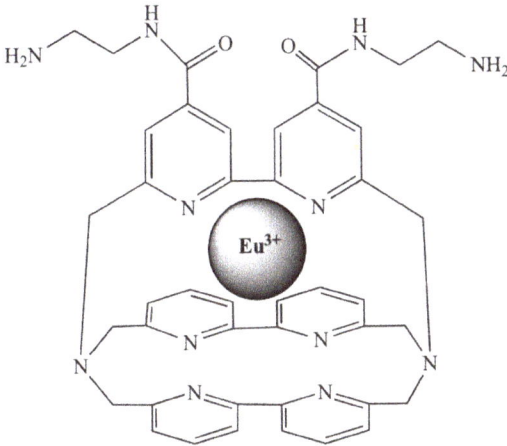

Figure 1.5: Structure of Eu^{3+} cryptate.

A different strategy for incrementing laboratory throughput is multiplexing. Multiplexing consists of simultaneous detection of several analytes in the same sample. The most extended multiplexing technology is based on the use of microspheres with different characteristics of fluorescence and size that can be identified using flow cytometers or similar detectors. Each microsphere class is functionalized for the specific detection of a target analyte. Quantification of the analyte is also performed by fluorescence, using a fluorophore with excitation and emission wavelengths that do not overlap with the microsphere signal (Figure 1.6). Multiplexing is achieved by addition of a mixture of several classes of microspheres to the same sample, and posterior separation of the microspheres in the reader by a cytometer-like fluidic system (Figure 1.6). The number of analytes could be higher than 50. This technology is widely used in the clinical analysis laboratory setup. Plate-based designs have been adapted to multiplexed assays for the detection of environmental contaminants, such as marine toxins, cyanotoxins, and mycotoxins [10–12]. Planar arrays for multiplexing have also been reported [13].

In the last decades, flow cytometers have been optimized to provide high-throughput workflow rates [14, 15]. Although they do not reach the speed of the multimode spectrophotometers discussed above, there are flow cytometers that can process 384-well plates in less than 1 h. This technology allows classification of events attending to their characteristics of size, light dispersion, and fluorescence intensity at different excitation and emission wavelengths. Therefore, flow cytometers can combine high-throughput speed and multiplexing for simultaneous detection of several analytes in the same sample.

Nowadays, electrophysiology methods have also been automated. Electrophysiology is used to measure plasma membrane voltage or currents to detect ion fluxes through plasma membrane ion channels. Although some automated electrophysiology instruments would perform parallel recordings from 96- or 384-well plates, they do not allow workflow rates as high as the other technologies yet; moreover, they are fairly expen-

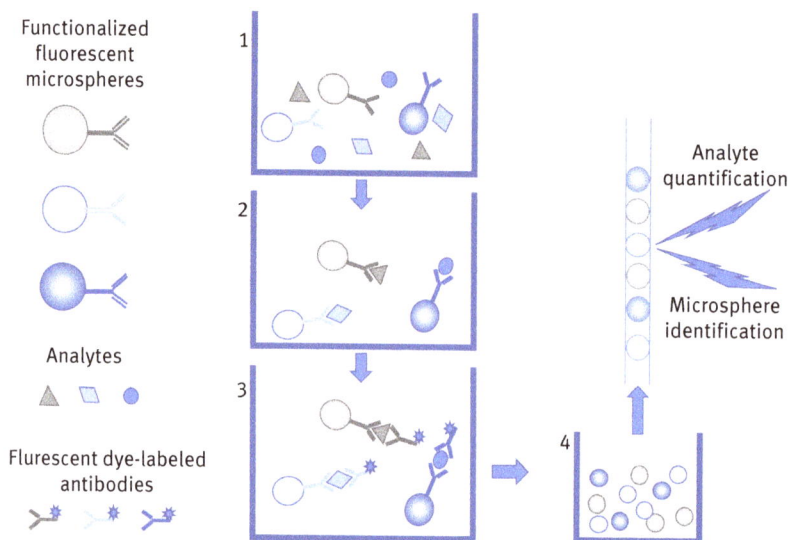

Figure 1.6: Diagram of a multiplexed immunoassay with a sandwich format. 1. Sample with different target analytes is combined with a mixture of several classes of microspheres, each class functionalized for the specific detection of an analyte. 2. Each microsphere class binds the specific analyte. 3. The amount of analyte bound to the microsphere surface is quantified using a specific fluorophore-labeled antibody. 4. The microspheres present in the sample are separated by the fluidic system of the reader. The internal fluorescence of the microsphere and the surface-labeled antibody signal are recorded for analyte class identification and quantification, respectively.

sive and require highly qualified personnel [16, 17]. Therefore, electrophysiology is not the best option for routine high-throughput contaminant detection at the moment.

Although still under development to provide HTS, one of the more promising fields in this area is microfluidic paper-based analytical devices (μPADs), which include lateral flow assays, discussed later in this chapter, and paper-based continuous-flow microfluidic devices. Currently, μPADs provide multiplexing for several analytes and have been adapted for on-site environmental monitoring because they are portable, rapid, and easy to perform. The readouts more commonly used in these devices are colorimetric, electrochemical, chemiluminescence, and electrochemiluminescence detection [18].

1.4 Assay design

Assay design is critical for efficient, robust, high-throughput detection. Several detection strategies can be miniaturized and adapted to reach a high-throughput workflow. Most of these techniques relay on the specific binding of the analyte to a detecting molecule. The nature of this detecting molecule will greatly determine the characteristics of the detection method. None of these techniques should be considered an analytical method because they would not allow the unequivocal identification and quantification of the analyte. However, all these assay designs are suitable for HTS.

Methods based on the interaction of the analyte with antibodies, receptors, enzymes, aptamers, and cells will be discussed.

1.4.1 Immunodetection

Immunodetection is based on binding of the analyte to a specific antibody. It offers multiple possible designs that can be implemented in high-throughput assays. ELISA (enzyme-linked immunosorbent assay) has been the most widely used immunoassay. There are several detection strategies, such as direct, indirect, sandwich, or competition ELISA (Figure 1.7), depending on the characteristics of the antigen and the available antibodies. Independently of the assay format, in all ELISAs, the reporter antibody is covalently bound to an enzyme and the readout is a molecule generated by the enzymatic reaction catalyzed by this enzyme that can be quantified by absorbance, luminescence, or fluorescence (Figure 1.7). The commonly used enzymes are horseradish peroxidase, alkaline phosphatase, and β-D-galactosidase. The substrate depends on the final readout. The enzymatic reaction provides significant signal amplification with a remarkable increase of sensitivity. Chemiluminescence is considered to provide the highest sensitivity and the widest detection range. ELISAs are robust, inexpensive, sensitive assays performed in plates, and they can be easily miniaturized for high throughput.

Other immunodetection possibilities amenable for high-throughput detection are labeling of the reporting antibody directly with a fluorescent dye or substitution of the microplate well flat solid phase by microspheres for multiplexing. Besides fluorescence intensity, TRF, FRET, and RT-FRET have also been used for immunodetection techniques.

Immunoassays are specific, easy to perform, reliable, and robust. However, antibody cross-reactivity with other molecules might be a problem. Frequently, antibodies bind to the molecules that have served as an immunogen, but also to other molecules with a similar structure. This characteristic might be a disadvantage if specific detection of a single molecule is required, and the presence of similar compounds in the sample is possible. In other cases, it might be advantageous if there are several toxic analogs in the same sample that should be detected. Marine toxin groups, for instance, have multiple analogs with varying toxicities. Ideally, a good antibody for a marine toxin detection immunoassay should detect efficiently the highly toxic analogs and should have lower binding affinity for the less toxic molecules. Unfortunately, this perfect cross-reactivity balance is very difficult to achieve [19, 20].

Therefore, immunoassays can be considered quantitative if the sample cannot contain any other molecule that binds the antibody except for the target analyte. On the contrary, if the sample may contain several analogs of the target group that bind the antibody, the immunoassay cannot be used to accurately quantify the content of contaminants and should be considered a qualitative screening tool.

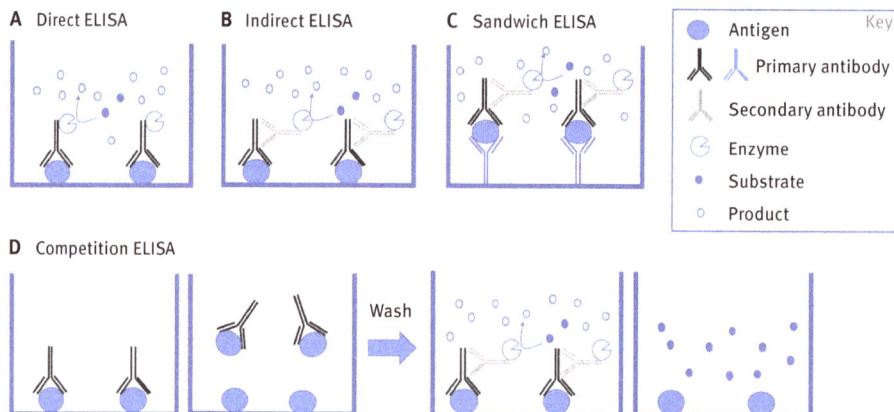

Figure 1.7: ELISA assay formats. (A) Direct ELISA: The analyte in the sample is immobilized onto the well surface, and its presence is detected by an enzyme-labeled primary antibody. A chemical reaction catalyzed by the enzyme produces a colored, fluorescent, or luminescent product. (B) Indirect ELISA: Similar to direct ELISA, expect that in this case the enzyme is bound to a secondary antibody that interacts with the primary antibody. (C) Sandwich ELISA: An analyte-specific antibody is immobilized onto the well surface. The analyte in the sample binds to this antibody, and then it is detected by the addition of another primary antibody. Direct or indirect detection is possible as in (A) and (B) if the enzyme is bound to a primary or secondary antibody, respectively. (D) Competition ELISA: The analyte present in the sample competes for binding to a specific antibody with the analyte previously immobilized on the surface. Positive samples will cause a reduction of the amount of antibody attached to the surface, and therefore, a reduction of the signal. Direct and indirect designs are also possible.

Another limitation of immunoassays is the challenge of producing antibodies against small molecules, which do not behave as antigens, and therefore have to be linked to bigger carriers to trigger the immune response. Many contaminants are quite small, and consequently, good-quality antibodies may not be available for assay development. In addition, the toxic potency of some compounds hinders the appearance of an adequate immune response in the hostage animal before death occurs [19, 21].

Although not strictly considered high throughput, rapid screening immunoassays for in situ detection of contaminants should be mentioned in this section. Parallel or simultaneous testing of many samples in short periods of time is possible with these techniques that can be also used in the field. The most commonly used methods are lateral flow immunoassays (LFIA, Figure 1.8A). The system consists of a combination of absorbent materials and nitrocellulose membrane that forms a path for continuous flow of liquid sample driven by capillary forces (Figure 1.8B) [22, 23]. They usually follow a sandwich or competition assay design (Figure 1.6), depending on the size of the antigen and the antibodies [23]. A "deposit" of a specific antibody (antibody conjugate pad) usually labeled with color nanoparticles is located in the path of the flow (Figure 1.8B). The sample enters in contact with the labeled, analyte-specific antibody and flows toward a test zone where another analyte-specific antibody is immobilized (sandwich assay design,

Figure 1.8C). If the sample contains the analyte, the first antibody–analyte complex will bind to the immobilized antibody, causing a concentration of color nanoparticles and revealing a color band (Figure 1.8A and C). Sample flow moves forward and reaches the control zone where an anti-immunoglobulin binds the excess nanoparticle-labeled antibody, which serves a positive control of assay performance. For inhibition assays, the analyte is immobilized in the test zone, and the absence of the color band is the positive test result. These assays were initially developed for visual score, but colorimetric quantification or fluorescent readout using fluorescent labeling of the antibody are also possible using adequate readers [24].

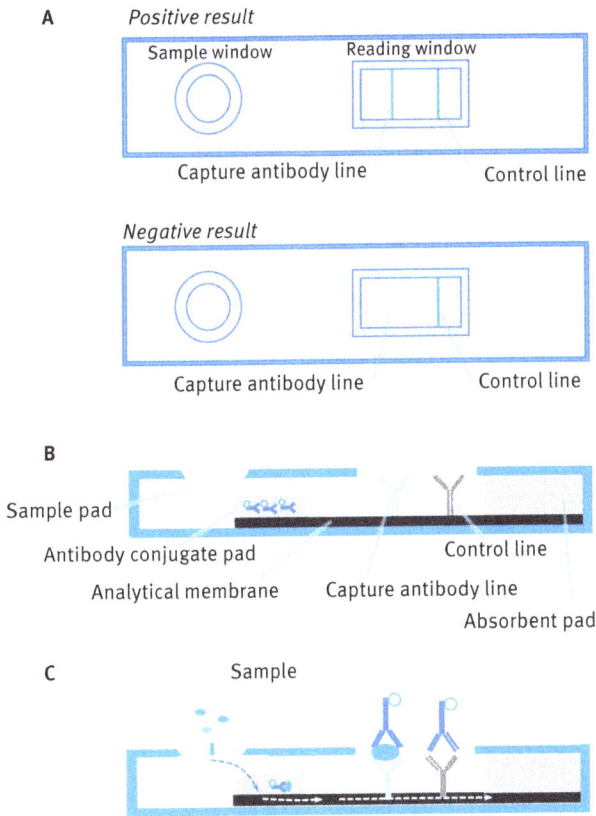

Figure 1.8: Lateral flow immunoassay (sandwich format). (A) Schematic representation of positive and negative results by visual score. (B) Internal design of the lateral flow device. (C) Diagram of sample application, flow (dashed lines) and capture antibody–analyte–nanoparticle-labeled antibody complex for a positive sample, and control line.

In spite of the limitations, there are immunoassays available for the detection of multiple contaminants, such as marine toxins, mycotoxins, or pesticides, using different approaches that include ELISA, multiplexing, or LFIA [13, 25, 26].

1.4.2 Receptor-based methods

Some techniques use binding of the analyte to a biological target as a detection strategy. These techniques are not so different from immunodetection in design and may use similar formats to quantify binding to the receptor, although the competition assay is probably the most common.

An important advantage of receptor-based assays versus immunoassays is that the affinity of the receptor for the different analogs of a toxin group correlates better with toxic activity than antibody cross-reactivity [27]. On the other hand, receptor-based assays have also some limitations. Membrane receptors or ion channels are often the natural targets of toxicants. These proteins lose their activity and binding properties if they are extracted from the plasma membrane. Therefore, working with purified receptors is not always possible, and membrane fragments or vesicles containing the receptors should be used. Although there are some receptor-based methods available for toxin detection, working with plasma membrane adds a degree of complexity and they are not so robust and commonly used as immunoassays. Another important disadvantage is that many of the receptor-based methods use radioactive labels, which are not well accepted for routine detection.

An example of target-based methods is the nicotinic receptor-based assay for cyclic imines that has been developed for detection in microplates with colorimetric, fluorescence, fluorescence polarization, or chemiluminescence readouts and in multiplexed platforms using microspheres [28–31]. This method consists in a competition format, where cyclic imines compete for binding to the nicotinic receptor with fluorescent bungarotoxin. This assay also performed adequately after miniaturization in 384-well plates [29]. Receptor-based assays usually quantify the interaction of the analyte with the receptor. A better evaluation of toxicity would be provided by quantification of receptor functionality, which is usually performed with cells, instead of isolated receptors. Recently, in the pharmacology field, some high-throughput assays have been optimized to screen libraries of compounds for modification of receptor function, mainly G-protein-coupled receptors, using receptor-rich membranes and TR-FRET readouts.

1.4.3 Enzymatic activity

Modulation of enzymatic activity can also be used to measure or detect contaminants. Frequently, the enzyme is a biological target of the analyte (or analyte class), providing as a result an evaluation of sample toxicity in relation to the contaminant class. Inhibition of enzymatic activity is most commonly used. In order to quantify the enzyme activity, the transformation of a substrate into a reporter product is measured by colorimetric, fluorescent, or luminescent detection. These assays usually require data acquisition in kinetic mode, rather than endpoint measurements, to assess enzymatic reaction rates. An enzymatic activity-based assay is the protein phosphatase inhibition assay for the detection of okadaic acid and its analogs. The inhibition of protein phosphatase activity is quantified by the reduction of a dephosphorylated product that can be measured by colorimetric or fluorescent techniques [32, 33]. Microcystins can also be detected with a similar assay [34]. Enzymatic assays are easily transferred to high-throughput platforms.

Optimization of enzymatic assays for sample analysis is complicated due to the complexity of biological matrices. Many enzymatic reactions require specific conditions to take place, and therefore they are easily affected by modifications of the composition of the reaction medium after sample addition [35].

1.4.4 Aptamer-based detection

Aptamers are short sequences of single-stranded DNA or RNA that bind the analyte [21]. They are selected from synthetic libraries of oligonucleotides for high-affinity binding to the analyte to increase selectivity and specificity, in a process called SELEX (systematic evolution of ligands by exponential enrichment). Aptamers offer advantages of high stability and chemical flexibility for labeling and addition of functional groups, when compared with antibodies. Several assay designs are also possible, including sandwich and competitive formats, and can be used as binding molecules in HTTs. However, most aptamer-based detection methods published to date are sensors, often called aptasensors, oriented to fast, low-cost, portable detection. Aptamer-based techniques are relatively recent and have not reached the extended use of immunoassays, although they have received a lot of attention and undergone remarkable development in the last decades. Therefore, interaction of analytes and aptamers might be used to optimize high-throughput methods in the future. Currently, aptamers for detection of mycotoxins, aquatic toxins, antibiotics, and pesticides have been reported [21].

1.4.5 Cell-based assays

Cell-based assays use the response of the cell to a substance to detect its presence in a sample. The nature of the measured response may be diverse. Probably, the most commonly used cell-based assay is quantification of cell viability, which serves as an indicator of cytotoxicity, reflecting a reduction of cell proliferation or cell death. There are multiple techniques to quantify cell viability with colorimetric, fluorescent, or luminescent readouts (see Table 1.1), and many have been used in HTS of cytotoxic compounds [36].

Besides cell viability assays, other cellular events can also be used as markers of toxicity for the detection of hazardous substances, such as intracellular calcium con-

Table 1.1: Techniques to measure cell viability in well plates [36].

Technique	Reagent	Measured product	Cell function	Readout	Advantages/ disadvantages
Tetrazolium reduction assay	Tetrazolium compounds (MTT[1], MTS, XTT, WST)	Formazan	Viable cell metabolism	Colorimetric	Lower sensitivity Cytotoxic Endpoint assay[5] 1–4 h for signal Less expensive
Resazurin reduction assay	Resazurin	Resorufin	Viable cell metabolism	Fluorescent	More sensitive than tetrazolium assays Cytotoxic in the long term 1–4 h for signal Continuous assay[5] Less expensive
Protease viability marker assay	GF-AFC[2]	AFC	Constitutive protease activity	Fluorescent	Nontoxic 30 min to 1 h for signal
ATP assay	Luciferin Firefly luciferase	Light[3]	Cellular ATP	Luminescent	Very sensitive Fast (10 min) Endpoint assay[5] Less handling Expensive
Real-time luciferase assay	Shrimp luciferase Pro-substrate	Light[4]	Viable cell metabolism	Luminescent	Continuous assay[5] 10 min for readout Real time: rapid decay of signal after death Expensive

[1]3-(4,5-*Dimethylthiazol*-2-yl)-2,5-di*phenyl*tetrazolium bromide.
[2]Glycylphenylalanyl-aminofluorocoumarin.
[3]Luciferin + ATP + O_2 → Oxyluciferin + PP_i + AMP + CO_2 + light. Reaction catalyzed by luciferase.
[4]Pro-substrate enters the cell and undergoes intracellular reduction to the luciferase substrate. The substrate exits the cell and luciferase catalyzes the enzymatic reaction with the production of light.
[5]Endpoint assay indicates that it requires cell lysis. Continuous assay allows monitoring of cell viability for hours without the need of cell lysis, and therefore may also be amenable for multiplexing.

centration, membrane potential or intracellular levels of cyclic adenosine monophosphate. Plasma membrane proteins, mainly receptors and ion channels, are often targets of toxic compounds, and as mentioned above, they must be inserted in the membrane to preserve their binding properties and activity. At the moment, testing of an effect on the activity of these biomolecules is usually performed in the whole cell. Measurements of intracellular ion concentration and ion flux through the plasma membrane or membrane potential have been optimized for the detection of molecules with activity on G-protein-coupled receptors or ion channels. HTS has been reported for these assay designs using ion- or voltage-sensitive fluorescent dyes and fluorescence plate readers with fluorescence intensity, FRET, or fluorescence imaging readouts [16, 37]. Automated electrophysiology is also available, although it is considered medium throughput by some authors [16]. Most electrophysiological screening methods measure current under voltage clamp conditions, providing an indication of ion fluxes through a specific channel type [16].

Cell-based assays often use cell models that have been genetically manipulated to express high levels of a desired receptor, to express recombinant proteins involved in signaling or readout signal, or even for organelle-directed protein location. Recombinant cell lines have been widely used in high-throughput pharmaceutical screening for new drugs, and offer many possibilities for detection of environmental toxicants. In addition, recombinant cell lines are also critical to produce reagents for receptor-based assays.

One of the main features of cell-based assays is the lack of specificity. The further down in the signaling cascade the less specific. For example, for a toxin that alters the function of an ion channel, a cell viability assay would be less specific than membrane voltage detection. However, specificity can be improved with an adequate assay design. A strategy used for specific detection of palytoxin in hemolysis and cytotoxicity assays consists of adding the drug ouabain that inhibits palytoxin effect on the Na^+/K^+ ATPase pump [38, 39]. Within the same assay, palytoxin is tested alone to screen its presence in the sample, and in combination with ouabain as a positive control of the effect being caused through alteration of the Na^+/K^+ ATPase pump. Similarly, the presence of ciguatoxins is confirmed by potentiation of the response by veratridine in membrane voltage or cytotoxicity cell-based assays [40, 41].

In spite of the great potential of cell-based assays for contaminant detection, they have some important disadvantages that preclude their extended use for routine lab work. The maintenance of cell cultures requires significant resources in terms of equipment, material, and personnel. In addition, cell-based assays entail special challenges for method validation due to variability of cell culture conditions and genotypic and phenotypic instability of cell lines [16, 42], among others.

Acknowledgments: This research, leading to these results, has received funding from the following grants: Ministerio de Ciencia e Innovación PID 2020-11262RB-C21, IISCIII/PI19/001248, Grant CPP2021-008447 funded by MCIN/AEI/10.13039/501100011033 and by the European Union NextGenerationEU/PRT; Campus Terra (USC), BreveRiesgo (2022-PU011) CLIMIGAL (2022-PU016); Conselleria de Cultura, Educacion e Ordenación Universitaria, Xunta de Galicia, GRC (ED431C 2021/01); and European Union, Interreg EAPA-0032/2022 – BEAP-MAR, HORIZON-MSCA-2022-DN-01-MSCA Doctoral Networks 2022 101119901-BIOTOXDoc, and HORIZON-CL6-2023-CIRCBIO-01 COMBO-101135438.

Keywords: Automation, screening, fluorescence, luminescence, absorbance, laboratory robot, immunoassay, ELISA, multiplexing, lateral flow immunoassay, cytotoxicity

Abbreviations: ELISA, enzyme-linked immunosorbent assay; FRET, fluorescence resonance energy transfer; HTS, high-throughput screening; LFIA, lateral flow immunoassays; SPE, solid-phase extraction; TRF, time-resolved fluorescence.

References

[1]   Burr A, Bogart K, Conaty J, Andrews J. Automated liquid handling and high-throughput preparation of polymerase chain reaction-amplified DNA for microarray fabrication. Methods Enzymol. 2006;410:99–120.

[2]   Kong F, Yuan L, Zheng YF, Chen W. Automatic liquid handling for life science: A critical review of the current state of the art. J Lab Autom. 2012;17:169–85.

[3]   Kuklenyik Z, Ye X, Needham LL, Calafat AM. Automated solid-phase extraction approaches for large scale biomonitoring studies. J Chromatogr Sci. 2009;47:12–18.

[4]   Holland I, Davies JA. Automation in the life science research laboratory. Front Bioeng Biotechnol. 2020;8:571777.

[5]   Thurow K. Liquid handling in laboratory automation. Wiley Analytical Science Magazine, 2023.

[6]   Klostermeier D, Millar DP. Time-resolved fluorescence resonance energy transfer: A versatile tool for the analysis of nucleic acids. Biopolymers. 2001;61:159–79.

[7]   Grundmann M, Kostenis E. Label-free biosensor assays in GPCR screening. Methods Mol Biol. 2015;1272:199–213.

[8]   Ke N, Wang X, Xu X, Abassi YA. The xCELLigence system for real-time and label-free monitoring of cell viability. Methods Mol Biol. 2011;740:33–43.

[9]   Blay V, Tolani B, Ho SP, Arkin MR. High-throughput screening: Today's biochemical and cell-based approaches. Drug Discov Today. 2020;25:1807–21.

[10]   Banati H, Darvas B, Feher-Toth S, Czeh A, Szekacs A. Determination of mycotoxin production of Fusarium species in genetically modified maize varieties by quantitative flow immunocytometry. Toxins (Basel). 2017;9:70.

[11]   Fraga M, Vilarino N, Louzao MC, Rodriguez LP, Alfonso A, Campbell K, Elliott CT, Taylor P, Ramos V, Vasconcelos V, Botana LM. Multi-detection method for five common microalgal toxins based on the use of microspheres coupled to a flow-cytometry system. Anal Chim Acta. 2014;850:57–64.

[12]   Fraga M, Vilarino N, Louzao MC, Rodriguez P, Campbell K, Elliott CT, Botana LM. Multidetection of paralytic, diarrheic, and amnesic shellfish toxins by an inhibition immunoassay using a microsphere-flow cytometry system. Anal Chem. 2013;85:7794–802.

[13]   Maragos CM. Multiplexed biosensors for mycotoxins. J AOAC Int. 2016;99:849–60.

[14]   Collins AR, Annangi B, Rubio L, Marcos R, Dorn M, Merker C, Estrela-Lopis I, Cimpan MR, Ibrahim M, Cimpan E, Ostermann M, Sauter A, Yamani NE, Shaposhnikov S, Chevillard S, Paget V, Grall R, Delic J, de-Cerio FG, Suarez-Merino B, Fessard V, Hogeveen KN, Fjellsbo LM, Pran ER, Brzicova T, Topinka J, Silva MJ, Leite PE, Ribeiro AR, Granjeiro JM, Grafstrom R, Prina-Mello A, Dusinska M. High throughput toxicity screening and intracellular detection of nanomaterials. Wiley Interdiscip Rev Nanomed Nanobiotechnol. 2017;9:1413.

[15]   Ding M, Kaspersson K, Murray D, Bardelle C. High-throughput flow cytometry for drug discovery: Principles, applications, and case studies. Drug Discov Today. 2017;22:1844–50.

[16]   McManus OB, Garcia ML, Weaver D, Bryant M, Titus S, Herrington JB. Ion channel screening. In: Sittampalam GS, Coussens NP, Brimacombe K, Grossman A, Arkin M, Auld D, Austin C, Baell J, Bejcek B, Chung TDY, Dahlin JL, Devanaryan V, Foley TL, Glicksman M, Hall MD, Hass JV, Inglese J, Iversen PW, Kahl SD, Kales SC, Lal-Nag M, Li Z, McGee J, McManus O, Riss T, Trask OJ, Jr, Weidner JR, Xia M, Xu X, eds. Assay guidance manual. Eli Lilly & Company and the National Center for Advancing Translational Sciences, Bethesda (MD), 2012.

[17]   Priest BT, Cerne R, Krambis MJ, Schmalhofer WA, Wakulchik M, Wilenkin B, Burris KD. Automated electrophysiology assays. In: Sittampalam GS, Coussens NP, Brimacombe K, Grossman A, Arkin M, Auld D, Austin C, Baell J, Bejcek B, Chung TDY, Dahlin JL, Devanaryan V, Foley TL, Glicksman M, Hall MD, Hass JV, Inglese J, Iversen PW, Kahl SD, Kales SC, Lal-Nag M, Li Z, McGee J, McManus O, Riss T, Trask OJ, Jr, Weidner JR, Xia M, Xu X, eds. Assay guidance manual. Eli Lilly & Company and the National Center for Advancing Translational Sciences, Bethesda (MD), 2017.

[18]   Boobphahom S, Ly MN, Soum V, Pyun N, Kwon OS, Rodthongkum N, Shin K. Recent advances in microfluidic paper-based analytical devices toward high-throughput screening. Molecules. 2020; 25(13):2970.

[19]   Campbell K, Rawn DF, Niedzwiadek B, Elliott CT. Paralytic shellfish poisoning (PSP) toxin binders for optical biosensor technology: Problems and possibilities for the future: A review. Food Addit Contam Part A Chem Anal Control Expo Risk Assess. 2011;28:711–25.

[20]   Vilarino N, Louzao MC, Vieytes MR, Botana LM. Biological methods for marine toxin detection. Anal Bioanal Chem. 2010;397:1673–81.

[21]   Rapini R, Marrazza G. Electrochemical aptasensors for contaminants detection in food and environment: Recent advances. Bioelectrochemistry. 2017;118:47–61.

[22]   Posthuma-Trumpie GA, Korf J, van Amerongen A. Lateral flow (immuno)assay: Its strengths, weaknesses, opportunities and threats. A literature survey. Anal Bioanal Chem. 2009;393:569–82.

[23]   Tripathi P, Upadhyay N, Nara S. Recent advancements in lateral flow immunoassays: A journey for toxin detection in food. Crit Rev Food Sci Nutr. 2017;58(10):1715–1734.

[24]   Hsieh HV, Dantzler JL, Weigl BH. Analytical tools to improve optimization procedures for lateral flow assays. Diagnostics (Basel). 2017;7(2):29.

[25]   Vilarino N, Louzao MC, Fraga M, Rodriguez LP, Botana LM. Innovative detection methods for aquatic algal toxins and their presence in the food chain. Anal Bioanal Chem. 2013;405:7719–32.

[26]   Andreu V, Pico Y. Determination of currently used pesticides in biota. Anal Bioanal Chem. 2012;404:2659–81.

[27]   Botana LM, Alfonso A, Botana AM, Vieytes MR, Vale C, Vilariño N, Louzao MC. Functional assays for marine toxins as an alternative high throughput screening solution to animal tests. Trends Anal Chem. 2009;28:603–11.

[28]   Rodriguez LP, Vilarino N, Molgo J, Araoz R, Antelo A, Vieytes MR, Botana LM. Solid-phase receptor-based assay for the detection of cyclic imines by chemiluminescence, fluorescence, or colorimetry. Anal Chem. 2011;83:5857–63.

[29]   Rodriguez LP, Vilarino N, Molgo J, Araoz R, Botana LM. High-throughput receptor-based assay for the detection of spirolides by chemiluminescence. Toxicon. 2013;75:35–43.

[30]   Vilarino N, Fonfria ES, Molgo J, Araoz R, Botana LM. Detection of gymnodimine-A and 13-desmethyl C spirolide phycotoxins by fluorescence polarization. Anal Chem. 2009;81:2708–14.

[31]   Rodriguez LP, Vilarino N, Molgo J, Araoz R, Louzao MC, Taylor P, Talley T, Botana LM. Development of a solid-phase receptor-based assay for the detection of cyclic imines using a microsphere-flow cytometry system. Anal Chem. 2013;85:2340–47.

[32]   Tubaro A, Florio C, Luxich E, Sosa S, Della Loggia R, Yasumoto T. A protein phosphatase 2A inhibition assay for a fast and sensitive assessment of okadaic acid contamination in mussels. Toxicon. 1996;34:743–52.

[33] Vieytes MR, Fontal OI, Leira F, Baptista de Sousa JM, Botana LM. A fluorescent microplate assay for diarrheic shellfish toxins. Anal Biochem. 1997;248:258–64.

[34] Fontal OI, Vieytes MR, Baptistade Sousa JM, Louzao MC, Botana LM. A fluorescent microplate assay for microcystin-LR. Anal Biochem. 1999;269:289–96.

[35] Serres MH, Fladmark KE, Doskeland SO. An ultrasensitive competitive binding assay for the detection of toxins affecting protein phosphatases. Toxicon. 2000;38:347–60.

[36] Riss TL, Moravec RA, Niles AL, Duellman S, Benink HA, Worzella TJ, Minor L. Cell viability assays. In: Sittampalam GS, Coussens NP, Brimacombe K, Grossman A, Arkin M, Auld D, Austin C, Baell J, Bejcek B, Chung TDY, Dahlin JL, Devanaryan V, Foley TL, Glicksman M, Hall MD, Hass JV, Inglese J, Iversen PW, Kahl SD, Kales SC, Lal-Nag M, Li Z, McGee J, McManus O, Riss T, Trask OJ, Jr, Weidner JR, Xia M, Xu X, eds. Assay guidance manual. Eli Lilly & Company and the National Center for Advancing Translational Sciences, Bethesda (MD), 2016.

[37] Arkin MR, Connor PR, Emkey R, Garbison KE, Heinz BA, Wiernicki TR, Johnston PA, Kandasamy RA, Rankl NB, Sittampalam S. FLIPR assays for GPCR and ion channel targets. In: Sittampalam GS, Coussens NP, Brimacombe K, Grossman A, Arkin M, Auld D, Austin C, Baell J, Bejcek B, Chung TDY, Dahlin JL, Devanaryan V, Foley TL, Glicksman M, Hall MD, Hass JV, Inglese J, Iversen PW, Kahl SD, Kales SC, Lal-Nag M, Li Z, McGee J, McManus O, Riss T, Trask OJ, Jr, Weidner JR, Xia M, Xu X, eds. Assay guidance manual. Eli Lilly & Company and the National Center for Advancing Translational Sciences, Bethesda (MD), 2012.

[38] Espina B, Cagide E, Louzao MC, Fernandez MM, Vieytes MR, Katikou P, Villar A, Jaen D, Maman L, Botana LM. Specific and dynamic detection of palytoxins by in vitro microplate assay with human neuroblastoma cells. Biosci Rep. 2009;29:13–23.

[39] Riobó P, Paz B, Franco JM, Vázquez JA, Murado García MA. Proposal for a simple and sensitive haemolytic assay for palytoxin. Toxicological dynamics, kinetics, ouabain inhibition and thermal stability. Harmful Algae. 2008;7:415–29.

[40] Louzao MC, Vieytes MR, Yasumoto T, Botana LM. Detection of sodium channel activators by a rapid fluorimetric microplate assay. Chem Res Toxicol. 2004;17:572–78.

[41] Manger R, Woodle D, Berger A, Dickey RW, Jester E, Yasumoto T, Lewis R, Hawryluk T, Hungerford J. Flow cytometric-membrane potential detection of sodium channel active marine toxins: Application to ciguatoxins in fish muscle and feasibility of automating saxitoxin detection. J AOAC Int. 2014;97:299–306.

[42] Geraghty RJ, Capes-Davis A, Davis JM, Downward J, Freshney RI, Knezevic I, Lovell-Badge R, Masters JR, Meredith J, Stacey GN, Thraves P, Vias M, Cancer Research UK. Guidelines for the use of cell lines in biomedical research. Br J Cancer. 2014;111:1021–46.

Jesús M. González-Jartín and Amparo Alfonso

2 Analytical instrumentation and principles

2.1 Introduction

Analytical instrumentation provides information about chemical and physical properties of substances and about individual components of samples. Data obtained in this way can be used either for quantitative or qualitative purposes. Nowadays, devices used as analytical instruments have advanced substantively, increasing not only their accuracy but also their versatility and easiness of use. An instrument for analysis converts the information about chemical or physical properties of some analyte to information that can be interpreted. In this context, several definitions should be considered:

- Analytical technique: This is a chemical or physical principle used to study an analyte.
- Analytical method: This is the application of a technique or a set of techniques to analyze a specific analyte in a specific matrix or used in a sample to know qualitatively and/or quantitatively the composition and the chemical state in which it is located.
- Analytical process: This process has instructions to use an analytical method.
- Analytical protocol: This protocol has specific guidelines with all steps necessary to develop a specific analytical method.

Analytical instrumentation includes traditional titrimetric and volumetric techniques and many other different procedures useful for the determination of active ingredients in a sample and to quantify related compounds and impurities associated in order to know the composition of complex matrices. These techniques have the advantage of using small amounts of sample, reagents, and time.

The classification of methods included in analytical instrumentation can be done in two ways:

1. According to the information obtained:
   a. Qualitative methods
   b. Quantitative methods
2. According to the property measured:
   a. Physicochemical methods: spectroscopy, including colorimetry, spectrophotometry and fluorimetry, nephelometry, turbidimetry, nuclear magnetic resonance, and mass spectrometry
   b. Electro-analytical methods: potentiometry, amperometry, voltammetry, electrophoresis, and polarography

https://doi.org/10.1515/9783111014449-002

c. Separation-based methods: used to separate individual components from a mixture. This includes methods such as extraction, filtration, distillation, chromatography, centrifugation, and crystallization. Electrophoretic techniques can also be considered as separation methods.

In this way, a wide range of old and new techniques, tools, and instruments can be included under the term "analytical instrumentation." This chapter focuses on separation-based methods, particularly liquid chromatography (LC) coupled to different devices to identify and quantify environmental toxicants. A summary of separation and chromatographic techniques is provided in Figure 2.1. To select a separation technique is important to consider physical and chemical properties of analyte, the size of the sample, the concentration of analyte, the composition of matrix, and the number of samples.

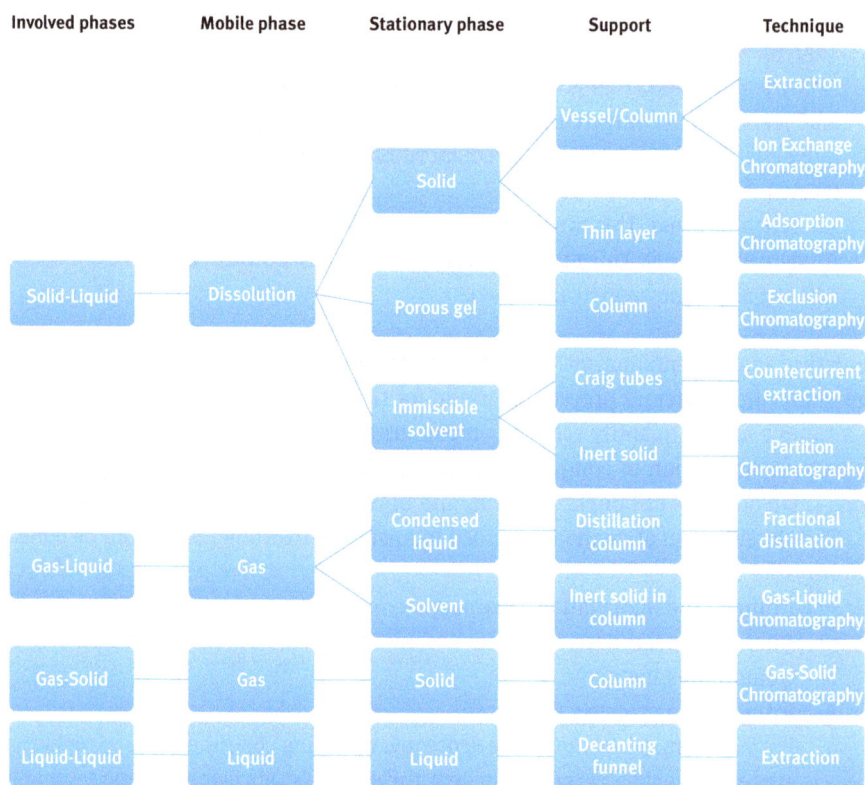

Figure 2.1: Separation and chromatographic techniques.

Environmental toxicants such as marine toxins, cyanotoxins, or mycotoxins are small molecules with different chemical characteristics (Table 2.1, see Chapters 6–8). These compounds are present in complex matrices, and therefore, for their analysis and individual identification, the combination of several separation methods, mainly extraction procedures and chromatographic techniques, is necessary (Figure 2.2).

Table 2.1: Solubility and polarity characteristic of main marine toxins, cyanotoxins, and mycotoxins groups.

Toxicant	Lipophilic/apolar	Hydrosoluble/polar
Marine toxins	Okadaic acid, dinophysistoxin I and dinophysistoxin II (diarrhetic shellfish toxins)	Saxitoxin group (paralytic shellfish toxins)
	Azaspiracids	Domoic acid
	Yesotoxins	Tetrodotoxins
	Pectenotoxins	
	Cyclic imines	
	Ciguatoxins	
	Brevetoxins	
Cyanotoxins	Microcystins	Anatoxins
	Nodularins	Cylindrospermopsin
		Saxitoxin group
Mycotoxins	Fumonisins	Deoxinivalenol
	Zearalenone	Patulin
	Aflatoxins	
	Beauvericin	
	Enniatins	

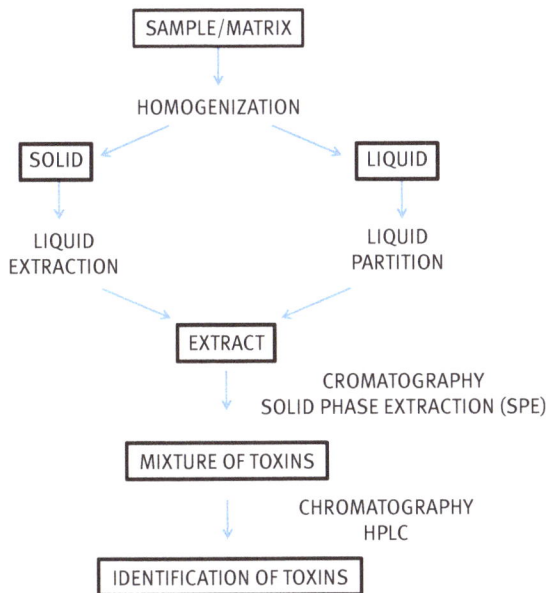

Figure 2.2: Scheme of separation methods used to analyze toxins.

2.2 Sample/matrix extraction

The first essential step in the analysis of toxic compounds is the adequate extraction of the compound of interest from the matrix. This step includes sample preparation and extraction procedures.

2.2.1 Sample preparation

The sample or matrix should be homogeneous in order to achieve good extraction efficiency. To homogenize the sample, if solid (e.g., shellfish, fish, or grains), a blender or grinder is often used. In addition, in case of official analysis, the sample should be representative of the lot to be analyzed; therefore, the sampling procedure should be carried out according to specific rules [1–3]. In any case, the sample should be as representative as possible. Table 2.2 shows the amount of sample that should be processed and homogenized and the amount necessary from this homogenate for the extraction of each group of toxins following official standard procedures.

Table 2.2: Amount of sample processed for representative sampling and amount necessary for the extraction following official standard operating procedures to identify EU-regulated toxins.

Toxicant	Sample processed	Amount used for extraction	Official method
Lipophilic marine toxins: diarrhetic shellfish poison, azaspiracids, yesotoxins, pectenotoxins, cyclic imines	100–150 g shellfish tissue	2.00 ± 0.05 g	[4]
Paralytic shellfish poisons	100–150 g shellfish tissue	5.00 ± 0.1 g	[5, 6]
Domoic acid	100–150 g shellfish tissue Scallops: at least 10 specimens	4.00 ± 0.1 g	[7] [8]

AOAC, Association of Official Analytical Chemists.

2.2.2 Extraction procedure

The extraction procedure for each compound is different depending on the matrix state (solid or liquid):

2.2.2.1 Extraction procedures from solid matrices (solid–liquid extraction)

In this case, the matrix homogenate is mixed with a solvent. The polarity of the solvent used should be based on the nature of the compound to be extracted. Organic solvents such as acetone, methanol, or acetonitrile are used to extract nonpolar/lipophilic compounds as in the case of lipophilic marine toxins, whereas aqueous solutions are used to extract polar/hydrophilic compounds such as paralytic shellfish poisons (PSPs), domoic acid (DA), or tetrodotoxins (TTXs). Sometimes a mixture of solvents in different proportions is used to extract compounds with different nature from the same sample, which is the case of mycotoxins. In any case, the solvent (pure or combined) is mixed with the matrix, and the separation between phases is achieved by gravity, filtration, or centrifugation. The compound of interest will be in the liquid phase. A summary of solvents used to extract toxins from different solid matrices is given in Table 2.3.

Table 2.3: Solvents used to extract toxins from solid matrices.

Toxicant	Solvent for extraction
Lipophilic marine toxins: diarrhetic shellfish poison, azaspiracids, yesotoxins, pectenotoxins, cyclic imines	Methanol
Ciguatoxins	Acetone or methanol
Brevetoxins	Acetone
Paralytic shellfish poison	HCl 0.1 N or acetic acid 1%
Domoic acid	Methanol (50%)
Tetrodotoxins	Acetic acid 1%
Mycotoxins	Water:acetic acid (49:1) – acetonitrile (50)

2.2.2.2 Extraction procedures from liquid matrices (liquid–liquid extraction)

In this case, the extraction of toxins from the liquid is based on the toxin distribution between two phases and in its ability to migrate from one to the other. In a simple liquid–liquid extraction, two immiscible phases are used, one aqueous and the other an organic solvent such as diethyl ether, chloroform, or hexane. These phases are immiscible and form two layers, with the denser phase on the bottom. The extraction efficiency is the percentage of toxin moving from one phase to the other, and it is determined by the partition coefficient of the toxin between both phases. In some cases, for example, mycotoxin extracts, the partition between phases is forced using a mixture of salts to saturate the aqueous phase and to force mycotoxins to migrate to the organic phase.

This is called dispersive liquid–liquid extraction [9]. In addition, to extract some toxins from liquid matrices, resins or polymers are also used to attach these compounds [10]. The binding toxin resin is specific and saturable. The resin is introduced in mesh bags to be easily extracted from liquids. In the case of cyanotoxin extraction from fresh water, since these compounds are located inside cells, the water sample should be filtrated and concentrated, and some step to lysate cells should be incorporated [11].

2.3 Chromatography

Chromatography is a commonly used technique to analyze and identify the extract obtained from matrix. In addition, chromatography is also used to clean and prepare the sample before analysis.

Chromatography is defined as an analytical technique of separation in which a chemical mixture (in solution or suspension) carried by a liquid or gas is separated into its components by passing through a stationary phase. Several methods included under this term are LC, thin-layer chromatography (TLC), gas chromatography, and supercritical fluid chromatography (SFC). The method is based on the distribution of components between a mobile phase and a stationary phase. Sample components interact with the stationary phase with different affinities and are dragged by the mobile phase. Their ability and rate of migration are based on their interaction with the stationary phase, and higher interaction gives rise to lower migration.

2.3.1 Classification

The classification of chromatographic methods can be done based on different criteria.

2.3.1.1 Based on phase combination

- Stationary phase: liquid or solid
- Mobile phase: liquid, supercritical fluid, or gas

Stationary phase	Mobile phase		
	Gas	**Supercritical fluid**	**Liquid**
Liquid	Gas–liquid chromatography	Supercritical fluid chromatography	Liquid–liquid chromatography
Solid	Gas–solid chromatography	Supercritical fluid chromatography	Liquid–solid chromatography (or LC)

2.3.1.2 Based on mechanism of separation

- Partition chromatography
- Adsorption chromatography
- Ion exchange chromatography
- Molecular exclusion, size-exclusion, or gel filtration chromatography
- Affinity chromatography
- Electrophoresis
- Chiral chromatography

2.3.1.3 Based on phase polarity

- Normal-phase chromatography: mobile-phase nonpolar and stationary-phase polar
- Reverse-phase chromatography: mobile-phase polar and stationary-phase nonpolar

2.3.1.4 Based on shape of chromatography bed

- Planar chromatography:
  - Paper chromatography
  - TLC
- Column chromatography:
  - Packed chromatography
  - Open tubular column chromatography

2.3.1.5 Based on the development procedure

- Frontal chromatography
- Displacement chromatography
- Elution chromatography

To separate toxins and to obtain qualitative and quantitative information, column chromatography, with different solid stationary phases and different mobile phases (liquid) depending on the analyte, is frequently used. The terms commonly used in the context of column chromatography are summarized in Table 2.4. In general, the mobile phase is a liquid, and the stationary phase is a solid held in a column.

Table 2.4: Terms used in column chromatography.

Term	Definition
Mobile phase or carrier	Solvent moving through the column
Stationary phase or adsorbent	Substance that stays fixed inside the column (solid or liquid)
Eluent	Fluid entering the column
Eluate	Fluid exiting the column (collected in fractions)
Elution	The process of washing out a compound through a column using a suitable solvent
Analyte	Mixture whose individual components (toxins) are going to be separated

2.4 Column chromatography–solid-phase extraction (SPE)–sample preparation

In the classic liquid column chromatography, the stationary phase is placed in a vertical glass or plastic column. The mobile phase, a liquid, is added to the top of the column and flows down (through the stationary phase) by either gravity (Figure 2.3) or an external low-pressure, flash chromatography (Figure 2.4). The sample, with compounds to be separate, is added on the top and the separation of components is achieved through varying interactions between phases. Fractions with components are separated by time units (minutes or seconds) or volume (drops or milliliters) and collected at the end of the column. The quality of separation depends on several factors. In this procedure, the absence of air bubbles in the stationary phase as well as correct packaging are particularly important.

This type of chromatography, currently applied to prepare and clean samples to avoid interferences in detection, is also called solid-phase extraction (SPE). The stationary phase is a solid adsorbent such as silica gel, alumina, dextran or agarose polymers, porous graphitic carbon, or synthetic resins like Biogel®, Diaion®, or Florisil®. The size of particles is usually given by the mesh value, which refers to the number of holes in the mesh used to sieve the adsorbent. Therefore, higher mesh values means more holes per area unit and correspondingly smaller particles. About 70–230 mesh particles are used for gravity columns and 230–400 mesh particles for flash columns.

Silica gel is usually modified with long hydrocarbon chains (8 or 18 carbon atoms), and the columns are called C18 octadecyl silica, or C8. Alumina is available in types I, II, and III (from lower to higher water content). The basis of separation in this chromatography is the polarity of molecules, and silica gel and alumina are both polar adsorbents. In this way, chromatographic techniques can be classified into two types:

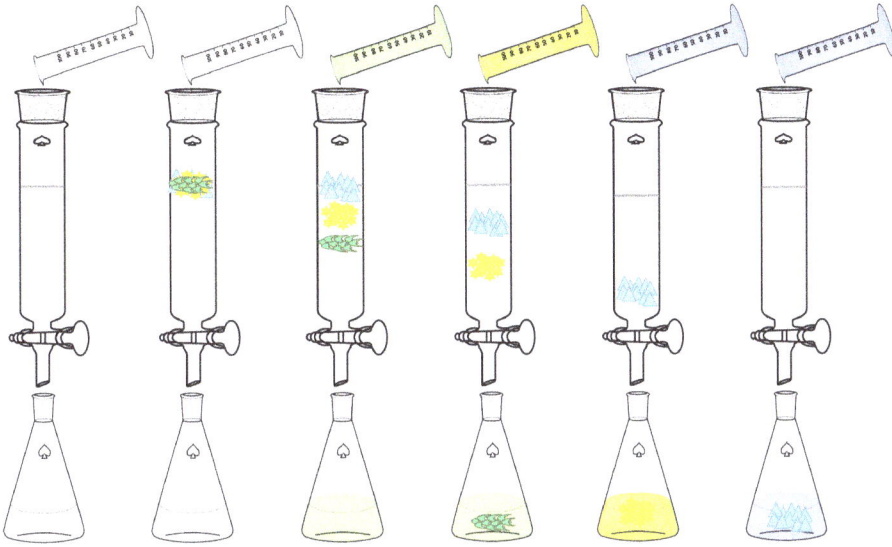

Figure 2.3: Separation of analytes by gravity column chromatography (solid stationary phase–liquid mobile phase).

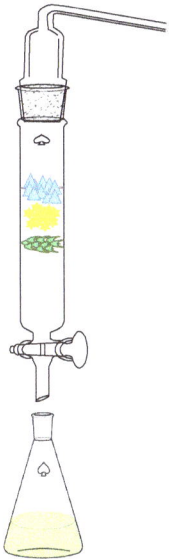

Figure 2.4: Separation of analytes by flash chromatography (solid stationary phase–liquid mobile phase).

– **Normal-phase chromatography**: when the stationary phase is more polar than the mobile phase. In this case, the more polar components in a sample will be strongly retained on the stationary phase, while nonpolar compounds will not be fixed by these adsorbents. Therefore, nonpolar compounds will be eluted faster than polar compounds. Thus, the components of a mixture can be adequately separated by increasing the polarity.
– **Reverse-phase chromatography**: when the mobile phase is more polar than the stationary phase. In this case, the mobile phase is a mixture of solvents such as water, methanol, and acetonitrile. Since many solid adsorbents used as stationary phases are polar by nature, nonpolar stationary phases are prepared by coating nonpolar molecules (silica gel coated with silicone or long hydrocarbon chains). With reverse phase, the most polar compounds will be eluted first and the components following will have decreasing polarities.

When the stationary phase is a resin, depending on its nature, different kinds of chromatography can be developed: ion-exchange, size-exclusion, exclusion, partition, or affinity chromatograph.

Flash chromatography is a modification of classic column chromatography in which the mobile phase moves faster through the column with the help of either vacuum or pressurized air. For this reason, it is considered as medium-pressure chromatography, whereas the chromatography is considered as low-pressure when only gravity is moving the mobile phase. The air used in medium-pressure chromatography should be inert to avoid interactions with the mobile or stationary phases or the sample.

These SPE protocols are currently used for preparative and cleanup purposes, to clean samples before analysis. As mentioned, the polarity of toxins is important to choose both the stationary and mobile phases. C8, C18, porous graphitic carbon, or Biogel® are used to clean and prepare samples with PSPs [12–14]. C8, C18, and porous graphitic carbon cartridges are used to process samples with tetrodotoxins (TTXs), whereas C18, silica gel, and Florisil® are used with ciguatoxin (CTX) samples [15–18]. The mobile phase and the elution protocol should be different in each case.

2.5 High-performance liquid chromatography (HPLC): analysis

High-performance liquid chromatography (HPLC) is the most common analytical technique used worldwide to analyze a sample containing toxins because it allows the separation, identification, and quantification of components in a mixture. HPLC is a modification of column chromatography. In this case, the stationary phase, a resin, is tightly packed into a column. The resin is a granular material made of solid small particles such as silica or polymers. A pump is necessary to force the mobile-phase elution through the column under high pressures of up to 400 atmospheres (404 bars). The pump makes the technique faster compared to other column chromatography, the efficiency is higher, and it provides continuous quantitative and qualitative information. Smaller particles in stationary phase have a much greater surface area for interactions with molecules passing through. This results in a much better separation of the components of a mixture. The necessary components for liquid chromatografy are shown in Figure 2.5.

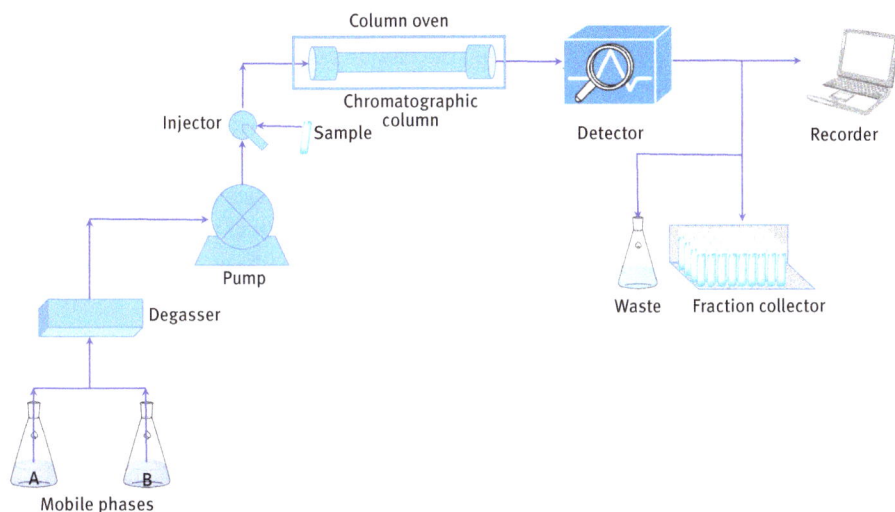

Figure 2.5: Liquid chromatography.

The response obtained after component separation in a column is called chromatogram, where the signal obtained from a detector is usually represented *versus* the time. Each component is represented as a peak, and the time when the peak appears is called retention time. The components of a sample are separated according to the time retained by the stationary phase, that is, according to the affinity for the stationary/mobile phases. These differences can be due to different chemical or physical properties including:
– Ionization state
– Polarity and polarizability
– Bindings (hydrogen bonding and Van del Waals forces)
– Hydrophobicity
– Hydrophilicity

2.5.1 Components for liquid chromatografy

2.5.1.1 Mobile phases and elution

Mobile phases should be dissolutions free of solids or particles in suspension, and therefore, should be filtrated or made with high-purity degree reagents. Reservoirs are glass containers (clear or amber). In general, a 2–5 μm filter is placed in the inlet of suction tube to avoid the pass of any particle in suspension. In addition to eliminate air bubbles solved in the mobile phases, a degasser is connected before the pump. Depending on the mobile-phase composition, the elution can be:
- Isocratic elution: the composition of the mobile phase is constant throughout the separation time.
- Gradient elution: the composition of mobile phase changes with time during separation in order to change the polarity and to increase efficiency. If the chromatography is in normal phase, the polarity of solvents in gradient would be increasing (from less to high). If the chromatography is in reverse phase, the polarity of solvents in gradient must be decreasing.

Methanol, water, and acetonitrile are the most common solvents used as mobile phase. When the sample contains ionizable compounds, some additives such as formic acid/ammonium formate or acetic acid/ammonium acetate are included in the mobile phase. In this way, the mobile phase pH is also controlled. This is important since the stability of the columns is also affected by mobile phase pH. The selection of additives is important, considering that a buffer is most effective at ±1 pH units of the target compound's pK_a. Besides additives can be appropriated or not depending on the detector. For example, citrate buffers are not suitable for ultraviolet (UV) detection below 220 nm or mass spectrometry detection.

2.5.1.2 Pumps

Pumping systems force to flow mobile phases through the column. To properly preform this function, some operating requisites are necessary:
- High range of pressures (0–1,200 bars)
- Variable fluxes (0.01–10 mL/min)
- Controlled and reproducible fluxes
- To be mechanically reliable
- Pulsation-free fluxes
- Inert materials in contact with fluids

Reciprocating piston pumps with two pistons are the pumping systems most frequently used. The basic elements in these pumps are a cylindrical pump chamber that holds a piston with a motor that operates a driving cam (this is called plunger because the motor rotates, and the piston is moved in and out of the pumper chamber), a pump seal, and a pair of check valves. Pistons are usually made in sapphire, stainless steel, or graphite. Check valves control the direction of flow through the pump. The pump seal avoids mobile-phase leaks around the piston. The systems with two plungers provide more uniform fluxes. In addition, a pulse damper is included to avoid signal fluctuations in detectors and to improve baseline noise. Nowadays, binary and quaternary pumps are commonly chosen in HPLC systems. Binary pumps have two channels. Each channel has a system with two plungers and inlet and outlet valves. Both channels are connected through a low-volume mixer chamber. Quaternary pumps are based in a system of one channel with two plungers and a valve to select the solvent line from four reservoirs.

2.5.1.3 Injection

The injection system, autosampler, is located between the pumping system and the column. This device introduces the sample (0.1–100 μL for analytical purposes or up to 5,000 μL for preparative purposes) at atmospheric pressure into the high-pressure system. This is a critical step in the chromatographic process. In this way, the requirement to stop the mobile-phase flow to inject the sample is eliminated. To allow this, sample injection valves (switching valves or rotor seal valves) are used. This system introduces reproducible amounts of sample in the mobile-phase stream without causing changes in pressure or flow. In addition, any disturbance in the chromatographic baseline is minimized. Autosamplers are the first contact of sample with HPLC instrumentation and therefore should be manufactured from inert materials. But contamination problems can still appear in the form of sample carryover due to sample adsorption in the injection system (rotor seal contamination). To avoid this, autosamplers are provided with washing systems able to eliminate any sample test. Washing solutions should be selected according to the solubility of the sample to clean the residual sample.

2.5.1.4 Columns

Columns are the essential part of HPLC system that will provide narrow peaks. Column efficiency and performance are measured by the number of theoretical plates (TPN). The bigger the TPN, the better the column. A typical HPLC column has an internal diameter (I.D.) of 4.6 mm or smaller and a variable length from 50 to 250 mm. Small I.D. columns increase sensitivity, but they may decrease efficiency and resolution. In semipreparative and preparative columns, I.D. is increased to 10 mm and 20 mm, respectively. These columns require flow rates of 10–20 mL/min while in analytical columns the optimum flow rate is 1–0.2 mL/min. Long columns increase efficiency and resolution over short columns, and also increase retention times and backpressure. Short columns can provide adequate resolution and high speed of analysis. The particle size of resin inside the column is also important. A particle size of 5 μm is the most common for speed and resolution in analytical columns. Smaller particles can increase resolution and decrease analysis time. In this way, when the size is lower than 3 μm, the pressure is increased. This is the case of ultra-HPLC (UHPLC) columns, with 1.5–3 μm particles. The pore size of internal particles is also important; the smaller the pore size, the larger the surface area. For small molecules, like environmental toxicants, the pore size particle should be 80–150 Å. Silica is the most popular resin useful for pH between 1.0 and 7.5. Other polymers are used for extreme pH. The resin is usually packed in stainless steel cartridges to support high pressures. Depending on the molecules to be separated, different chromatographic techniques will be chosen. Normal-phase columns are packed with polar resins like silica, amide, or hydrophilic interaction LC (HILIC) resin to separate small polar compounds by hydrophilic interactions [19]. Reverse-phase columns are the most widely used columns for samples with polar, nonpolar, ionic, or ionizable analytes. This phase is compatible with mobile phases containing methanol, water, or acetonitrile. C6, phenyl, C18, and C8 resins are often selected for reverse-phase columns, whereas for polar analytes, others like porous graphitic carbon are used. To separate toxins, either normal-phase or reverse-phase columns are selected [17, 20–22]. Figure 2.6 shows different stationary-phase resins for HPLC. Other chromatographic resins for ion chromatography, ion exchange chromatography, ligand exchange, or size exclusion, among others, can be selected depending on the analyte nature: inorganic compounds, amino acids, pesticides, carbohydrates, and so on.

2.5.1.5 Oven

In addition to the appropriate column, the separation temperature is also crucial. This should be as constant as possible; therefore, the oven is essential to keep constant column temperature, even when room temperature is selected in the chromatographic conditions.

2.5.1.6 Detectors

Detectors are devices to convert a physical or chemical change into a measurable signal and in this way to recognize an analyte. In LC detection, these devices are used to identify and determine the concentration of eluting compounds in the mobile phase coming from the column. Detectors should have several characteristics such as high sensitivity toward solute over mobile phase, reproducible responses, either specific or general response to compounds in a mixture, wide linear dynamic range, unaffected by changes in temperature or mobile phases, and fast response, among others [23]. In addition, detectors should have a low volume in detection cell and low noise detection

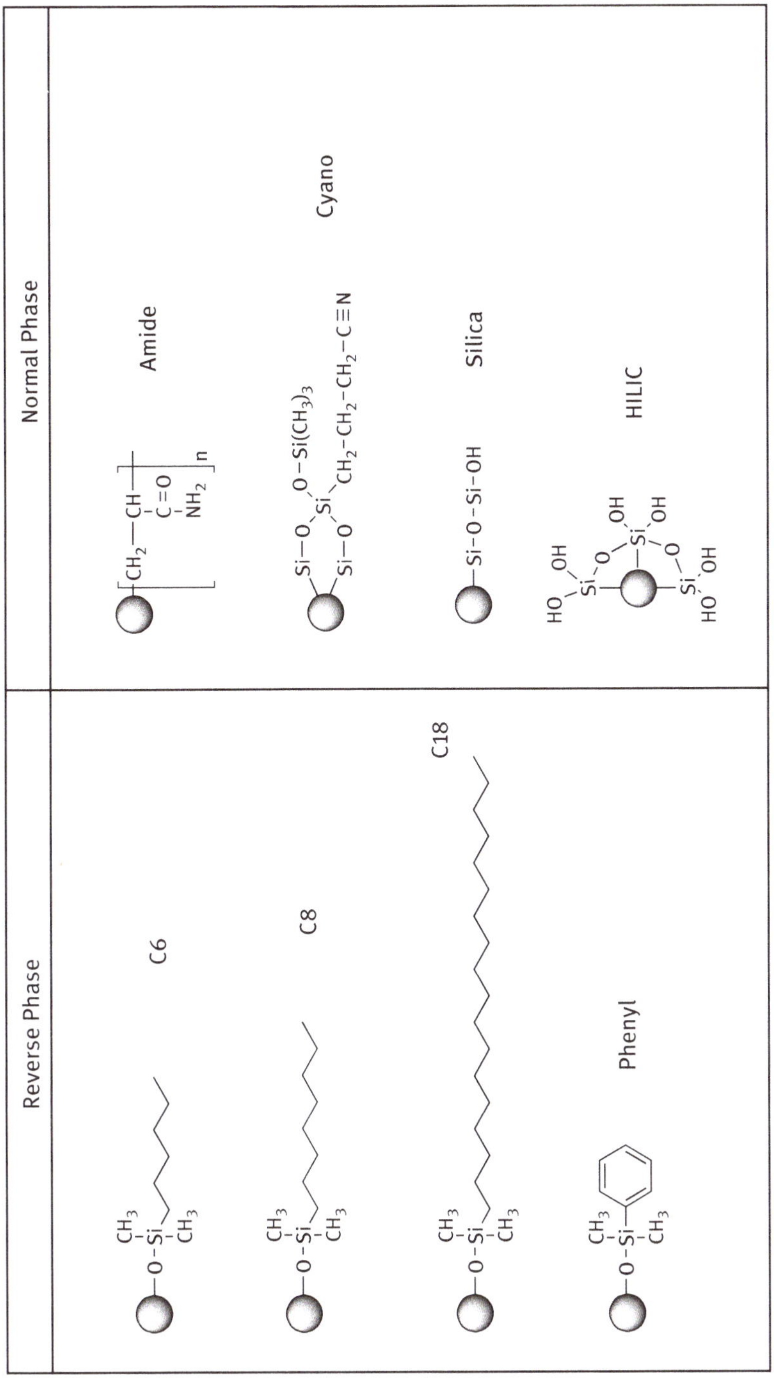

Figure 2.6: Resins for stationary phases (normal and reverse).

and limits. The detectors often used in LC systems can measure specific or bulk properties and, in this way, can be classified into two groups.

2.5.1.6.1 Bulk property detectors
– Electrochemical detectors:

These are used for compounds that can be oxidized or reduced and require the use of electrically conductive mobile phase. The system has three electrodes (a counter, a working, and a reference electrode), and the reaction takes place in the working electrode surface after the application of a fixed potential difference between working and reference electrodes [23]. These detectors are sensible and selective but aqueous mobile phases or polar solvents are required with electrolytes (oxygen free). In addition, flux and mobile-phase conditions are important.

– **Refractive index detectors:**

These are the oldest LC detectors. In this case, the property measured is the difference in optical reactive index between mobile phase and sample. This is a differential detector; therefore, the higher the optical reactive index, the bigger the signal.

– **Conductivity detection detectors:**

In this case, the conductivity of the mobile phase is measured. This property changes when analytes are passing through the detection cell.

– **Light scattering detectors:**

The property measured is the scattered light. These detectors are useful to measure large molecular weight compounds.

2.5.1.6.2 Specific property detectors
– **UV–visible detectors:**

The property measured is the absorbance of eluate. This is a common LC detector since many compounds of interest absorb in the UV or visible region. Sample concentration is a function of the fraction of light transmitted through the detection cell. Mobile-phase composition is important for optimum sensitivity and linearity detection. These detectors can have fixed or variable wavelengths (Figure 2.7A), or photodiode array that relies on one or more wavelengths generated by a broad-spectrum lamp. As shown in Figure 2.7B, in this case, the light passes through the flow cell prior to hitting the grating, allowing it to spread the spectrum across an array of photodiodes. UV detection is used in the official method of DA detection after HPLC separation and is also used to detect palytoxins, brevetoxins, and some mycotoxins [24, 25].

A

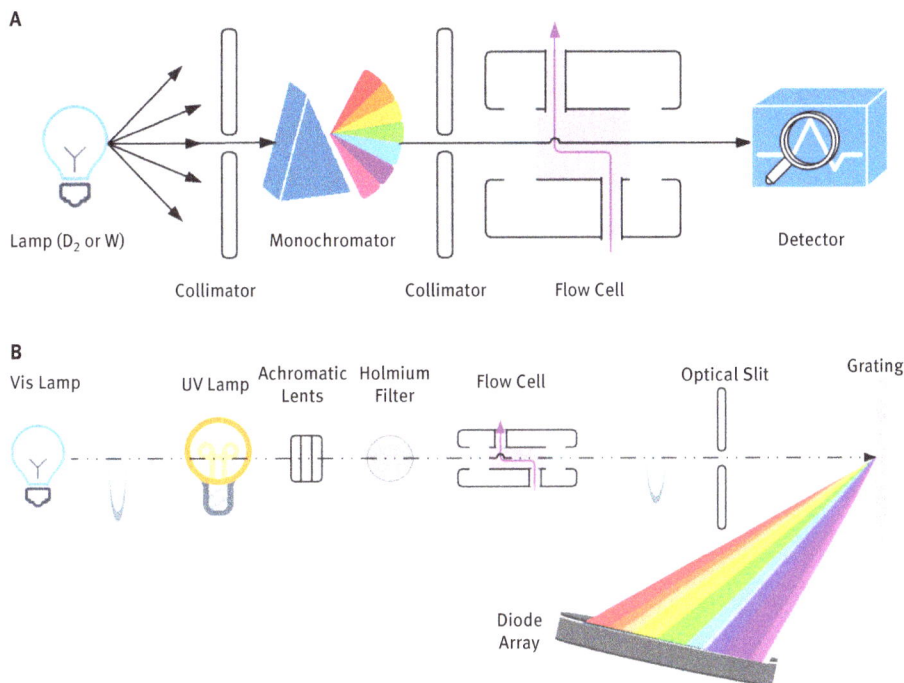

Lamp (D$_2$ or W)   Monochromator   Detector

Collimator   Collimator   Flow Cell

B

Vis Lamp   UV Lamp   Achromatic Holmium   Flow Cell   Optical Slit   Grating
Lents   Filter

Diode
Array

Figure 2.7: (A) Scheme of ultraviolet (UV) detector and (B) scheme of photodiode array (PDA) detector.

– Fluorescence detectors (FLDs):
The property measured is the optical light emitted by analytes after they have been excited to a higher wavelength. These detectors are most sensitive, specific, and selective than UV–visible detectors. The excitation light source is a lamp with broad spectrum (deuterium or xenon), and the excitation wavelength is selected with a filter or a monochromator (Figure 2.8). In some cases, a laser can be used instead of the lamp. Analytes should have native fluorescence or should be converted into derivatives with fluorescent properties. After sample excitation, the FLD should have a filter system or monochromator to select the most appropriate emission wavelength of an analyte. The dynamic range is high (change in fluorescence when the amount of analyte changes); however, the linear dynamic range is small. In addition, the composition of mobile phase and interferences of matrix should be taken into account. The FLD is often used to detect environmental toxicants after HPLC separation [25]. Several cyanotoxins and marine toxins can be detected by fluorescence after derivatization [25]. In the case of PSPs, the conversion in fluorescent molecules by oxidation can be done before or after HPLC separation [12, 26]. In the case of palytoxins, a pre-column derivatization method was also developed. FLDs are also used to detect microcystins (MCs) and anatoxins (ATXs) [11, 27].

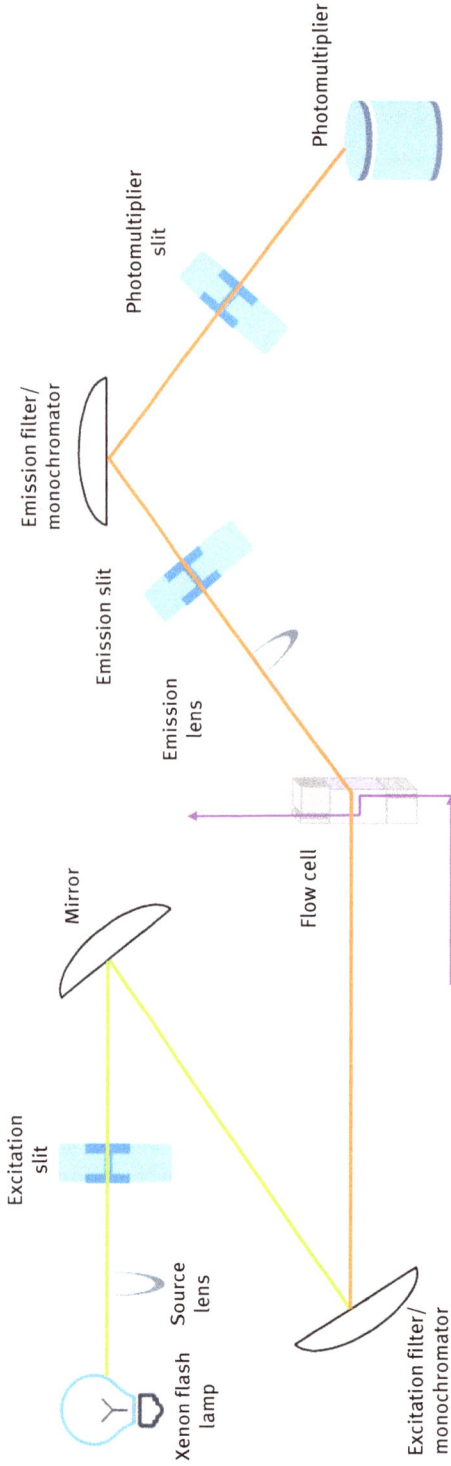

Figure 2.8: Scheme of fluorescence detector (FLD).

– **Mass spectroscopy (MS) detectors:**

In MS detectors, the property measured is the difference in mass-to-charge ratio (m/z) of ionized molecules to separate them. In addition, molecules can be fragmented by electrical fields. In this way, MS allows the quantification of analytes and provides structural information by identification of distinctive fragmentation pathways. This is currently the detection with higher applications because of advantages such as small sample size, high sensitivity, specificity, and resolution of time. However, when MS identification is used, samples are destroyed; therefore, for preparative purposes, a splitter should be included. In addition, this is a very costly technique in terms of technical and human resources.

MS is an analytical technique by which chemical substances are identified by the sorting of gaseous ions in electrical and magnetic fields. MS consists of four basic parts: first is the handling system to introduce the sample in the equipment; second is the gas phase where ions should be created; then the analyzer where ions will be separated, in space or time, based on their m/z; and finally a detector where detection and measure of the quantity of ions of each m/z will be done (Figure 2.9).

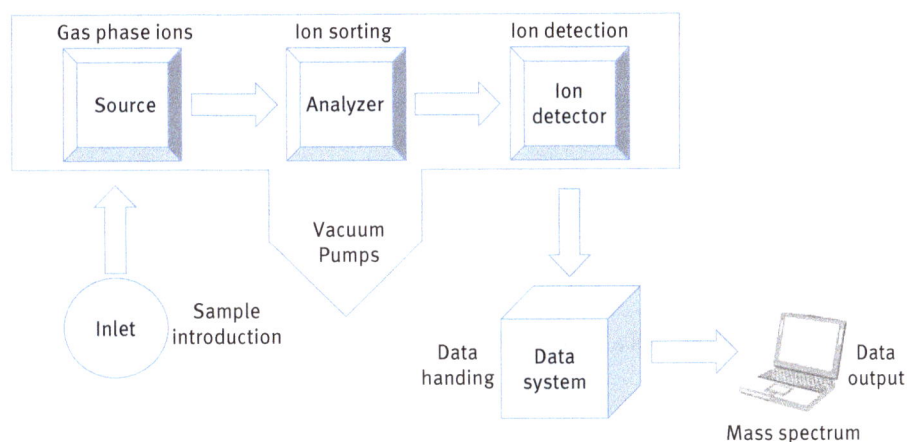

Figure 2.9: Scheme of mass spectrometer (MS) detector.

MS operates under vacuum conditions to have a collision-free path for ions. The sample inlet system is designed for minimal loss of vacuum when the sample coming from LC is introduced. From the inlet system, the sample is introduced into the ionization chamber. Ion fragments can be created in two different ways: gas phase (volatile substances and vaporized samples) or desorption techniques (samples in condensed phase inside the ionization chamber).

Gas phase: This ionization includes:
- Chemical ionization: the ionization is produced by collision of sample molecules with a gaseous reagent.
- Electroimpact ionization: the ionization is produced by an electron beam generated from a filament; ions are formed during collision of the electron beam and sample molecules.
- Field ionization: the ionization is produced because molecules can lose an electron when placed in a very high electric field created by applying high voltage.

Desorption: This ionization includes:
- Field desorption: this is useful for nonvolatile, thermally unstable compounds, and large lipophilic or polar molecules. The ionization is produced by high potential applied in an electrode.
- Fast atom bombardment: this is useful for large biological molecules difficult to get into gas phase. The ionization is produced when a high-energy beam of neutral atoms, xenon or argon, strikes a solid sample.
- Electrospray ionization (ESI): the ionization is produced at atmospheric pressure (API) and it is useful for both polar and nonpolar molecules. The sample is dispersed by an electrospray, nebulizing gas (nitrogen), into a fine aerosol. This aerosol is highly charged by high voltage and then dried by a heated gas (nitrogen).
- Matrix-assisted laser desorption ionization: sample and matrix are co-crystallized. The ionization is produced after crystallization by application of short pulses of laser in a high vacuum chamber, which causes the absorption of energy by the matrix. This energy is converted into excitation energy and used for sample ionization.

In mass analyzers, the ions produced in the ion source are separated according to their m/z. The most common analyzers are magnetic sector, quadrupole, and time of flight. Magnetic sector uses a magnetic field that causes ions to travel to a circular path. In quadrupole analyzers, the field is formed by four electrical conducting parallel rods (adjacent rods have opposite polarity), where ion circulation is dependent on applied voltage (for given voltages, only ions at certain m/z are allowed to pass through while others, such as uncharged molecules, are carried away). Time-of-flight analyzers use the differences in transit time through a drift region to separate ions of different masses previously accelerated by an electrical pulse. In addition, ion-trap analyzers use electrical fields to trap ions in a small space and to separate them by certain m/z values. Others like ion cyclotron resonance and electrostatic mass analyzers can also be included.

Tandem mass spectrometry (MS/MS) refers to the use of a second stage of analysis, that is, two mass analyzers in the same experiment. In this way, selected ions from a complex mixture can be studied. The first analyzer (Q1) is used to select the ion of interest, which is driven into a pressurized collision cell with an inert gas. In this collision cell, selected ions are dissociated after collision with the gas (collision-induced dissociation). Ions produced after dissociation, fragment ions, are analyzed in the second analyzer (Q3). In this way, product ions from a precursor ion are analyzed. This is called MS^n, where n is the number of mass analysis. An example of tandem equipment is formed by tandem quadrupoles. Originally, three tandem quadrupoles were used but nowadays devices are composed of two quadrupoles separated by the collision cell (Figure 2.10). In these devices, dissociation and fragmentation of precursor ion are produced in the collision cell where fragments are retained and then liberated to the second analysis stage.

Different experiments can be done with tandem MS detection:

– Full scan: Q1 analyzer is working in scan mode covering a specific mass range; in this way, a mass spectrum is registered for each analyte. The chromatogram obtained shows a line (total ion chromatogram) where the mass spectrum of each time interval can be observed (see Figure 2.11A).

– Selected ion monitoring (SIM): Q1 analyzer only detects selected ions. The chromatogram obtained shows peaks of selected ions.

– Multiple reaction monitoring (MRM): Q1 and Q3 working in SIM mode. Q1 analyzer selects precursor ions that are fragmented in the collision cell and Q3 monitors the corresponding product ions of each precursor ion. The chromatogram obtained shows a peak with all product ions (m/z with different intensity) for each precursor.

– Product ion scan: In Q1 analyzer, a precursor ion is selected, then this ion is fragmented in the collision cell and finally the product ions are detected in Q3 analyzer. The chromatogram obtained shows a precursor peak, and the mass spectrum of the peak shows m/z of product ions.

– Precursor ion scan: Q1 analyzer works in full scan mode and Q3 analyzer works in SIM mode. Like this, Q1 is scanning across a specific mass range. Ions in this mass range are passed into the collision cell where they are fragmented, thus giving the product ions of interest. The chromatogram obtained shows the peak of product ions, and the mass spectrum of peak shows m/z of the precursor ion.

– Neutral loss scan: Q1 and Q3 both work in full scan mode. In this way, Q1 is scanning across a specific mass range. The selected ions are passed into the collision cell where they are fragmented and Q3 is scanned over a similar mass range, offset by the neutral mass of the diagnostic fragment.

Examples of these chromatograms for TTX are shown in Figure 2.11.

Finally, ion beams, after passing through the mass analyzer, strike on the detector where ions separated according to their m/z in the mass analyzer can be electrically detected. In this way, a mass spectrum is obtained. The spectrum represents the relative abundance of ions of different m/z produced in an ion source and it contains molecular weight, and structural and quantitative information.

When a sample is analyzed, all MS parameters such as source temperature, voltage, or collision energy should be previously optimized with standard compounds in order to achieve the maximum level of sensitivity since these parameters are dependent both on equipment and toxin.

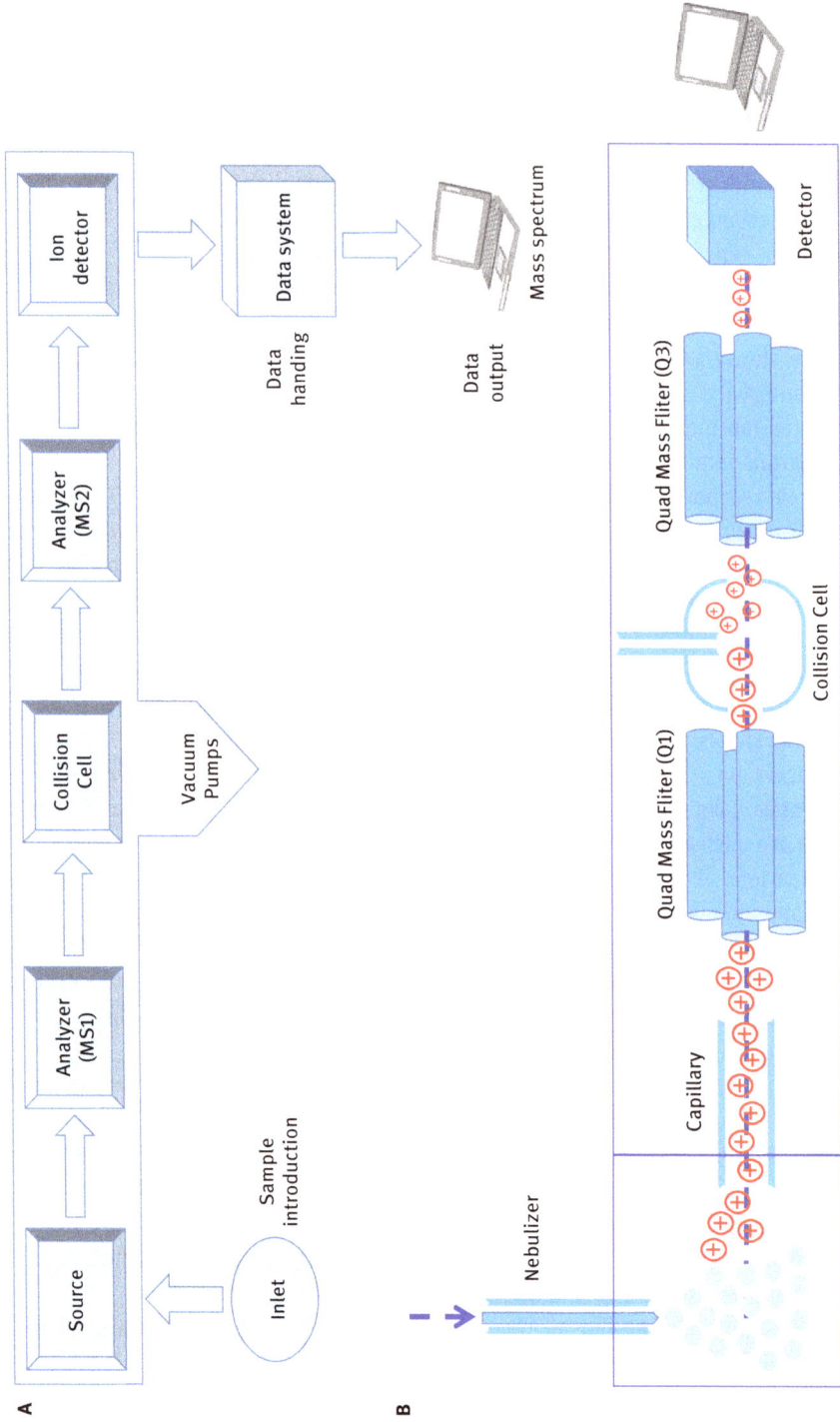

Figure 2.10: (A) Tandem mass spectrometer and (B) detail of ion fluxes and fragmentations through the mass spectrometer.

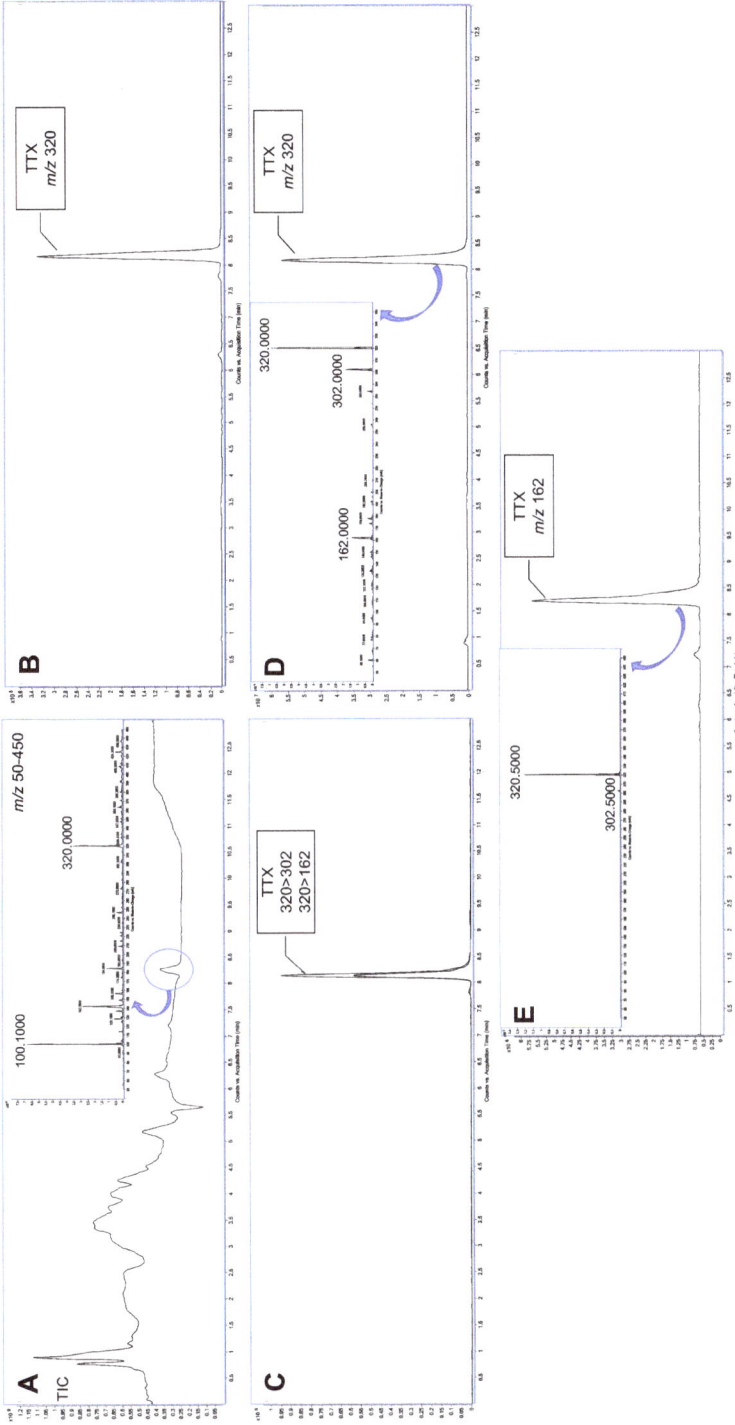

Figure 2.11: Chromatograms and mass spectrum of different experiments with TTX obtained by tandem MS: (A) full scan; (B) selected ion monitoring (SIM) mode; (C) multiple reaction monitoring (MRM) mode; (D) product ion scan; and (E) precursor ion scan mode.

MS/MS tandem detectors coupled to LC are widely used to identify and quantify all kinds of environmental toxicants. Calibration curves of signal versus concentration can be constructed (see below), and any toxin can be quantified. The calibration curve of one toxin standard is sometimes used to quantify other toxins from the same group, assuming an ionization-conversion factor 1:1. However, it is incorrect to assume that analogues from the same group provide an equimolar response by MS/MS tandem detection.

2.6 Liquid chromatography for toxin identification

The analysis of environmental toxicants is a challenge due to the wide range of structures and chemical properties of these molecules. The largest effort to detect these compounds is being done in HPLC with fluorescent detection (HPLC-FLD), UV detection (FPLC-UV), and MS detection (LC-MS) [28]. As it was discussed, these methods must be able to deal with compounds extracted from complex matrices and be capable of differentiating toxins of interest from other compounds. In addition, in all cases the use of certified toxin standards is necessary for method validation and quality control purposes. Although fluorescent and UV detection have been used for years to detect these compounds, and in some cases they are official methods, MS detection has proved to be a better technique [25, 28–30]. In all cases, after matrix extraction and cleanup, samples are injected in LC-MS/MS systems, and the chromatograms obtained are compared with standards. LC-MS detection methods can be used for monitoring the presence of toxicants in samples with official purposes in that case should be validated (see Chapter 3) or to detect and describe new compounds or unknown toxin metabolites. Several applications of UHPLC-MS to identify each group of toxins in samples are shown below as well as the use of high-resolution MS to identify new compounds is described. In MS detection, it is always important to bear in mind the matrix effect, that is, the deviation in MS signal when the toxin is solved in matrix. For this reason, matrix effect and matrix corrections should be always determined using spiked matrices with a known toxin concentration. In addition, the recovery of the extraction procedure should also be evaluated, that is, the amount of toxin extracted from a reference material with a certified amount of toxin.

2.6.1 Lipophilic toxins

LC-MS/MS is the official method to identify lipophilic toxins in shellfish for consumption according to the European Union (EU) regulations (EU 15/2011) since 2014 [31]. LC-MS/MS methodology is the reference method for identification of the following lipophilic compounds: diarrhetic shellfish poison (DSP) group (OA, DTX-1, DTX-2, and DTX-3), pectenotoxin (PTX) group (PTX-1 and PTX-2), yesotoxin (YTX) group (YTX, 45 OH YTX, homo

YTX, and 45 OH homo YTX), and azaspiracid (AZA) group (AZA-1, AZA-2, and AZA-3). In addition, spirolides (SPXs) (from the cyclic imine family) can also be detected, although this group of toxins is not regulated in Europe. In 2021, the PTX group was excluded from the group of regulated lipophilic toxins since there are no reports of human intoxications associated with this group of toxins [32]. To develop this analysis, a standard protocol currently available is harmonized between several national laboratories in Europe. Samples are first extracted with methanol, 2.00 ± 0.05 g of homogenate tissue/20 mL. The methanolic extract is filtered through a methanol-compatible 0.45-mm filter and injected in the LC-MS system to detect free toxins. To convert esterified forms of DSP group in free acids, an aliquot of extract is hydrolyzed in alkaline conditions. Toxins can be detected using acidic of basic chromatographic conditions, in both cases, employing two mobile phases in elution gradient and reversed-phase columns, C8 or C18 silica columns. For basic chromatographic conditions, mobile phases are composed of water (A) and acetonitrile–water (90:10) (B), both containing 0.05% ammonia. For acidic conditions, mobile phases are water (A) and acetonitrile–water (95:5), both containing 50 mM formic acid and 2 mM ammonium formate. Table 2.5 summarizes LC parameters for acidic conditions in a reverse-phase column and 6.5 min per injection. Gradient is done between 0 and 3 min starting in 30% of mobile phase B (organic) until 70%; then initial conditions are recovered, and the column is equilibrated for 2 min. MS/MS detection is performed in MRM mode using two transitions per toxin. The transition with higher intensity is used for quantification, while the lower transition is used for confirmatory purposes. PTX, AZA, and SPX groups are ionized in positive ion mode, while DSP and YTXs groups are ionized in negative mode (Table 2.5). MS parameters were previously optimized with toxin standards injected in mobile phases in order to achieve the maximum level of sensitivity. Figure 2.12 shows the chromatogram obtained (two peaks for toxin) when a mixture of lipophilic marine toxins solved in methanol was injected. The retention time order is a function of polarity: the higher polarity the first elution, that is, from spirolide 20 (SPX20), the most polar of lipophilic toxins and the first eluted, to AZA2, the most lipophilic (less polar) and the last eluted. A calibration curve with at least five concentrations of each toxin should be injected. In this way, a calibration curve representing the transition with higher intensity *versus* concentration is constructed to obtain a linear equation for each toxin (linear regression). When a sample is injected, the chromatogram can be compared with the standard one, and any peak (with the same transitions and retention time) can be identified as a toxin and quantified using the calibration curve [33].

Table 2.5: Main characteristics of UHPLC and MS/MS methods for detection of lipophilic marine toxins (Figure 2.12).

LC conditions	
Column	Acquity UPLC BEH C18, 100 mm × 2.1 mm, 1.7 μm particle size
Flow	0.4 mL/min
Injection volume	5 μL
Column temperature	40 °C
Mobile phase A	H_2O (2 mM ammonium formate and 50 mM formic acid)
Mobile phase B	CH_3CN H_2O (95:5) (2 mM ammonium formate and 50 mM formic acid)

	Time (min)	Mobile phase A (%)	Mobile phase B (%)
Gradient	0	70	30
	3	30	70
	4.5	30	70
	4.51	70	30
	6.5	70	30

MRM conditions						
Compound	Precursor ion *m/z*	Product ion *m/z*	CE	Fragmentor	CAV	Polarity
45-OH-homo-YTX	1,171.5	1,091.5	40	250	4	Negative
		869.5	88			
45-OH-YTX	1,157.5	1,077.5	38	240	4	Negative
		871.5	86			
Homo-YTX	1,155.5	1,075.5	40	250	4	Negative
		869.5	88			
YTX	1,141.2	1,061.3	38	240	4	Negative
		855.3	86			
PTX1	892.5	821.5	28	270	2	Positive
		213.2	44			
PTX2	876.5	823.5	28	270	2	Positive
		213.2	44			
AZA2	856.5	838.5	36	213	2	Positive
		820.5	40			
AZA1	842.5	824.5	32	206	4	Positive
		806.5	44			
AZA3	828.5	810.5	32	216	4	Positive
		792.5	44			
DTX1	817.5	255,2	52	320	7	Negative
		113.2	70			
OA/DTX2	803.5	255.2	50	320	7	Negative
		113.2	66			
SPX20G	706.5	688.3	32	233	4	Positive
		164.2	56			
SPX13	692.5	674.3	36	75	4	Positive
		164.2	60			
SPX13,19	678.5	660.3	36	135	4	Positive
		164.2	60			

CE, collision energies; CAV, cell accelerator voltage.

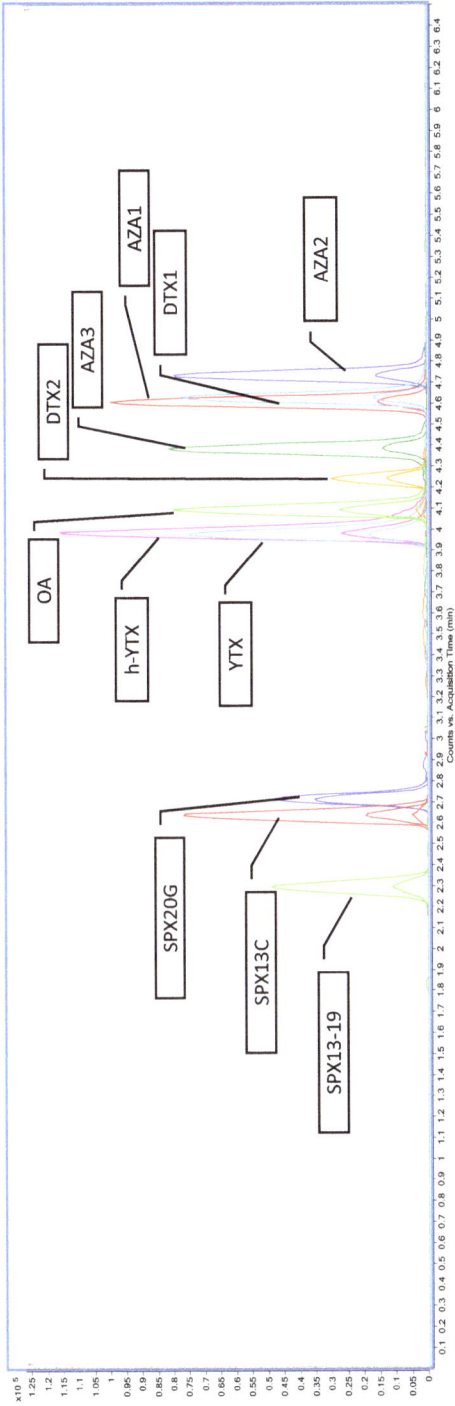

Figure 2.12: Characteristic chromatogram of lipophilic marine toxins. Chromatographic separation was carried out using a 1290 Infinity ultra-high-performance liquid chromatography system coupled to an Agilent G6460C Triple Quadrupole mass spectrometer equipped with an Agilent Jet Stream ESI source (Agilent Technologies, Waldbronn, Germany). The triple quadrupole was operated with the following optimized parameters: drying gas temperature of 350 °C and a flow of 8 L/min; nebulizer gas pressure of 45 psi; sheath gas temperature of 400 °C and a flow of 11 L/min; and the capillary voltage was set to 4,000 V in a negative mode with a nozzle voltage of 0 and 3,500 V in a positive mode with a nozzle voltage of 500 V. The collision energy (CE), cell accelerator voltage (CAV), and fragmentor were optimized using MassHunter Optimizer software.

2.6.2 Hydrosoluble marine toxins: PSPs, TTXs, DA

To officially detect and quantify the amount of PSPs and DA in shellfish for consumption, HPLC methods with fluorescent or UV detection should be used. In the case of TTXs, there is no official methods for detection although LC-MS is being used. The PSP group does not have natural UV adsorption or fluorescence. Therefore to detect these toxins, before (pre-column oxidation) or after (post-column oxidation) HPLC, this group of compounds should be converted into purine derivatives with fluorescent properties [12, 26]. This happens by the reaction in alkaline media at high temperature [34]. In this way, PSPs can be detected by fluorescence in any shellfish sample after acidic extraction. In the case of DA, HPLC with UV detection after an aqueous/methanol extraction of shellfish is the official method. To detect TTXs, several LC-MS-validated methods were developed [35–37]. To know the amount of all hydrophilic toxins in a sample, at least three different analyses with different equipment should be done. For this reason, the unification of the three groups into one method is very interesting. This is only possible by using LC-MS technology. However, the conventional reversed-phase columns used in LC are not useful to separate all compounds. In this regard, the use of normal-phase resins like ion-pair reagent for HILIC helps to solve this problem. With an amide column (normal phase) and a gradient starting in a high percentage of organic phase, the three groups can be analyzed in one injection [35]. Each toxin group should be extracted following the conditions in Table 2.2, that is, for PSP or TTX acidic conditions, 5.00 ± 0.01 g of homogenate tissue/5 mL AcOH 1%, and cleanup by SPE with carbon resins while DA is extracted with methanol:water (4.00 ± 0.01 g of homogenate tissue/ 16 mL AcOH 1% and methanol 50%). Then extracts obtained are injected in the LC-MS system. Conditions for this analysis are summarized in Table 2.6. In this case, normal-phase column, the gradient is done between 0 and 11 min starting with 95% of mobile phase B (organic) until 5% is reached, then initial conditions are recovered, and the column is equilibrated for 2 min. MS/MS detection is performed in the MRM mode using two transitions per toxin (Table 2.6). In these conditions, the three groups of toxins can be detected and quantified (Figure 2.13). In this case, DA with the lower polarity is eluted first while PSPs and TTX with higher polarity are eluted together at the end of the gradient. Like previous procedure, the calibration curve with different standard concentrations can be constructed, and any sample can be quantified.

Table 2.6: Main characteristics of UHPLC and MS/MS methods to detect hydrophilic marine toxins (Figure 2.13).

LC conditions	
Column	Acquity UPLC BEH Amide, 100 mm × 2.1 mm, 1.7 μm particle size
Flow	0.4 mL/min
Injection volume	5 μL

Table 2.6 (continued)

LC conditions

Column temperature	40 °C
Mobile phase A	H_2O (10 mM ammonium acetate and 0.1% formic acid)
Mobile phase B	98% CH_3CN + 2%H_2O (100 mM ammonium acetate and 0.1% formic acid)

	Time (min)	Mobile phase A (%)	Mobile phase B (%)
Gradient	0	5	95
	11	95	5
	12	95	5
	13	5	95
	15	5	95

MRM conditions

Toxins	Precursor ion *m/z*	Product ion *m/z*	CE	Fragmentor	CAV	Polarity
DA	312.1	74.2	24	116	4	Positive
		266	12			
C1 and C2	474	121.9	40	147	4	Negative
		351	24			
dcGTX2	351	164	32	156	2	Negative
		333	16			
dcGTX3	353	255	16	89	2	Positive
		335	2			
dcSTX	257	84.1	32	128	2	Positive
		126	20			
GTX1	410	349	20	142	2	Negative
		367	12			
GTX2	394	333	20	151	2	Negative
		351	12			
GTX3	396	110	56	108	2	Positive
		298	16			
GTX4	412	314	16	93	4	Positive
		332	12			
GTX5	378	121.9	24	147	2	Negative
		360	12			
GTX6	396.1	316.1	8	68	4	Positive
	394.2	122	20	129	1	Negative
NEO	316	110.1	52	112	2	Positive
		298.1	20			
dcNEO	273.1	110.1	44	111	2	Positive
		126	20			
STX	300	60.2	40	108	2	Positive
		204	24			
4,9-Anhydro-TTX	302	161.9	40	152	4	Positive
		256	28			
TTX	320	161.9	36	160	2	Positive
		302	24			

CE, collision energies; CAV, cell accelerator voltage.

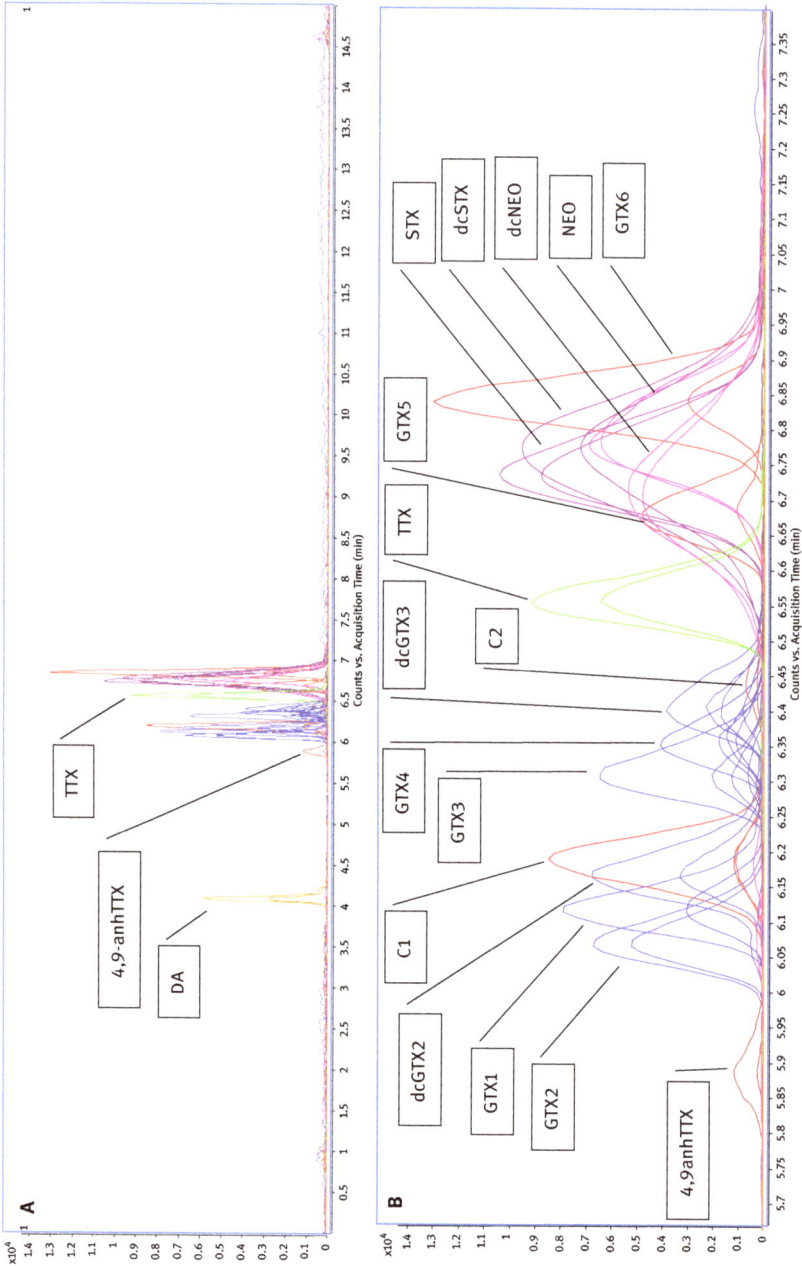

Figure 2.13: (A) Characteristic chromatogram of hydrophilic marine toxins. (B) Zoom-in of a chromatogram from 5.5 to 7.5 min. Chromatographic separation was carried out using a 1290 Infinity ultra-high-performance liquid chromatography system coupled to an Agilent G6460C Triple Quadrupole mass spectrometer equipped with an Agilent Jet Stream ESI source (Agilent Technologies, Waldbronn, Germany). The triple quadrupole was operated with the following optimized parameters: drying gas

Table 2.7: Main characteristics of UHPLC and MS/MS methods to detect hydrophilic and lipophilic toxins (Figure 2.14).

LC conditions

Column	Acquity UPLC BEH Amide, 100 mm × 2.1 mm, 1.7 µm particle size
Flow	0.4 mL/min
Injection volume	5 µL
Column temperature	40 °C
Mobile phase A	H_2O (10 mM ammonium formate and 0.1% formic acid)
Mobile phase B	98% CH_3CN + 2%H_2O (100 mM ammonium formate and 0.1% formic acid)
Mobile phase C	CH_3CN

	Time (min)	Mobile phase A (%)	Mobile phase B (%)	Mobile phase C (%)
Gradient	0	0	0	100
	1	0	0	100
	2	5	95	0
	3	5	95	0
	14	95	5	0
	16	95	5	0
	16.5	5	95	0
	17.5	5	95	0
	18	0	0	100
	19	0	0	100

MRM conditions

Toxins	Precursor ion *m/z*	Product ion *m/z*	CE	Fragmentor	CAV	Polarity
DA	312.1	74.2	24	116	4	Positive
		266	12			
C1 and C2	474	121.9	40	147	4	Negative
		351	24			
dcGTX2	351	164	32	156	2	Negative
		333	16			
dcGTX3	353	255	16	89	2	Positive
		335	2			
dcSTX	257	84.1	32	128	2	Positive
		126	20			
GTX1	410	349	20	142	2	Negative
		367	12			
GTX2	394	333	20	151	2	Negative
		351	12			
GTX3	396	110	56	108	2	Positive
		298	16			

━━━━

Figure 2.13 (continued)
temperature of 250 °C and a flow of 11 L/min; nebulizer gas pressure of 55 psi; sheath gas temperature of 400 °C and a flow of 12 L/min; and the capillary voltage was set to 3,000 V in a negative mode with a nozzle voltage of 0 and 3,000 V in a positive mode with a nozzle voltage of 0 V. The collision energy (CE), cell accelerator voltage (CAV), and fragmentor were optimized using MassHunter Optimizer software.

Table 2.7 (continued)

MRM conditions

Toxins	Precursor ion *m/z*	Product ion *m/z*	CE	Fragmentor	CAV	Polarity
GTX4	412	314	16	93	4	Positive
		332	12			
GTX5	378	121.9	24	147	2	Negative
		360	12			
GTX6	396.1	316.1	8	68	4	Positive
	394.2	122	20	129	1	Negative
NEO	316	110.1	52	112	2	Positive
		298.1	20			
dcNEO	273.1	110.1	44	111	2	Positive
		126	20			
STX	300	60.2	40	108	2	Positive
		204	24			
4,9-Anhydro-TTX	302	161.9	40	152	4	Positive
		256	28			
TTX	320	161.9	36	160	2	Positive
		302	24			
AZA1	842.5	806.5	44	206	4	Positive
		824.5	32			
AZA2	856.5	820.5	44	213	2	Positive
		838.5	36			
AZA3	828.5	792.5	44	216	4	Positive
		810.5	32			
45-OH-homoYTX	1171.5	869.5	88	250	4	Negative
		1091.5	40			
45-OH-YTX	1157.5	871.5	86	240	4	Negative
		1077.5	38			
HomoYTX	1155.5	869.4	88	250	4	Negative
		1075.5	40			
YTX	1141.5	855.4	86	240	4	Negative
		1061.5	38			
DTX1	817.5	113	70	350	7	Negative
		255.1	54			
OA DTX2	803.5	113.2	66	350	7	Negative
		255.1	50			
PTX1	892.5	213.2	44	175	2	Positive
		821.5	28			
PTX2	876.5	213.2	44	175	2	Positive
		823.5	28			
SPX13	692.5	164.2	60	75	4	Positive
		674.3	36			
SPX13,19	678.5	164.2	60	250	4	Positive
		660.2	36			
SPX20G	706.5	164.2	56	233	4	Positive
		688.2	32			

CE, collision energies; CAV, cell accelerator voltage.

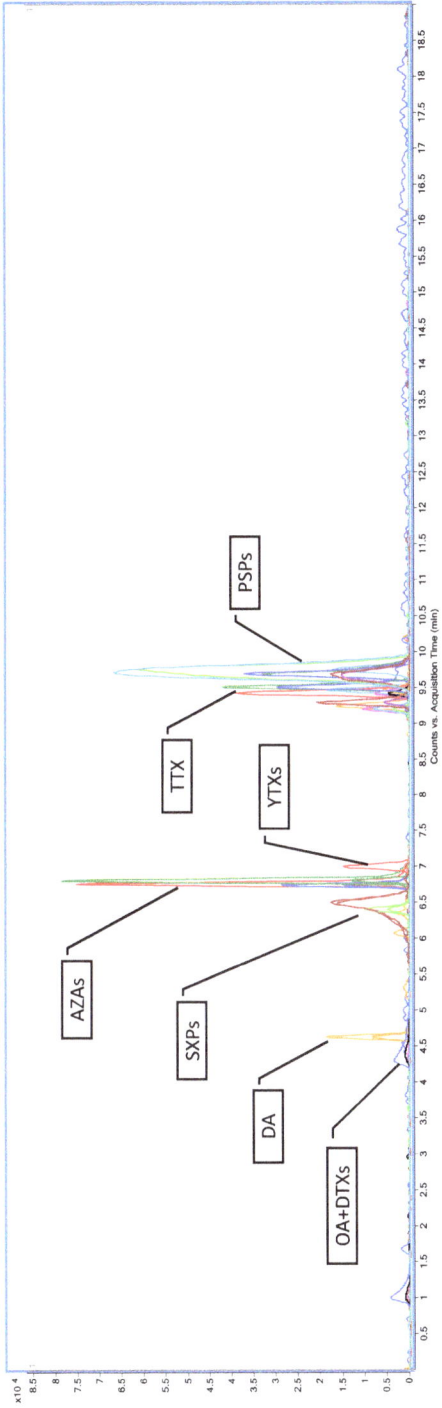

Figure 2.14: (A) Characteristic chromatogram of lipophilic and hydrophilic marine toxins. Chromatographic separation was carried out using a 1290 Infinity ultra-high-performance liquid chromatography system coupled to an Agilent G6460C Triple Quadrupole mass spectrometer equipped with an Agilent Jet Stream ESI source (Agilent Technologies, Waldbronn, Germany). The triple quadrupole was operated with the optimized parameters shown in Figure 2.13.

In addition, a method to detect all UE-regulated toxins by using UHPLC-MS/MS has been developed [35]. To separate lipophilic, PSPs, DA, and TTXs, a normal-phase column, three mobile phases, and a quaternary pump are necessary. Table 2.7 summarizes LC and gradient conditions for a 19-min separation in a normal-phase column. For the chromatographic separation, an isocratic period of 1 min with mobile phase C (100% acetonitrile) is followed by a 1-min gradient from 100% mobile phase C to 95% mobile phase B (98% acetonitrile) and 5% A (water). Then, a gradient between 3 and 14 min starting with 95% of mobile phase B until 5% is reached, and finally the initial conditions are recovered, and the column is equilibrated for 1 min. In these conditions, the three groups of regulated toxins and TTXs can be detected and quantified (Figure 2.14). Lipophilic toxins and DA are eluted first (organic phases) and then TTXs and PSPs. This method is an excellent tool to detect all toxins with the same equipment, and it is useful for routine test, although DTX2 and OA have the same retention time and transitions, and therefore cannot be separated.

2.6.3 Cyanotoxins

Several LC-MS/MS methods had been developed to quantify cyanotoxins; however, a long time, 70 min, was necessary to separate all toxins [38]. By using UHPLC, the time of analysis is notoriously reduced. In this case, instead of extraction from matrix, the sample should be in some way concentrated because toxins are solved in water [39]. Figure 2.15 shows a chromatogram of a mixture of six cyanotoxins eluted in LC and MS/MS conditions summarized in Table 2.8. To separate these toxins, a reverse-phase column is used with a gradient starting in 0% of mobile phase B at minute 4 until 70% at minute 8, then initial conditions are recovered, and the column is equilibrated for 2 min. The analysis is improved by adding an isocratic period (0–4 min) with 0% of organic phase [40]. With this period, phenylalanine can be easily separated from ATX. These two compounds have the same molecular weight and transitions and therefore can produce misidentifications. MS/MS detection is performed in the MRM mode in positive mode using two transitions (three for ATX-a) per toxin (Table 2.8). As it can be expected, the most polar is the first eluted, that is, cylindrospermopsins and ATXs are eluted first and then MCs and nodularins.

Table 2.8: Main characteristics of UHPLC and MS/MS methods to detect cyanotoxins (Figure 2.15).

LC conditions

Column	Acquity UPLC HSS T3, 100 mm × 2.1 mm, 1.8 μm particle size
Flow	0.4 mL/min
Injection volume	5 μL
Column temperature	35 °C
Mobile phase A	H_2O (0.05% formic acid)
Mobile phase B	CH_3CN (0.05% formic acid)

Gradient	Time (min)	Mobile phase A (%)	Mobile phase B (%)
	0	100	0
	4	100	0
	8	30	70
	10	30	70
	10.5	100	0
	13	100	0

MRM conditions

Compound	Precursor Ion *m/z*	Product Ion *m/z*	CE	Fragmentor	CAV	Polarity
Epoxy-homoATX-a	196	138	16	93	1	Positive
		178	12			
H2-homoATX-a	182	147	16	93	1	Positive
		164	12			
Epoxy-ATX-a	182	138	16	93	1	Positive
		164	12			
homoATX-a	180	145	16	93	1	Positive
		163	12			
H2-ATX-a	168	133	16	93	1	Positive
		150	12			
ATX-a	166,13	43,2	24	93	1	Positive
		131	16			
		149	12			
CYN	416,13	194	40	146	1	Positive
		336,1	20			
MC-YR	1046	103	120	151	1	Positive
		135	80			
MC-FR	1029	103	120	151	1	Positive
		135	80			
MC-LW	1026	103	120	151	1	Positive
		135	80			
MC-YM	1020	103	120	151	1	Positive
		135	80			
MC-LY	1003	103	120	151	1	Positive
		135	80			

Table 2.8 (continued)

MRM conditions						
Compound	**Precursor Ion *m/z***	**Product Ion *m/z***	**CE**	**Fragmentor**	**CAV**	**Polarity**
MC-LR	995,56	135	76	185	1	Positive
	498,28	861,4	8	105		
MC-LF	987	103	120	151	1	Positive
		135	80			
MC-AW	983	103	80	151	1	Positive
		135	120			
dmMC-LR	981,54	103	120	285	1	Positive
		135	80	215		
MC-VF	973	103	120	151	1	Positive
		135	80			
MC-YA	960	103	120	151	1	Positive
		135	80			
MC-AR	953	103	120	151	1	Positive
		135	80			
MC-LL	952	103	120	151	1	Positive
		135	80			
MC-LA	911	103	120	151	1	Positive
		135	80			
MC-RR	520,11	103,1	68	151	1	Positive
		135	32			

CE: collision energies; CAV: cell accelerator voltage.

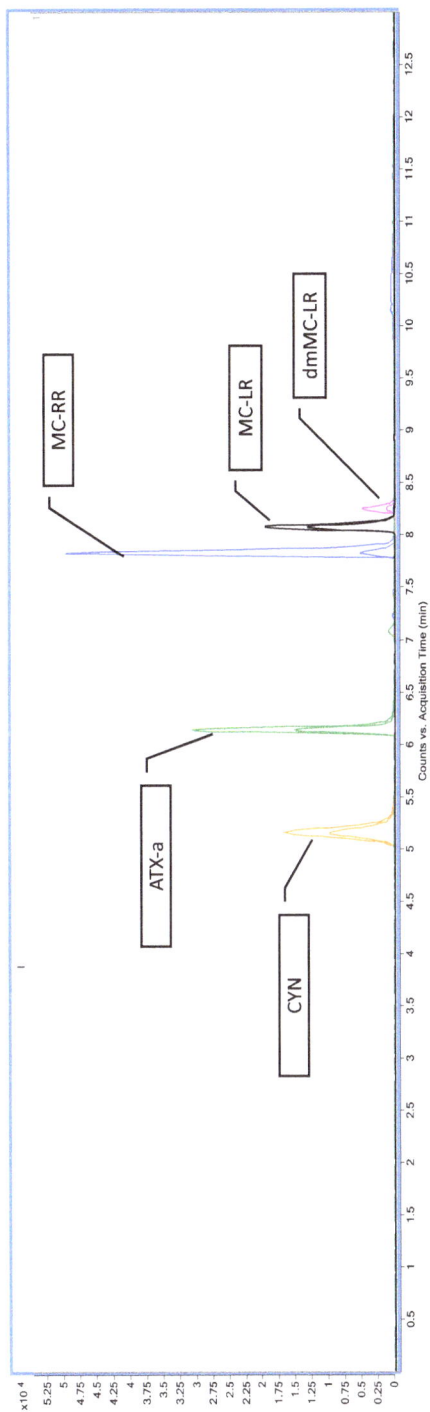

Figure 2.15: Characteristic chromatogram of a mixture of six cyanotoxins. Chromatographic separation was carried out using a 1290 Infinity ultra-high-performance liquid chromatography system coupled to an Agilent G6460C Triple Quadrupole mass spectrometer equipped with an Agilent Jet Stream ESI source (Agilent Technologies, Waldbronn, Germany). The triple quadrupole was operated with the following optimized parameters: drying gas temperature of 350 °C and a flow of 5 L/min; nebulizer gas pressure of 30 psi; sheath gas temperature of 400 °C and a flow of 12 L/min; and the capillary voltage was set to 3,500 V in a positive mode with a nozzle voltage of 1,500 V. The collision energy (CE), cell accelerator voltage (CAV), and fragmentor were optimized using MassHunter Optimizer software.

2.6.4 Mycotoxins

Mycotoxins are accumulated in solid matrices like cereals, figs, nuts, grapes, and coffee beans or liquid matrices like wine or bear. After toxin extraction from these matrices, additional cleanup processes developed for pesticide analysis, such as QuEChERS methodology (from quick, easy, cheap, effective, rugged, and safe) and dispersive liquid–liquid extractions should be done before analysis to avoid interferences. UHPLC-MS/MS is the technique more used, although HPLC-UV or HPLC-FLD have been traditionally used for mycotoxin analysis [41]. UHPLC-MS/MS allows the simultaneous detection of many mycotoxins with good sensibility and specificity. Table 2.9 summarizes LC and MS/MS conditions using a reverse-phase column for a fast mycotoxin separation (13 min). The gradient was done in several steps starting in 0% of organic phase (methanol) up to 100%. Acidic mobile phase allows the simultaneous detection of fumonisins and trichothecenes. MS is used in dynamic MRM mode, selecting transitions during a specified period of time, hence, providing higher sensitivity and reproducibility due to the dwell times for each transition that are maximized [42]. In this way, as shown in Figure 2.16, 22 mycotoxins can be easily identified in 13 min in several solid and liquid matrices [43–46].

Table 2.9: Main characteristics of UHPLC and MS/MS methods to detect mycotoxins (Figure 2.16).

LC conditions			
Column	Waters ACQUITY HSS T3, 100 mm × 2.1 mm, 1.8 μm particle size		
Flow	0.3 mL/min		
Injection volume	5 μL		
Column temperature	40 °C		
Mobile phase A	H_2O (5 mM ammonium formate and 0.1% formic acid)		
Mobile phase B	Methanol		

	Time (min)	Mobile phase A (%)	Mobile phase B (%)
Gradient	0	100	0
	0.5	86	14
	2	86	14
	3	40	60
	3.5	40	60
	6.5	0	100
	10	0	100
	10.5	100	0
	13	100	0

MRM conditions						
Compound	Precursor ion	Product ion	CE	Fragmentor	RT	Polarity
Aflatoxin B$_1$	313.07	285	24	142	6.45	Positive
		241	44			

Table 2.9 (continued)

MRM conditions

Compound	Precursor ion	Product ion	CE	Fragmentor	RT	Polarity
Aflatoxin B$_2$	315.07	287	28	147	6.25	Positive
		259	32			
Aflatoxin G$_1$	329.07	243	28	132	6.05	Positive
		200	48			
Aflatoxin G$_2$	331.1	245	32	75	5.9	Positive
		217	40			
Beauvericin	806.4	402.1	56	280	10.05	Positive
		302.1	68			
Deoxynivalenol	297.15	249	8	74	5	Positive
		203	12			
3 + 15-Acetyl-deoxynivalenol	339.15	137	8	65	5.72	Positive
		261	8			
Deoxynivalenol 3-glucoside	503.18	457.1	12	125	4.9	Negative
		427.0	20			
Enniatin A	704.4	557.2	52	255	10.4	Positive
		210	64			
Enniatin A$_1$	690.4	520.1	60	280	10.25	Positive
		232.1	64			
Enniatin B	662.4	467.2	52	240	9.95	Positive
		449.1	48			
Enniatin B$_1$	676.4	168.1	70	260	10.1	Positive
		236	66			
Fumonisin B$_1$	722.4	352.2	40	170	7.08	Positive
		334.3	40			
Fumonisin B$_2$	706.5	336.2	40	165	8	Positive
		318.2	48			
HT-2 toxin	447.1	345	16	108	7.14	Positive
		285	20			
Moliniformin	96.99	96.99	0	51	1.1	Negative
		41.1	12			
Neosolaniol	400.19	215	16	74	5.38	Positive
		169	28			
Ochratoxin A	404.09	239	22	84	8.0	Positive
		102	75			
T-2 toxin	484.25	245	12	84	7.64	Positive
		197.1	16			
Zearalenone	317.14	273	20	123	8.08	Negative
		131	28			
α-Zearalenol	319.1	275.1	20	155	7.9	Negative
		130	36			
β-Zearalenol	319.1	275.1	20	155	7.4	Negative
		130	36			

CE, collision energies; CAV, cell accelerator voltage.

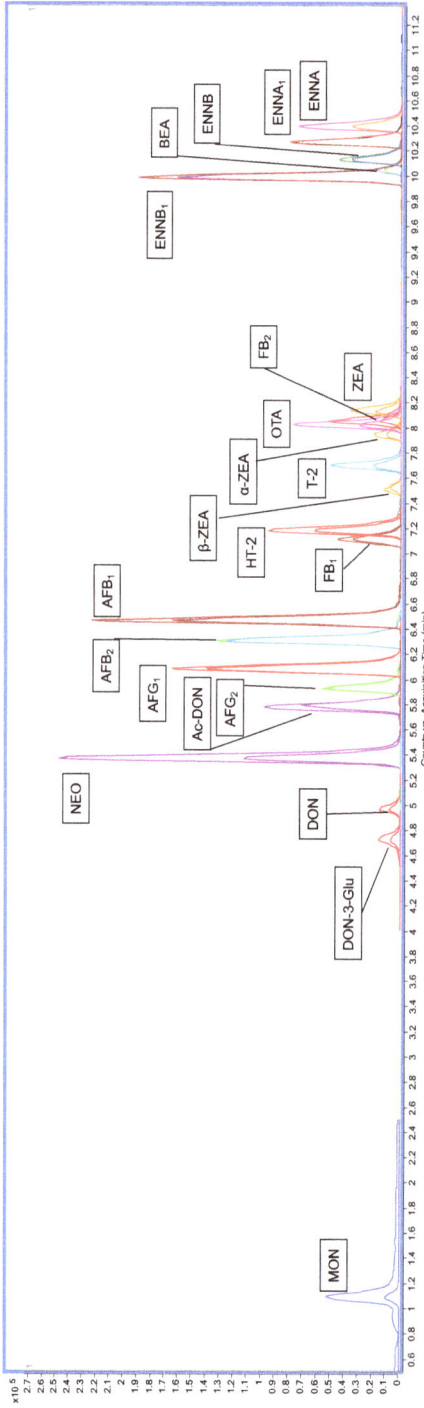

Figure 2.16: Characteristic chromatogram of a mixture of mycotoxins: deoxynivalenol (DON), deoxynivalenol 3-glucoside (DON-3-Glu), 3 + 15-acetyl-deoxynivalenol (Ac-DON), neosolaniol (NEO), aflatoxin G_2 (AFG$_2$), aflatoxin G_1 (AFG$_1$), aflatoxin B_2(AFB$_2$), aflatoxin B_1(AFB$_1$), fumonisin B_1(FB$_1$), HT-2 toxin (HT-2), T-2 toxin (T-2), fumonisin B_2 (FB$_2$), ochratoxin A(OTA), enniatin B_1 (ENB$_1$), beauvericin (BEA), enniatin B (ENB), enniatin A_1 (ENA$_1$) and enniatin A (ENA), moniliformin (MON), zearalenone (ZEA), α-zearalenol (α-ZEA), and β-zearalenol (β-ZEA). Chromatographic separation was carried out by using a 1290 Infinity ultra-high-performance liquid chromatography system coupled to an Agilent G6460C Triple Quadrupole mass spectrometer equipped with an Agilent Jet Stream ESI source (Agilent Technologies, Waldbronn, Germany). The triple quadrupole was operated with the following optimized parameters: sheath gas temperature, 400 °C; sheath gas flow, 12 L/min; gas temperature, 350 °C; gas flow, 8 L/min; nebulizer, 45 psi; capillary voltage, 4,000 V; and nozzle voltage, 0 V. Cell voltage accelerator (CAV) employed for all analytes was 2 V. The collision energy (CE) and fragmentor were optimized using MassHunter Optimizer software.

2.6.5 Ciguatoxins

The presence of CTX in fish is not regularly monitored due to the high variability of toxins, the variability of fish species, and due to large inter-species variations [29]. In addition, the lack of standards hinders correct identifications. In any case, UHPLC-MS is a methodology regularly used for these toxins. The extraction from fish matrices is carried out with acetone or methanol followed by several SPEs and liquid–liquid extractions to avoid matrix interferences and then LC-MS analysis [16, 18, 20]. Table 2.10 summarizes LC and MS conditions to separate and identify several CTXs and other analogues like gambierol and gambierone. To separate these toxins, a reverse-phase column is used with a 4.5-min gradient starting in 50% of mobile phase B at minute 2.5 until 100% at minute 7. Then the column is stabilized for 5 min. MS identification is done in two steps. First, MS is used in SIM mode, positive mode, selecting m/z of compounds described in bibliopraphic references. Then, each peak is confirmed using scan mode and checking the typical fragmentation pathway of CTX, that is, ammonium, sodium, and potassium adducts and several water losses. In this way, as shown in Figure 2.17, CTX can be identified even when no standards are available. In this case, the calibration curve of one toxin standard will be used to quantify other toxins, assuming ionization-conversion factor 1:1, which is not exact.

Table 2.10: Main characteristics of UPLC and MS/MS methods for detection of Ciguatoxins (Figure 2.17).

LC conditions			
Column	Acquity UPLC BEH C18, 100 mm × 2.1 mm, 1.7 µm particle size		
Flow	0.4 mL/min		
Injection volume	5 µL		
Column temperature	35 °C		
Mobile phase A	H_2O (2 mM ammonium formate and 50 mM formic acid)		
Mobile phase B	CH_3CN H_2O (95:5) (2 mM ammonium formate and 50 mM formic acid)		

	Time (min)	Mobile phase A (%)	Mobile phase B (%)
Gradient	0	50	50
	2.5	50	50
	7	0	100
	11.5	0	100
	11.6	50	50
	14	50	50

SIM conditions				
Compound	Precursor ion m/z	Fragmentor	CAV	Polarity
I-CTX3 and I-CTX4	1,157.6	240	7	Positive
CTX-1143	1,143.6	240	7	Positive
Caribbean-CTX	1,141.7	240	7	Positive

Table 2.10 (continued)

SIM conditions

Compound	Precursor ion *m/z*	Fragmentor	CAV	Polarity
CTX-1127	1,127.6	240	7	Positive
CTX-1123	1,123.6	240	7	Positive
CTX1B	1,111.6	240	7	Positive
CTX-1109	1,109.5	240	7	Positive
54-Deoxy-CTX/1B and 52-epi-CTX1B	1,095.6	240	7	Positive
M-CTX-4A/4B	1,079.6	240	7	Positive
CTX-4A/4B	1,061.6	240	7	Positive
2,3-OH-CTX3C and M-CTX3C	1,055.6	240	7	Positive
2-OH-CTX3C and M-CTX3C	1,041.6	240	7	Positive
CTX-1,040	1,040.6	240	7	Positive
51-OH-CTX3C	1,039.6	240	7	Positive
Gambierone	1,025.5	180	7	Positive
49-Epo-CTX3C and CTX3C	1,023.6	240	7	Positive
Gambierol	757.9	240	7	Positive
CTX-1159	1,159.6	240	7	Positive

CAV, cell accelerator voltage.

2.6.6 Identification of unknown and novel toxins

MS/MS analysis is based on a target methodology of finding predetermined compounds while missing other nonpredetermined molecules that could be present in a sample. When no standards are available, but the compounds to be identified are described in bibliography and the fragmentation pathway is known, they can be as it was described for CTXs. However, when unknown compounds should be identified, this cannot be applied. The term SWATH from "sequential windowed acquisition of all theoretical fragment ion mass spectra" is sometimes applied for the total search of compounds in a sample, eliminating the targeted nature of MS/MS analysis. However, this is not easy when the number of compounds to look for is high, with wide molecular weight range and complex structures, in addition to the variability of many different matrices [47]. Therefore, the use of a high-resolution MS (time-of-flight or Orbitrap MS) can be a solution to identify unknown compounds. With this technology, the accurate mass data as well as the fragmentation pathway can be obtained, and the formula of any new analogue can be predicted [22, 48–51].

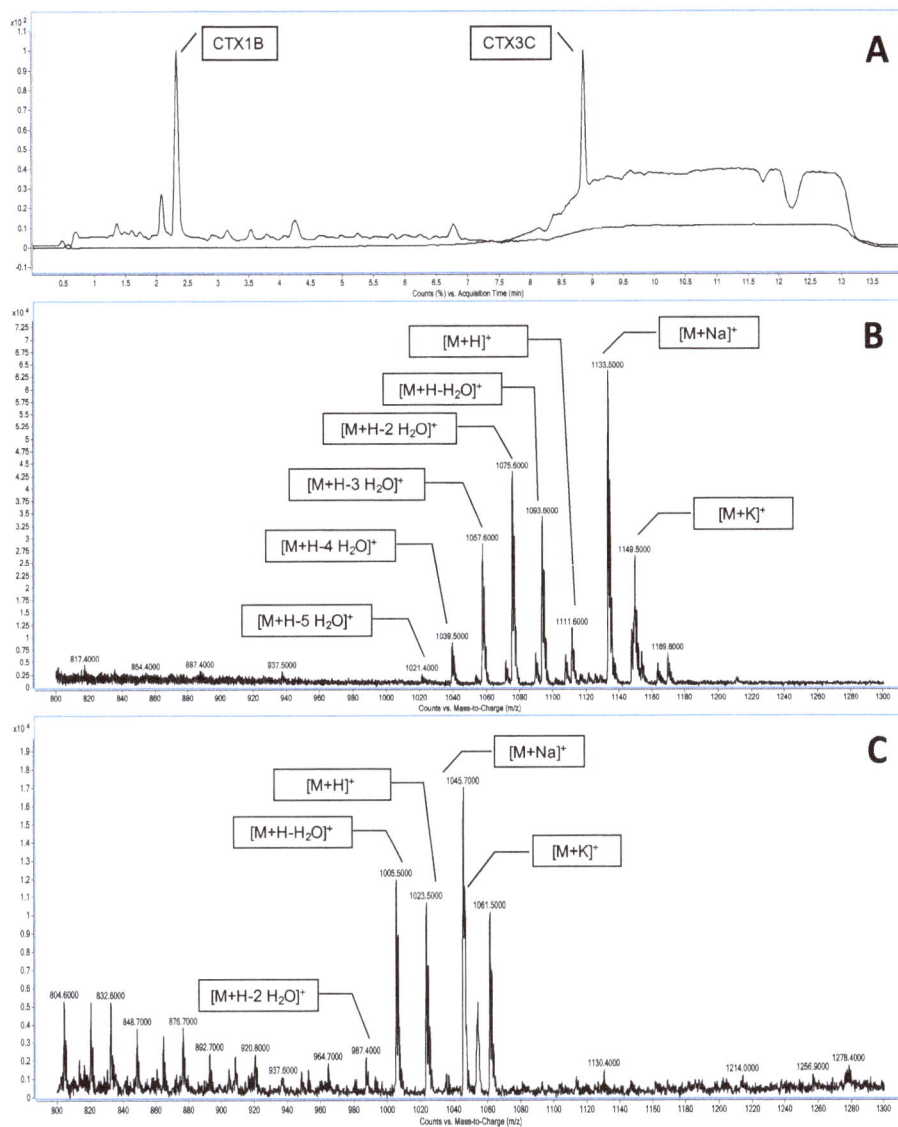

Figure 2.17: (A) Characteristic chromatogram of a mixture of two ciguatoxins obtained in SIM mode. (B) Mass spectrum of CTX1B peak obtained in scan mode. (C) Mass spectrum of CTX3C peak obtained in scan mode. Chromatographic separation was carried out using a 1290 Infinity ultra-high-performance liquid chromatography system coupled to an Agilent G6460C Triple Quadrupole mass spectrometer equipped with an Agilent Jet Stream ESI source (Agilent Technologies, Waldbronn, Germany). The triple quadrupole was operated with the following optimized parameters: drying gas temperature of 350 °C and a flow of 8 L/min; nebulizer gas pressure of 45 psi; sheath gas temperature of 400 °C and a flow of 11 L/min; and the capillary voltage was set to 4,000 V in a positive mode with a nozzle voltage of 0 V. The collision energy (CE), cell accelerator voltage (CAV), and fragmentor were optimized using MassHunter Optimizer software.

References

[1] Commission Regulation (EC) No 401/2006 of 23 February 2006 laying down the methods of sampling and analysis for the official control of the levels of mycotoxins in foodstuffs, 2006.

[2] Commission Regulation (EU) No 519/2014 of 16 May 2014 amending Regulation (EC) No 401/2006 as regards methods of sampling of large lots, spices and food supplements, performance criteria for T-2, HT-2 toxin and citrinin and screening methods of analysis, 2014.

[3] Sainz MJ, Alfonso A, Botana LM. Considerations about international mycotoxin legislation, food security, an climate change. In: Botana LM, Sainz MJ, eds. Climate change and mycotoxins. Berlin, Germany: De Gruyter, 2015;153–73.

[4] Biotoxins EURLfM. EU-Harmonised Standard Operating Procedure for determination of Lipophilic marine biotoxins in molluscs by LC-MS/MS, 2015;1–20.

[5] AOAC. Official method 959.08. Paralytic shellfish poison. Biological method. Official methods of analysis of AOAC International. 18th ed. Gaithersburg, MD, USA: AOAC International, 2005. http://www.eoma.aoac.org/methods/info.asp?ID=28589.

[6] AOAC. Official method 2005.06. Quantitative determination of Paralytic shellfish poisoning toxins in shellfish using pre-chromatographic oxidation and liquid chromatography with fluorescence detection. Official methods of analysis of AOAC International. 18th ed. Gaithersburg, MD, USA: AOAC International, 2006. http://www.eoma.aoac.org/methods/info.asp?ID=48717.

[7] Biotoxins EURLfM. EU-Harmonised Standard Operating Procedure for determination of Domoic Acid in Shellfish and Finfish by RP-HPLC using UV detection, 2008;1–20.

[8] AOAC. Official method 991.26. Domoic acid in mussels, liquid chromatography method. Official methods of analysis of AOAC International. Gaithersburg, MD, USA: AOAC International, 2000. http://www.eoma.aoac.org/methods/info.asp?ID=28606.

[9] Pereira VL, Fernandes JO, Cunha SC. Mycotoxins in cereals and related foodstuffs: A review on occurrence and recent methods of analysis. Trends Food Sci Technol. 2014;36:96–136.

[10] Rodríguez P, Alfonso A, Turrell E, Lacaze J-P, Botana LM. Study of solid phase adsorption of paralytic shellfish poisoning toxins (PSP) onto different resins. Harmful Algae. 2011;10:447–55.

[11] Merel S, Walker D, Chicana R, Snyder S, Baures E, Thomas O. State of knowledge and concerns on cyanobacterial blooms and cyanotoxins. Environ Int. 2013;59:303–27.

[12] Lawrence JF, Niedzwiadek B, Menard C. Quantitative determination of paralytic shellfish poisoning toxins in shellfish using prechromatographic oxidation and liquid chromatography with fluorescence detection: Collaborative study. J AOAC Int. 2005;88(6):1714–32.

[13] Rey V, Botana AM, Botana LM. Quantification of PSP toxins in toxic shellfish matrices using post-column oxidation liquid chromatography and pre-column oxidation liquid chromatography methods suggests post-column oxidation liquid chromatography as a good monitoring method of choice. Toxicon. 2017;129:28–35.

[14] Turner AD, McNabb PS, Harwood DT, Selwood AI, Boundy MJ. Single-laboratory validation of a multitoxin ultra-performance LC-hydrophilic interaction LC-MS/MS method for quantitation of paralytic shellfish toxins in bivalve shellfish. J AOAC Int. 2015;98(3):609–21.

[15] Lewis RJ, Yang A, Jones A. Rapid extraction combined with LC-tandem mass spectrometry (CREM-LC/MS/MS) for the determination of ciguatoxins in ciguateric fish flesh. Toxicon. 2009;54(1):62–66.

[16] Otero P, Perez S, Alfonso A, Vale C, Rodriguez P, Gouveia NN, Gouveia N, Delgado J, Vale P, Hirama M, Ishihara Y, Molgo J, Botana LM. First toxin profile of ciguateric fish in Madeira Arquipelago (Europe). Anal Chem. 2010;82(14):6032–39.

[17] Rodríguez P, Alfonso A, Otero P, Katikou P, Georgantelis D, Botana LM. Liquid chromatography–mass spectrometry method to detect Tetrodotoxin and its analogues in the puffer fish Lagocephalus sceleratus (Gmelin, 1789) from European waters. Food Chem. 2012;132:1103–11.

[18]   Yogi K, Oshiro N, Inafuku Y, Hirama M, Yasumoto T. Detailed LC-MS/MS analysis of ciguatoxins revealing distinct regional and species characteristics in fish and causative alga from the Pacific. Anal Chem. 2011;83(23):8886–91.

[19]   Buszewski B, Noga S. Hydrophilic interaction liquid chromatography (HILIC) – A powerful separation technique. Anal Bioanal Chem. 2012;402(1):231–47.

[20]   Silva M, Rodriguez I, Barreiro A, Kaufmann M, Isabel Neto A, Hassouani M, Sabour B, Alfonso A, Botana LM, Vasconcelos V. First report of ciguatoxins in two starfish species: Ophidiaster ophidianus and Marthasterias glacialis. Toxins. 2015;7(9):3740–57.

[21]   Rodriguez I, Alfonso A, Antelo A, Alvarez M, Botana LM. Evaluation of the impact of mild steaming and heat treatment on the concentration of okadaic acid, dinophysistoxin-2 and dinophysistoxin-3 in mussels. Toxins. 2016;8(6):1–12.

[22]   Rodriguez I, Alfonso A, Alonso E, Rubiolo JA, Roel M, Vlamis A, Katikou P, Jackson SA, Menon ML, Dobson A, Botana LM. The association of bacterial C9-based TTX-like compounds with Prorocentrum minimum opens new uncertainties about shellfish seafood safety. Sci Rep. 2017;7:40880.

[23]   Swartz M. HPLC detectors: A brief review. J Liquid Chromatogr Related Technol. 2010;33:1130–50. Taylor & Francis Group, LLC.

[24]   Riobo P, Franco JM. Palytoxins: Biological and chemical determination. Toxicon. 2011;57(3):368–75.

[25]   Rodriguez I, González JM, Botana AM, Sainz MJ, Rodriguez MR, Alfonso A, Botana LM. Analysis of Natural Toxins by Liquid Chromatography. In: Fanali S, Haddad PR, Poole CF, Riekkola ML, eds. Liquid chromatography applications. 2nd ed. Elsevier, 2017;479–514.

[26]   Oshima Y. Postcolumn derivatization liquid chromatographic method for paralytic shellfish toxins. J AOAC Int. 1995;78(2):528–32.

[27]   Moreira C, Ramos V, Azevedo J, Vasconcelos V. Methods to detect cyanobacteria and their toxins in the environment. Appl Microbiol Biotechnol. 2014;98(19):8073–82.

[28]   Botana LM. Guide to phycotoxins monitoring of bivalve mollusk-harvesting areas. In: Botana LM, ed. Seafood and freshwater toxins: Pharmacology, physiology and detection. 3rd ed. Boca Ratón: CRC Press, 2014;39–58.

[29]   Gerssen A. Analysis of marine biotoxins by liquid chromatography mass spectrometry detection. In: Botana LM, ed. Seafood and freshwater toxins: Pharmacology, physiology and detection. 3rd ed. Boca Ratón: CRC Press, 2014;409–27.

[30]   Rodriguez I, Vieytes MR, Alfonso A. Analytical challenges for regulated marine toxins. Detection methods. Curr Opin Food Sci. 2017;18:29–36.

[31]   15/2011 CRE. Commission Regulation (EU) 15/2011 of 10 January 2011 amending Regulation (EC) 2074/2005 as regards recognised testing methods for detecting marine biotoxins in live bivalve molluscs. Official Journal of the European Union. 2011;L6/3–L6/.

[32]   Commission delegated regulation (EU) 2021/1374 of 12 April 2021 amending Annex III to Regulation (EC) No 853/2004 of the European Parliament and of the Council on specific hygiene requirements for food of animal origin, 2021.

[33]   Silva M, Rodriguez I, Barreiro A, Kaufmann M, Neto AI, Hassouani M, Sabour B, Alfonso A, Botana LM, Vasconcelos V. Lipophilic toxins occurrence in non-traditional invertebrate vectors from North Atlantic Waters (Azores, Madeira, and Morocco): Update on geographical tendencies and new challenges for monitoring routines. Mar Pollut Bull. 2020;161(Pt B):111725.

[34]   Vale P. Saxitoxin and analogs: Ecobiology, origin, chemistry and detection. In: Botana LM, ed. Seafood and freshwater toxins: Pharmacology, physiology and detection. 3rd ed. Boca Ratón: CRC Press, 2014;991–1011.

[35]   Rodriguez I, Alfonso A, Gonzalez-Jartin JM, Vieytes MR, Botana LM. A single run UPLC-MS/MS method for detection of all EU-regulated marine toxins. Talanta. 2018;189:622–28.

[36]   Turner AD, Dhanji-Rapkova M, Fong SYT, Hungerford J, McNabb PS, Boundy MJ, Harwood DT, Collaborators. Ultrahigh-performance hydrophilic interaction liquid chromatography with tandem

mass spectrometry method for the determination of paralytic shellfish toxins and tetrodotoxin in mussels, oysters, clams, cockles, and scallops: collaborative study. J AOAC Int. 2020;103(2):533–62.

[37] Turner AD, Dean KJ, Dhanji-Rapkova M, Dall'Ara S, Pino F, McVey C, Haughey S, Logan N, Elliott C, Gago-Martinez A, Leao JM, Giraldez J, Gibbs R, Thomas K, Perez-Calderon R, Faulkner D, McEneny H, Savar V, Reveillon D, Hess P, Arevalo F, Lamas JP, Cagide E, Alvarez M, Antelo A, Klijnstra MD, Oplatowska-Stachowiak M, Kleintjens T, Sajic N, Boundy MJ, Maskrey BH, Harwood DT, Gonzalez Jartin JM, Alfonso A, Botana L. Interlaboratory evaluation of multiple LC-MS/MS methods and a commercial ELISA method for determination of tetrodotoxin in oysters and mussels. J AOAC Int. 2023;106(2):356–69.

[38] Yen HK, Lin TF, Liao PC. Simultaneous detection of nine cyanotoxins in drinking water using dual solid-phase extraction and liquid chromatography-mass spectrometry. Toxicon. 2011;58(2):209–18.

[39] Rodriguez I, Fraga M, Alfonso A, Guillebault D, Medlin L, Baudart J, Jacob P, Helmi K, Meyer T, Breitenbach U, Holden NM, Boots B, Spurio R, Cimarelli L, Mancini L, Marcheggiani S, Albay M, Akcaalan R, Koker L, Botana LM. Monitoring of freshwater toxins in European environmental waters by using novel multi-detection methods. Environ Toxicol Chem. 2017;36(3):645–54.

[40] Rodriguez I, Alfonso C, Alfonso A, Otero P, Meyer T, Breitenbach U, Botana LM. Toxin profile in samples collected in fresh and brackish water in Germany. Toxicon. 2014;91(0):35–44.

[41] Rahmani A, Jinap S, Soleimany F. Qualitative and quantitative analysis of mycotoxins. Compr Rev Food Sci Food Saf. 2009;8:202–51.

[42] Dong Y, Yan K, Ma Y, Wang S, He G, Deng J, Yang Z. A sensitive dilute-and-shoot approach for the simultaneous screening of 71 stimulants and 7 metabolites in human urine by LC-MS-MS with dynamic MRM. J Chromatogr Sci. 2015;53(9):1528–35.

[43] Gonzalez-Jartin JM, Alfonso A, Rodriguez I, Sainz MJ, Vieytes MR, Botana LM. A QuEChERS based extraction procedure coupled to UPLC-MS/MS detection for mycotoxins analysis in beer. Food Chem. 2019;275:703–10.

[44] Gonzalez-Jartin JM, Rodriguez-Canas I, Alfonso A, Sainz MJ, Vieytes MR, Gomes A, Ramos I, Botana LM. Multianalyte method for the determination of regulated, emerging and modified mycotoxins in milk: QuEChERS extraction followed by UHPLC-MS/MS analysis. Food Chem. 2021;356:129647.

[45] Gonzalez-Jartin JM, Ferreiroa V, Rodriguez-Canas I, Alfonso A, Sainz MJ, Aguin O, Vieytes MR, Gomes A, Ramos I, Botana LM. Occurrence of mycotoxins and mycotoxigenic fungi in silage from the north of Portugal at feed-out. Int J Food Microbiol. 2022;365:109556.

[46] Rodriguez-Canas I, Gonzalez-Jartin JM, Alvarino R, Alfonso A, Vieytes MR, Botana LM. Detection of mycotoxins in cheese using an optimized analytical method based on a QuEChERS extraction and UHPLC-MS/MS quantification. Food Chem. 2023;408:135182.

[47] Botana LM, Alfonso A, Rodríguez I, Botana AM, Louzao MC, Vieytes MR. How safe is safe for marine toxins monitoring? Toxins. 2016;8(208):1–8.

[48] Gonzalez-Jartin JM, Alfonso A, Sainz MJ, Vieytes MR, Botana LM. UPLC-MS-IT-TOF identification of circumdatins produced by Aspergillus ochraceus. J Agric Food Chem. 2017;65(23):4843–52.

[49] Gonzalez-Jartin JM, Alfonso A, Sainz MJ, Vieytes MR, Botana LM. Detection of new emerging type-A trichothecenes by untargeted mass spectrometry. Talanta. 2018;178:37–42.

[50] Rodriguez I, Genta-Jouve G, Alfonso C, Calabro K, Alonso E, Sanchez JA, Alfonso A, Thomas OP, Botana LM. Gambierone, a Ladder-Shaped Polyether from the Dinoflagellate Gambierdiscus belizeanus. Org Lett. 2015;17(10):2392–95.

[51] Gonzalez-Jartin JM, Alfonso A, Sainz MJ, Vieytes MR, Aguin O, Ferreiroa V, Botana LM. First report of Fusarium foetens as a mycotoxin producer. Mycotoxin Res. 2019;35(2):177–86.

Carmen Alfonso

3 Quantitative and qualitative methods, and primary methods

3.1 Quantitative and qualitative methods: different objectives with the same quality

A qualitative method can be defined [1] as an analytical method that identifies a substance based on its chemical, biological, or physical properties, whereas a quantitative method is an analytical method that determines the amount of a substance and expresses it as a numerical value and an appropriate unit.

A qualitative method tries to determine the presence or absence of one analyte in a sample (Figure 3.1). This measure is determined by a specified level; below it the analyte is considered insignificant and above it the analyte is considered present. A quantitative method serves us to determine the concentration of that analyte in the sample. At the middle of these two types of methods, semiquantitative methods are situated [2], analytical methods that assign the test samples to a given class, for example, high, medium, low, or very low. These methods provide some quantification of the analyte in the sample, useful when the accuracy of the measurement can be low.

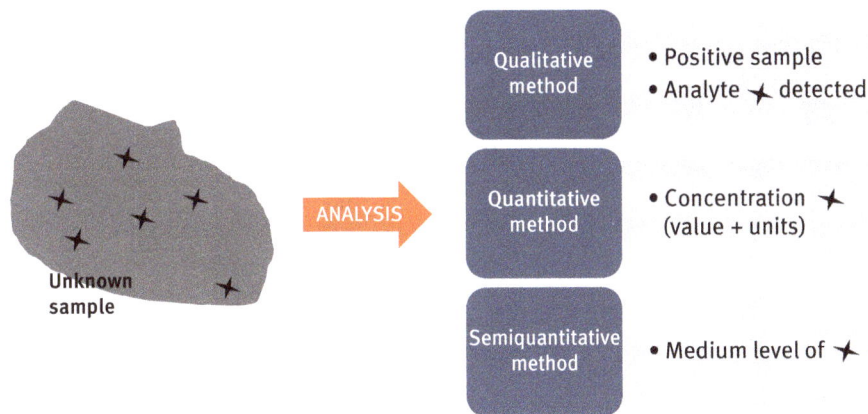

Figure 3.1: Differences in results from qualitative, quantitative, and semiquantitative methods.

We must select a qualitative method if our main objective is to know if a compound is present in a sample, whereas a quantitative method will be the correct option if the information needed is how much compound contains the sample.

Nowadays other category of methods with increasing applications is nontargeted methods. These methods try to identify and/or quantitate all chemicals present in the

https://doi.org/10.1515/9783111014449-003

sample; they are not focused in one or several analytes and want to know its complete composition. Nontargeted methods allow the detection of known and unknown analytes and the quantification of some of them (depending on their characteristics and known properties) and produce huge amounts of data. In practice they do not detect everything present in the sample, but allow its characterization and the exact comparison between multiple samples analyzed in different moments. Usage of these methods is becoming popular in several fields, like food fraud [3], contaminants in environmental [4] and food samples [5], plastics in food packaging [6], or biomarkers in human samples [7].

An essential part in all analytical methods, qualitative and quantitative, is the sample. It is the object which will be analyzed, and its representativeness will determine the applicability of the obtained results. Usually a sample is a portion of a larger material, which is collected to be analyzed (Figure 3.2), and the final result of the analysis will depend on the reliability of the sampling step [8]. There are several standardized methods to collect representative samples, described in national or international standards [9, 10] and regulations [11, 12]. These methods vary depending on the characteristics of the initial material and the needs of the measurement process, showing different sample size, parts of material to be collected, or number of subsamples. During sampling process attention must be paid to sources of variability, like different particle size, evaporation or sedimentation of components or heterogeneous distribution. Also handling of the sample from the collection to the analysis can determine the final method result because degradation due to inadequate conservation conditions can lead to an underestimation or an overestimation of the analyte.

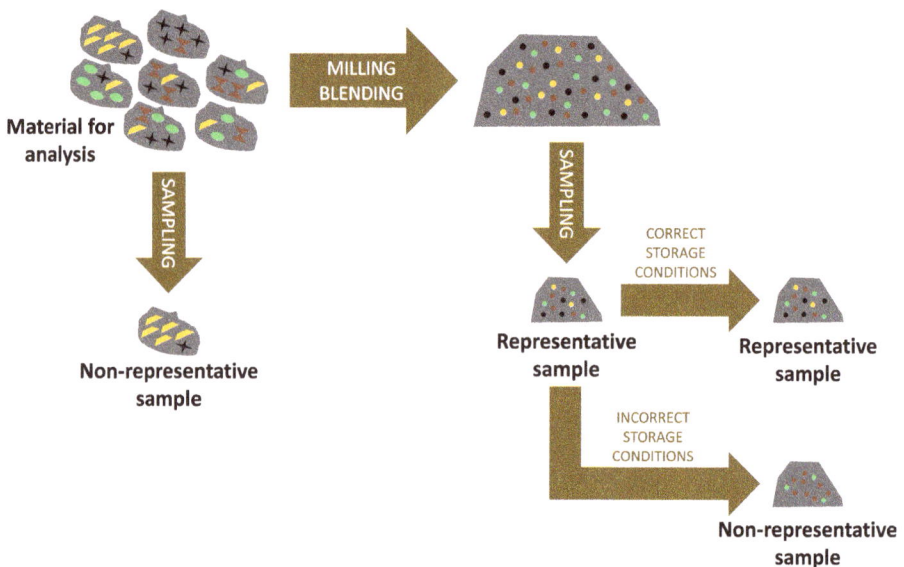

Figure 3.2: Sampling process, examples of representative, and non-representative samples.

3.2 Qualitative methods: a new focus on analytical chemistry

Qualitative methods are becoming more important in recent years due to their rapid and reliable response, which allows a high sample throughput in short time periods, and their adaptation to instantaneous requirements. In certain cases, the most important issue is determining if a sample complies with a specific legislation and the accurate concentration result is not a key point. An example of this situation is the screening of a set of samples to determine the presence or absence of contaminants above or below the maximum permitted level (specification limit or threshold value) fixed by certain authority. In this situation customers need to know the classification of each sample and the quantification of each contaminant is a useless activity. Other example is determining the presence or absence of a biomarker in a human sample, as blood or plasma, which can be related with the prevalence of a disease. In COVID-19 pandemic situation, classification of patients' nasal swabs as positive or negative for SARS-CoV-2 RNA was essential for managing populations, whereas the exact quantification of samples was not required.

Different objectives can be listed for qualitative methods [13]: obtain information required for a specific purpose in a rapid procedure, eliminate as far as possible the pretreatment of the samples prior to their analysis (time consuming and source of variability), or reduce the use of instruments (principally to avoid their costs and the qualified personnel involved). These methods are sometimes used to filter a set of samples and determine which ones must be analyzed by quantitative processes, so they are also called screening systems.

There are various classifications of sample screening systems, according to different criteria [13]. The relation between the analyte and the signal can be direct or due to a chemical, biochemical, or immunological reaction, the binary response obtained can discriminate one analyte or can be the same for various analytes from a chemical family, the state of the sample can be solid, liquid, or gaseous, the system can be used inside or outside the laboratory, and the screening can be independent or implemented during an analytical quantitative process.

According to the sample treatment, screening systems can be classified in [13] direct screening without sample treatment, screening after a simple sample treatment, and screening after a full sample treatment. In the last case screening methods are preferred to minimize the use of expensive analytical equipment or to avoid another more complicated sample pretreatment.

Considering the type of detection, screening systems can be divided into sensorial and instrumental (Figure 3.3) [2]. In sensorial detection human senses are used to interpret the response of the method, the most usual one is the vision. Responses are obtained due to the reaction between the analyte in the sample and a specific reagent involved in the procedure and the concentration of the analyte can be related directly or indirectly with the magnitude of the response. Different reaction concepts are used (chemical, immunological) and the final color can be determined either by visual inspection or by comparison with a color card or similar. In instrumental detection the comparison is performed between the unknown sample and a reference sample which contains a specified level of the analyte, obtaining an instrumental response, for example, absorbance. The result of this kind of methods is a binary response (yes/no) as in the sensorial detection, no calibration curves are done, and no quantification is obtained.

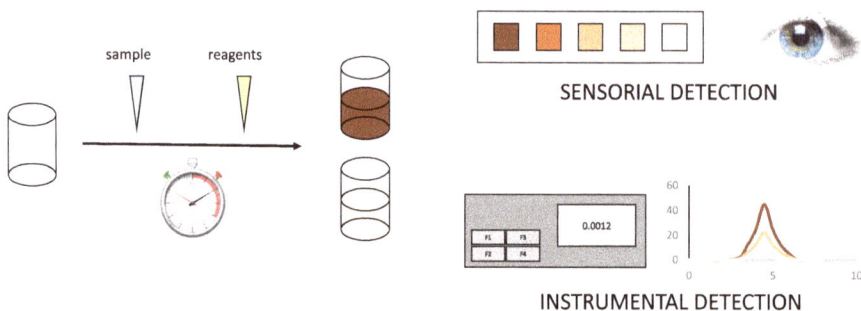

Figure 3.3: Sensorial and instrumental detection in qualitative methods.

According to the fitness for purpose [14, 15], there are two main types of qualitative analysis. The first one is performed to classify the samples according to a specific criterion, provided by a regulation, a customer, or a laboratory specification and related to a single measurand or to a group of them. The second type pursues the identification of analytes or groups of analytes using techniques of separation and detection, for example, chromatography and spectroscopy.

Considering the practical implementation of the qualitative method [15], different types of systems can be used: test kits and sensors which show the binary response directly (usually for sample classification) and conventional instrumentation which need a treatment of the obtained data to convert it into a binary response (usually for identification of measurands). Maybe test kits are the most extended alternative, including a chemical reaction and an evaluation system, usually based on color.

The need to demonstrate traceability of results is essential either in quantitative as in qualitative methods. This property assures that results can be related to a reference through a documented unbroken chain of calibrations [16]. This unbroken chain of calibrations relates the measurement result with a reference, which can be a measurement unit (e.g., kg), a value obtained from a reference measurement procedure or a measurement standard (e.g., a certified reference material (certified RM) produced by an organization accredited according to ISO 17034). Each step of this chain must be documented and its contribution to the overall measurement uncertainty must be assigned. Usually in qualitative methods the final reference can be established through three different approaches [14]: comparing the results of the qualitative method with those obtained by a confirmatory, reference, or primary method, analyzing aliquots of certified RMs or participating in interlaboratory exercises (in this case one of the laboratories has established the traceability of the tested method).

Validation of qualitative methods is needed to apply them in solving analytical problems. For example, in food chemistry these methods are used to detect contaminants (e.g., pesticides or mycotoxins) and the results produce consequences for producers, consumers and authorities. In these applications assuring that an analyte is present in a tested sample or that its concentration is above or below a specified level is a very important issue and performance characteristics of the methods shall be determined.

A general procedure for validation of qualitative methods (Figure 3.4) includes different steps [14]. In the first one the analytical information that must be obtained by the method is fixed, according to customer's needs, objectives established by the laboratory, quantities provided by legislation or any other input, and the most appropriate method to solve the analytical problem is selected. In the following step the performance characteristics of the method are determined, and the experimental processes needed to obtain them are designed. Finally, experimental results are obtained and compared with analytical information initially fixed as necessary, and decision about validation of the method is established. To perform a complete validation of qualitative methods three types of samples must be available [14]: certified RMs in an appropriate range of concentration, reliable blank samples, and unknown target samples.

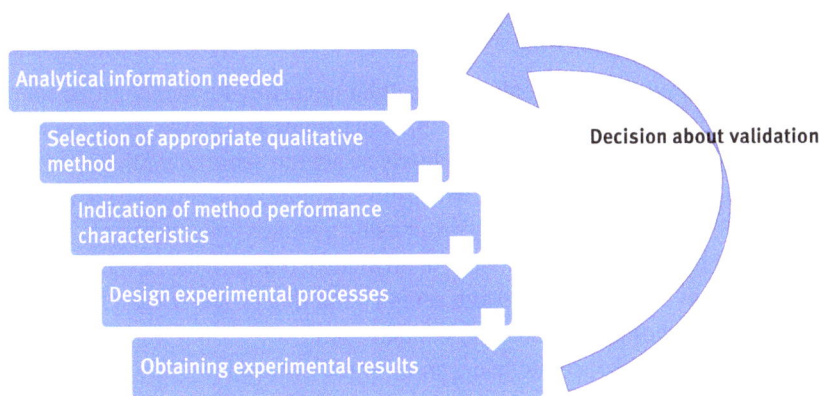

Figure 3.4: Steps of validation procedure for qualitative methods.

The main performance characteristics for qualitative methods (Figure 3.5) are specificity, limit of detection, reliability (false-positive and false-negative rates), and recovery. From a practical point of view the productivity-related properties (speed, costs, and risks) are also very important for a real implementation of qualitative methods [14] since a very expensive and slow method will not be selected by common users and it will be desirable to handle low risk reagents.

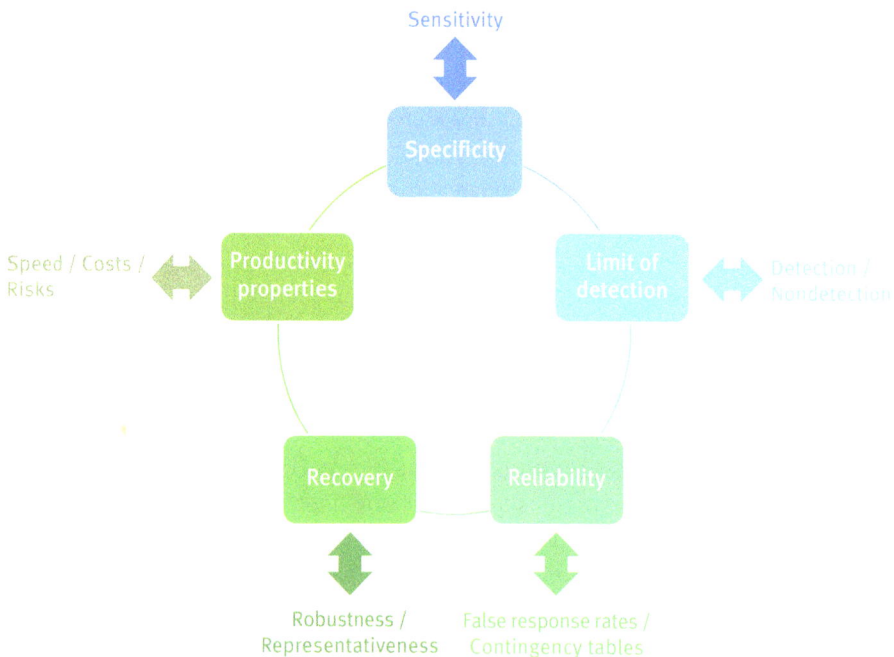

Figure 3.5: Main performance characteristics for qualitative methods.

Specificity [2] is the capacity of a qualitative method to classify known negative samples as negative, so it is the probability [17, 18] of achieving a negative result using a known negative sample. This parameter is clearly related to sensitivity [2], the capacity of a qualitative method to classify known positive samples as positive, which is the probability [17, 18] of achieving a positive result using a known positive sample.

The limit of detection [14] is the amount of an analyte that provides a signal statistically different from the signal of a blank sample, and this property establishes the separation between zones of detection and nondetection.

In validation of quantitative methods, an essential parameter is the uncertainty [15], expressed as the concentration range within which the results can be expected, so it is a measure of the dispersion of the results [19]. In qualitative testing uncertainty is an indication [19, 20] of the probability of a sample classification being wrong. In this context, reliability [13, 14] is defined as the percentage of correct results provided by many independent tests using aliquots of the same sample (usually a certified RM). This characteristic can be considered as a combination of accuracy and precision, both basic properties of quantitative analysis. Considering this, unreliability [2, 15] is the response range where the false responses (positive or negative) are produced. False negative and positive rates measure the probability that a known sample (positive or negative) is incorrectly classified (negative or positive, respectively) by the qualitative method. Bibliography refers to these terms with different definitions according to the specific author. For example [17], false-negative rate can be the fraction of observed negatives that are false, the fraction of true positives that are considered negatives or the fraction of all results that are false negatives. Different definitions correspond to different quantities [21], so the term must be correctly specified prior to the experimental design of the validation study. Importance of false responses depends on the application performed. For example, the identification of a positive sample as negative is especially serious in the detection of toxic chemical in food or environmental samples [22] since no confirmation analysis will usually be performed.

Although false-response rates are the most used form to express uncertainty in qualitative methods, other alternatives are possible [19, 20]:

– Contingency tables: In these tables the results of qualitative analysis are classified in categories in the simplest case positive or negative. Some parameters, as sensitivity and specificity, are calculated from the table and provide an overall idea of the performance of the assay and its reliability. They are also used to compare different assays, but cannot estimate the probability of obtaining a wrong result in an individual sample. The performance of this alternative depends on the sample size, but it is frequently used due to its easy application to a lot of different qualitative assays.

– Bayes' method: In this approximation the probabilities are estimated according to similar measurements under similar circumstances, considering historical and conditional probabilities, obtaining a more complete measure of uncertainty than in the previous approach. This method is more complicated but provides an estimation of the uncertainty associated with a new measurement.

– Statistical intervals based on the normal or Gaussian distribution and its properties: A relationship is established between the specification limit (usually a concentration) and its instrumental response. Then the response of the unknown sample is compared to the one at the specification limit and the result is defined with a calculated probability of committing errors. This approach must be used with instrumental responses; it cannot be applied to binary approaches.

– Performance curves: These curves are established after the analysis of several samples with different concentrations using a screening method. The percentage of positive results above the specification limit is represented versus the concentration level and a cut-off concentration can be calculated. This is the value from which the response of the method is above the specification limit with a specified probability of error.

The recovery of a method needs to be determined if it includes extraction steps. One variable which must be considered to define these steps is the robustness of the qualitative method [13, 14], defined as the invariability of the sample results using slightly different experimental conditions (pH, temperature, etc.). For example, one parameter involved in robustness is the stability of the reagents used in the tests, which can be affected by variables as the temperature or the exact composition of the samples. As it has been mentioned earlier, the representativeness [14] of the sample is essential for the result of the method, so correct sampling protocols must be established. Finally, depending on the selectivity of the method, the sample treatment will involve different number of steps, which will affect the recovery achieved.

Quality control procedures must be established to monitor the validity of analytical methods either quantitative or qualitative. Important issues are the complete description of the parameter related to quality, the definition of the influences in quality and the features which determine the optimal quality. Typical approaches [23] are the analysis of RMs and blind samples, the use of control charts, the analysis of blank, duplicate and spiked samples, and the proficiency testing. In qualitative analysis quality control focuses on identifying false positives and false negatives. Internal quality control is related to the maintenance of validation conditions in the laboratory for a long time, whereas external quality control is used to compare results between laboratories.

Control charts [23] can be used for binary response, using sample A that must produce a 100% positive response to detect false negative and sample B that must produce a 100% negative response to detect false positive. In the simplest approach individual experimental data are represented and out-of-control situations are detected. If several analyses have been done, false rates can be represented in the control chart; for example, one point for each experimental day. In this approach reference lines of the chart can be established, as an acceptance error level at 5% of false response, the warning limit at 10% of false response, and the control limit at 15% of false response, and false responses include false positive and false negative. A more complex approach is used to represent false positives and false negatives independently, using a control chart with two axes and four regions: situation under control, false positive response, false negative response, and double false response (Figure 3.6). As in the previous case results of each day must be plotted to detect out-of-control situations. In all approaches, the frequency of quality control must be established by the analyst, based on the number of samples used, the reliability of the method, or the application of the results, but it is always recommended to use at least two control samples: one positive and one negative.

Proficiency testing programs have been adapted to qualitative analysis, showing its capacity to evaluate the comparability of the results produced by different laboratories. In these programs specificity and sensitivity of the method are estimated [17], employing samples at a representative range of concentrations and blank samples, all completely characterized by quantitative methods.

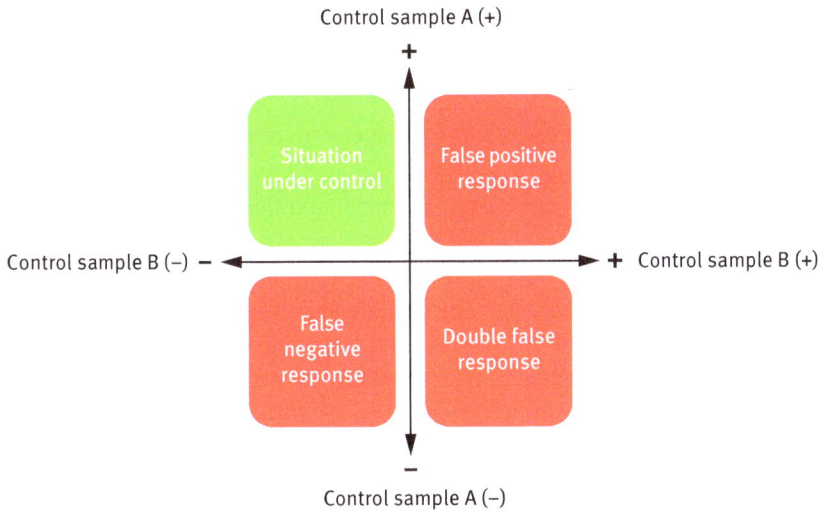

Figure 3.6: Control chart for qualitative analysis representing four regions: situation under control, false positive response, false negative response, and double false response.

Certified RMs are rather complicated to use in qualitative methods [2] because they must contain the analyte at a concentration level close to the limit of the method. If the concentration level is either far below or far above this limit, its use will only indicate if the method correctly classifies samples as negative or positive. For concentrations close to this level, the probabilities of false positive and false negative responses must be computed, so the comparison with a certified RM must be in terms of probabilities and cannot be in terms of concentrations.

Pure RMs are analyzed by the instrumental techniques in qualitative methods used for identification [15] and their results are stored in libraries. After that, the results of the unknown sample are compared with these libraries and the identification of the compounds is performed due to probabilistic techniques. Usually for classification methods a matrix RM must be used since the interferences of the samples can modify the obtained results. Screening and confirmatory methods must be related to reference methods to demonstrate traceability, comparing the proportions of positive results.

In summary, qualitative methods are an often-forgotten part of analytical chemistry with very interesting properties to fulfill customer requirements since they can provide a speed answer to solve an analytical problem, without using expensive equipment and specialized personnel. Due to these reasons it is quite probable that in next years they will be developed and expanded to every field of analytical measurements.

3.3 Quantitative methods: amounts, units, and uncertainties

Quantitative methods measure one or more analytes in a sample, obtaining their amount, expressed in certain units. In a simplified manner, in these methods the analyte generates a signal in an instrument or equipment and this signal is traduced in a quantity (Figure 3.7). The first step for the development of a quantitative method is the complete definition of the analyte to be measured: its name, its formula, and the amount of sample used for the analysis. The formula and the name of the analyte are needed to avoid confusions between similar compounds, like cations with different oxidation state or various salts of the same compound, whereas the amount of sample is needed to calculate the concentration or the quantity of the analyte. As it has been stated earlier, the representativeness of the sample is essential, and the pretreatment procedure for the measurement can be more or less complex, including several different steps depending on sample nature, analytes characteristics, and signal measured.

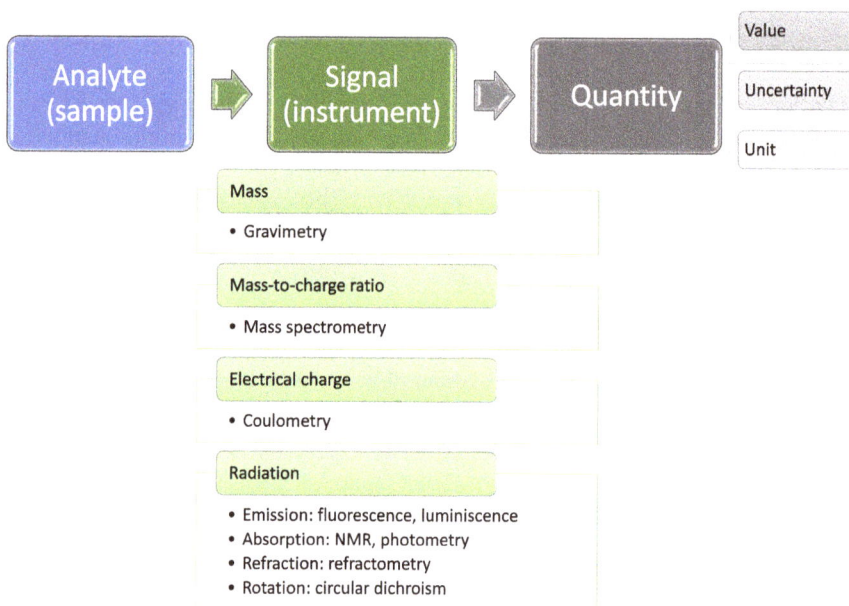

Figure 3.7: Overall scheme of a quantitative method.

Once the analyte is defined, the quantitative method can be chosen or developed. The first option can be an internationally accepted method, completely defined for the analyte and the sample. If this method does not exist, or if the necessary equipment is not available, a new method can be developed by a particular laboratory. A previous similar method can be used as an initial point in some cases, whereas in others a completely novel method is designed, involving much more work, with several initial ideas and multiple experiments to achieve a final result. The fitness for purpose of a quantitative method, either an internationally accepted or a novel one, is providing the right quantification for the analyte in the sample [24], which must be demonstrated through its validation. The extent of this validation varies if the method is an internationally accepted or a novel one and also depends on its future use or the risks associated with the obtained results. Usually an internationally accepted method is verified previously to its use in a laboratory since a complete characterization is available. On the other hand, performance characteristics of a novel method must be calculated to determine if its results comply with fixed specifications. Quality of validation is affected by factors like personnel qualification, equipment calibration, suitability of chemical or physical principle, and quality management of the entire procedure [8]. Costs and risks also affect the process since improving performance characteristics suppose an increase in the inversion, which must be recovered during the routine use of the method, and a decrease in risks associated with incorrect results, so both parameters must be balanced. Finally, periodical reviews of the obtained results are essential to guarantee that the method is operating as expected in routine measurements.

Validation process begins with the elaboration of the validation plan, which contains the identification of the method, the extent of the programmed validation (number of experiments, characteristics of samples, and schedule of activities), and the performance characteristics to be met. Once the experiments have been performed and analyzed, results are obtained and a validation report is written. This document must contain these results, a summary, and a conclusion about the fitness for purpose of the method, which will depend on the achievement of the performance characteristics initially fixed (Figure 3.8).

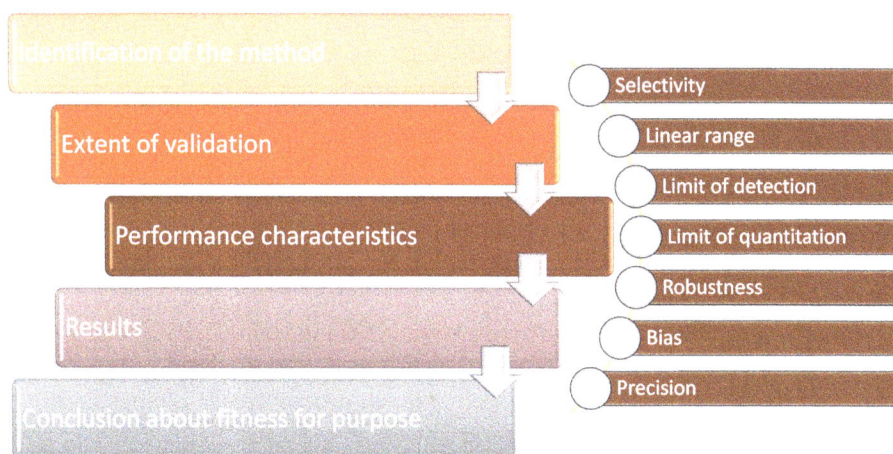

Figure 3.8: Validation procedure and performance characteristics for quantitative methods.

Several performance characteristics can be checked during the process of quantitative method validation [25], for example, selectivity, linear range, limit of detection and limit of quantitation, robustness, bias, and precision (repeatability and reproducibility). In several practical applications, method performance characteristics have to comply with regulatory requirements, for example, in the European Union methods for the determination of the levels of pesticide residues in matrices like soil, water, air, or food [26]. In most common situations these characteristics can be initially set in line with method capability, based on previous experiences, estimated from mathematical models or based in proficiency testing with similar methods. If performance characteristics obtained during validation do not meet the requirements initially fixed, method can be improved, but costs associated with this improvement have to be analyzed, as it has been mentioned earlier.

Selectivity can be defined as the capacity of a method to determine different analytes in mixtures without mutual interference. In practice, there is no quantitative method totally free of interferences, so their value must be established during method validation. These interferences can produce a modification in the measured signal (an enhancement or a decrease), which will affect the calculated analyte quantity [24]. In some particular areas, like pharmaceutical sector, the term selectivity is often substituted by specificity.

Linear range is the range of quantities of each analyte that can be determined in a sample with a linear relation between signal and quantity. To determine this range samples with different concentrations of analyte inside the range (usually 6–10 values) and blanks must be available, and linear relation between analyte concentration and signal obtained must be calculated. Each different matrix used in the same method has to be checked since interferences can produce a nonlinear response in some cases [24].

Limit of detection and limit of quantitation represent the smallest amount of analyte which can be detected or quantified by the method, with an acceptable performance in both limits. Their precise estimation must be defined previously to method validation and can be related to signal/noise ratios (e.g., in chromatographic procedures) or blank values (calculated in many software programs). For example, experimental determination of these limits can be done after a series of measurements of blank samples (matrices with no detectable analyte), test samples with low concentration of analyte, reagent blanks, or reagent blanks spiked with low concentration of analyte [24], applying different mathematical approaches.

Robustness is the capacity of the measuring system to maintain its characteristics when there are changes in environmental conditions, like temperature, minor components of reagents, or slight changes in preparation procedures. In a robustness test these changes are implemented and results are analyzed, establishing the effect and importance of each change and the critical variables which have to be controlled during the routine use of the method [24].

Quantitative methods determine a measured value for an analyte in a sample, usually different from the true value for the analyte in the sample (by definition, this true value cannot be known). Difference between the true value and the measured value is the error associated with the measurement [27] with two components: systematic and random. Bias is an estimate of the systematic component and precision an estimate of the size of random components [27]. Bias determination can be done analyzing certified RMs. If they are not available, bias can be calculated from recovery experiments with spiked samples or comparing method results with results from a previously characterized method [24], preferably a reference method or a method in routine use in the laboratory.

Precision indicates the spread of the obtained results under specified conditions. Considering these conditions different variations can be calculated:
- Repeatability: results obtained by a single analyst with the same equipment in a short timescale.
- Reproducibility: results obtained by different laboratories.
- Intermediate precision: results obtained in a single laboratory, by different analysts, in an extended timescale, and/or using different equipment.

Precision estimation is expressed using standard deviation or relative standard deviation, calculated from replicate measurements of suitable samples. Usually 6–15 replicates of each sample are preferred, although number can be adapted depending on the particular method [24].

Precision and bias are considered in method accuracy, a qualitative property which indicates how close is the measured value to the true value of the analyte in the sample [16] (Figure 3.9). A method will be more or less accurate, depending on the measurement errors associated, either random or systematic. This qualitative concept is substituted by uncertainty to provide a quantitative indication of the quality of method results. Uncertainty associated with a measurement indicates the range of values which can be reasonably attributed to the quantity being measured. It is estimated considering all known effects influencing the measurement: precision, bias, uncertainty associated with certified RM used in bias measurements, equipment calibration, steps performed, unknown interferences, etc. [24].

A complete result obtained in a quantitative method must contain a quantity value and an uncertainty. This component is usually expressed as expanded uncertainty, in the same units than the quantity value, and the final result is indicated with the format: quantity value ± expanded uncertainty (Figure 3.9), which means that the true value is contained in the indicated interval with a high level of confidence. Estimation of uncertainty associated with a quantitative measurement is a complex activity: all sources of uncertainty must be detected, their standard uncertainties must be determined, combined standard uncertainty must be calculated (considering previous standard uncertainties), and finally, expanded uncertainty must be obtained, applying a coverage factor to the combined standard uncertainty [16]. Although the coverage factor depends

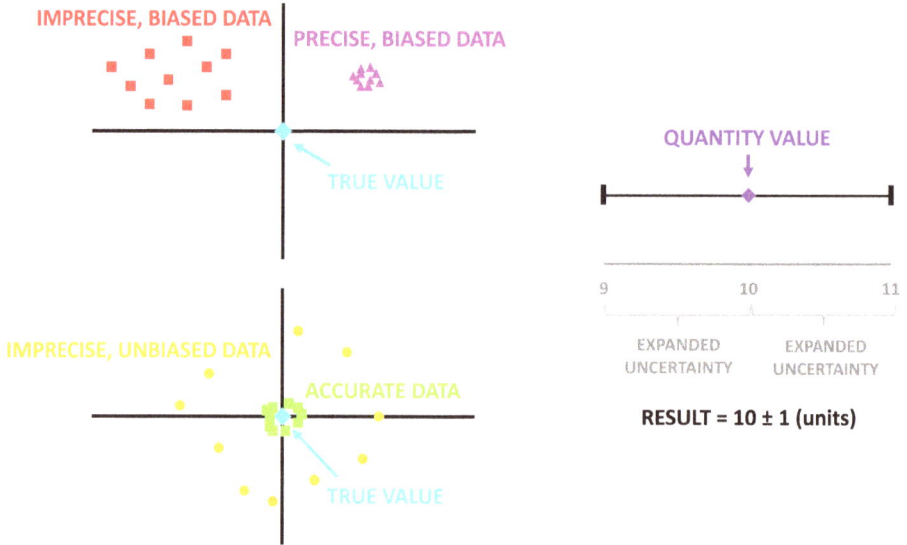

Figure 3.9: Left: Precision, bias, and accuracy. Right: uncertainty.

on the level of confidence desired, a factor of 2 is commonly used, associated with a 95% of confidence. For validated methods, combined standard uncertainty is usually calculated from measurement precision and uncertainty of bias using the law of propagation. Measurement uncertainty must always be accompanied by an explanation of how it was estimated since there can be different correct approaches [16].

One of the widespread applications of quantitative measurements is the decision about the compliance with a specification. For example, deciding if a batch of food contains a contaminant above or below a regulated limit (or specification limit). Uncertainty associated with the quantity value must be considered in these applications to correctly decide the compliance. For an upper limit of a contaminant which cannot be exceeded, after a quantitative measurement of the contaminant, five situations can occur [28] (Figure 3.10):

1. Quantity value above upper limit, uncertainty interval does not contain upper limit.
2. Quantity value above upper limit, uncertainty interval contains upper limit.
3. Quantity value at limit, uncertainty interval contains upper limit.
4. Quantity value below upper limit, uncertainty interval contains upper limit.
5. Quantity value below upper limit, uncertainty interval does not contain upper limit.

In cases 1 and 5 decision about compliance is quite simple since considering quantity value and uncertainty interval, contaminant is above (fail decision) or below (pass decision), respectively, of the upper limit. Cases 2, 3, and 4 represent more complex situations, where the probability of an incorrect decision may be or may be not sufficiently small to justify a decision about compliance, depending on the risks associated with a wrong decision or on other circumstances [28]. For example, for contaminants in a batch of food, a wrong decision can produce effects in human health or, in the other hand, unaffordable costs for the providers. Uncertainty interval is a key point in this analysis since a reduced interval will suppose a smaller number of complex situations. Due to this, in several legislations with specification limits, there are rules related to the maximum uncertainty associated with the quantitative method employed to reduce associated risks [11, 29].

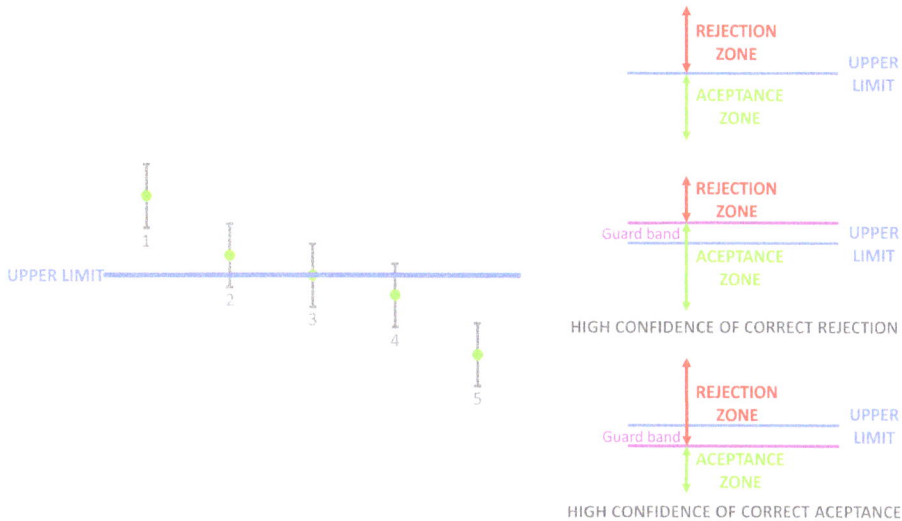

Figure 3.10: Left: Decision rules, different situations related to compliance. Right: Effect of guard band on confidence about acceptance and rejection zones.

Decision rules are established for a routine management of compliance. These rules fix an acceptance zone and a rejection zone so that measurement results contained in the first one are considered compliant and measurement results contained in the second one are considered noncompliant. Depending on the risks associated with wrong decisions, it can be set a guard band to increase one of these zones and provide more confidence in one of the decisions (Figure 3.10). Decision rules can include the possibility of other results (conditional or inconclusive) when the specification limit lays inside the expanded uncertainty interval and can also establish the need of additional analysis [28]. In the example previously described, decision rules can not only consider situations 2 and 3 as conditional fail and situation 4 as conditional pass but also can consider situations 2, 3, and 4 as inconclusive and order a new analysis.

As it has been mentioned earlier in this chapter, results from quantitative methods must be traceable. Traceability to the SI means traceability to reference values obtained by agreed realizations of SI units. For example, in the case of a quantitative method with results expressed in units of grams, these results must be traceable to the realization of the kilogram [30]. References used to demonstrate traceability will depend on the activity performed. For physical measurements, equipment and measurement standards are available for several magnitudes (like mass, length, or temperature), with low calibration uncertainties. Materials used must be accompanied by a certificate of calibration, where the traceability to adequate references is stated [30]. For other activities, like calibration measurements, choosing an adequate certified RM would be the best option. This material is accompanied by a certificate, where the traceability to SI units or other reference is indicated. If this option is not possible (e.g., for activities related to an in-house synthesized new material), in-house RMs can be prepared. Finally, reference data can be used in several measurements, provided that they are traceable to appropriate references and they are applied in the same conditions as acquired. For example, mass spectrum of a compound in certain chromatographic conditions must be compared with samples in the same conditions, not using different columns, mobile phases, or gradients. There are fields where traceability to a universal reference is still being defined. An example is the quantification of genetically modified organisms present in food and feed, which must be expressed (as stated in EU regulations) as genetically modified mass fraction and is usually measured by PCR, obtaining the ratio of DNA copy numbers (from biological species and genetically modified ones) [31].

Quality control of validated quantitative methods is necessary to assure that capabilities and limitations observed during validation are maintained during routine use.

Schedule of internal quality control must be established by laboratory management, based on parameters like risk assessment, frequency of the analysis, criticality of results, complexity of method, or other factors considered essential during method validation. A level of 5% of quality control samples in routine use is often considered as acceptable, but for analysis performed infrequently a complete validation could be required in each batch.

Control charts are frequently used to evaluate the performance of a quantitative method. In them the mean value obtained for the quality control sample during method validation is represented, with warning limits and action limits at values corresponding to twice the standard deviation of the mean and thrice that standard deviation. Results obtained for quality control samples during routine use are plotted in these charts, their tendencies are analyzed applying previous defined rules (e.g., a maximum of six values decreasing or increasing or a maximum of nine values at one side of the mean) and abnormal results are identified (Figure 3.11). Stated values for mean and warning and action limits must be reassessed when enough routine values are collected or even in a timeline manner since values obtained during validation could not reflect all the variations which appear in operating conditions [32].

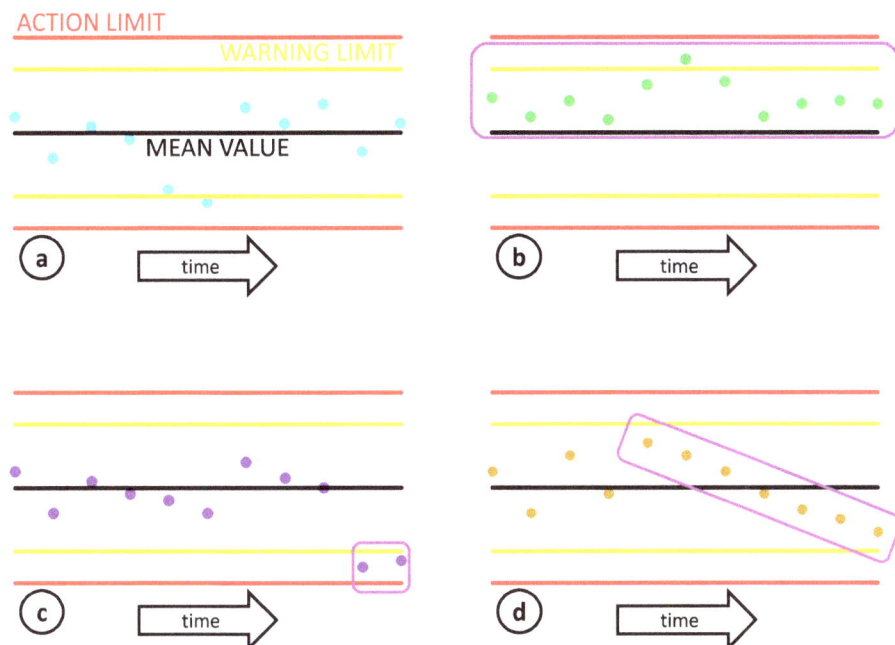

Figure 3.11: Control charts: (a) controlled situation; (b) abnormal results: more than nine values at the same side of the mean value; (c) abnormal results: two consecutive values outside the warning limits; (d) abnormal results: more than six values decreasing.

Also blank samples, replicate samples, and blind samples are part of the internal quality control: blank samples are used to eliminate contributions to the result which are not related to the analyte, replicate analysis of unknown samples allows checking the precision of the results, and blind samples avoid preconceived ideas in the analyst, which can affect results obtained [24].

External quality control is an option to demonstrate that performance characteristics of the method are maintained and is used to compare results with other laboratories. This external control consists of participation in proficiency testing schemes or other types of interlaboratory comparisons. Accreditation bodies require that laboratories participate in these schemes to assure its competence, and there are international standards which describe the competence of proficiency testing providers [33] and the selection and interpretation of these schemes [34].

Environmental protection is an important issue for the future of quantitative analysis, with several strategies to be implemented, like the replacement of toxic reagents, the miniaturization of measurements, or the online treatment of wastes [35]. The next challenge is the maintaining of the characteristics of current quantitative methods adapting them to these new strategies, with the main task of prioritizing direct measurement of untreated samples, which will suppose a greater reduction of energy and reagents consumed and waste generated.

In summary, quantitative methods must be completely defined and correctly validated and controlled in order to produce accurate results, essential for world trade, fraud detection, environmental control, or clinical diagnosis. Future developments of these methods are related to the protection of the environment and to the quantitation of new natural or synthetized compounds.

3.4 Primary methods: characteristics and applications

The Consultative Committee for Amount of Substance (CCQM), created by the International Committee of Weights and Measures (CIPM) defined a primary method of measurement as a method with the highest metrological qualities, with a complete uncertainty statement in terms of SI units and whose operation can be completely described and understood [36, 37], differentiating:
– Primary direct methods: measure the quantity of an analyte without reference to a standard of the same quantity.
– Primary ratio methods: measure the ratio of an analyte to a standard of the same quantity, with a measurement equation describing method operation [38].

Other definition which can be cited says that a primary reference measurement procedure (or primary method of measurement) is a reference measurement procedure which allows the determination of a measurement result without reference to a standard of the same quantity, with reference to the definition of its measurement unit or to fundamental constants. In this definition, a reference measurement procedure is a well-characterized method, which provides results fit for its intended use, usually with small uncertainties completely described. This use can be the characterization of RMs, the calibration, or the assessment of trueness of quantity results obtained in other procedures [16].

Today, it is recognized that there are methods with the potential of being primary methods for certain analytes in specific circumstances since a method or technology usually cannot be a primary method for all its applications. These potential primary methods have to be validated for each specific use and their scope must be perfectly described [39].

Briefly, a primary direct method provides a result using data obtained by an instrument or equipment and mathematical formulae. For example, gravimetry provides the quantity of each molecule present in a reaction of precipitation, using the mass of the precipitate obtained in a balance and the molecular formulae, molecular weights, and stoichiometric coefficients of the reaction. On the other hand, a primary ratio method needs the use of a standard to provide the same result. For example, isotope dilution mass spectrometry (IDMS) calculates the quantity of an analyte using data obtained in a mass spectrometer, an enriched isotope standard of the analyte, and mathematical formulae.

There are several methods classically recognized as primary methods or with the potential of being primary methods. Some of them are described below, together with novel methods which need further research to be internationally accepted (Figure 3.12).

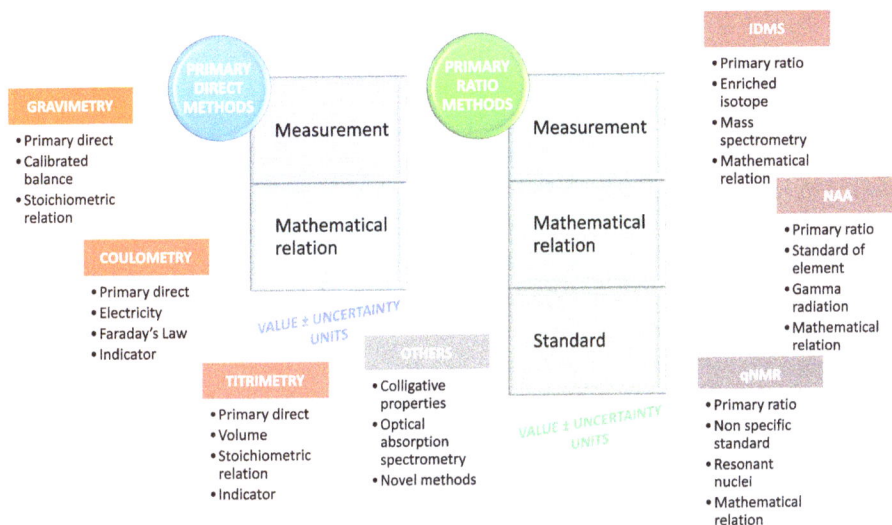

Figure 3.12: Primary direct methods and primary ratio methods: examples.

Gravimetry is a classically recognized primary direct method. It uses a calibrated balance to weight an analyte, a stoichiometric reaction product, or a derivative [40]. The calibration of the balance must be made using transfer weights traceable to the kilogram prototype, so the traceability chain for the final result is usually short and the obtained accuracy is high. In gravimetry, treatment of the initial sample produces a solid, which is weighted. This treatment can be a precipitation, after the addition to the sample of an appropriate reagent, specific for each measurement. Also a volatilization can be done, heating the sample and weighting before and after heating, or weighting the volatile retained product [40]. The analyte is calculated from the weight, by a stoichiometric relationship which defines the reaction performed, usually precipitation or volatilization. The uncertainty associated with gravimetry can be completely determined since all of its contributions are known. Some of them are related to the separation of the solid to be weighted from the rest of the sample or to the complete realization of the reaction [41]. Gravimetry is sometimes employed in the certification of RMs, mixing known amounts of substances, for example, gases [42].

Coulometry is an electrochemical technique classically recognized as primary direct method. It measures the electricity needed to transform an analyte at an electrode (converting it from one oxidation state to another) and uses a mathematical expression, based on Faraday's law, to calculate the amount of analyte [40]. The endpoint of the reaction (when all the conversion has been performed) is detected by the change of an indicator also present in the sample, for example, a colorimetric change in a solution. This technique is traceable to the SI but shows low sensitivity so its use is limited. Uncertainty of the results can be related to the practical realization of the technique, for example, the side reactions that can contribute to the process [41]. This technique is used in some applications for the direct determination of purity of substances [43].

Titrimetry is a volumetric technique classically recognized as primary direct method. It measures the volume of a standard solution (named titrant solution) needed to react with the analyte in a sample. The endpoint of the reaction, when the volume of standard solution is enough to react with all the analyte in the sample, is detected by the change of an indicator as in coulometry. Different reactions can be used; for example, acid-base, precipitation or redox, and an appropriate indicator must be selected in each case. The quantity of analyte is calculated using the volume of standard solution used and the stoichiometric relation in the reaction performed [40]. Uncertainty associated with results comes from the practical realization of the technique, for example, from the determination of the endpoint of the reaction [41]. Titrimetry is used for the certification of SI traceable RMs [44].

IDMS is a classically recognized primary ratio method. It uses an enriched isotope of an analyte as standard and mass spectrometry to measure the isotope abundance ratios in the sample, in the standard, and in the sample spiked with the standard [40]. The concentration or the amount of the analyte is obtained using these measures, a mathematical expression, obtained from the relation of the isotopes, and the quantity and isotopic composition of the standard. Uncertainty associated with the method is principally related to the steps of sample treatment performed previously to the measurement [41]. Different approaches of IDMS have been established (double IDMS and triple IDMS) when there is a lack of metrological information about the concentration or the isotopic composition of the standard [45, 46]. Several applications of this technique have been described, like the quantification of mercury in environmental samples [47], the quantification of ochratoxin A in roasted coffee [48], validated and fit for purpose to be used in the certification studies of a RM, or the absolute protein quantification, using acid hydrolysis and IDMS [49].

Neutron activation analysis (NAA) is a recognized primary method [50] for the quantification of several elements in a sample. It measures, using a gamma ray detector, the gamma radiation emitted by radioactive nuclei, generated in a sample due to the irradiation with neutrons of its stable atomic nuclei. This measure and the mathematical equations that describe the process are used to quantify the elements in the sample that have radioactive daughter products. This technique is usually used as a ratio method, with a standard of the element which is going to be quantified. In this case, the ratio of the gamma radiation emitted by the sample and by the standard is a direct measure of the ratio of the masses of the element in the sample and in the standard. All the contributions to the uncertainty of the process can be determined and the results are traceable to the SI. This technique has been used for the quantification of trace elements in complex matrices, obtaining results comparable to IDMS; for example, cadmium in rice [51] or chromium, copper, iron, manganese, and zinc in aluminum alloy [52]. This technique has advantages for solid materials which cannot be dissolved easily, are easily contaminated, and/or are irreplaceable [50].

Quantitative nuclear magnetic resonance (qNMR) spectroscopy is considered a potential primary method due to the fact that the integrated signal area obtained by NMR is directly proportional to the number of resonant nuclei which contribute to the signal. It is a ratio method because a certified RM of known purity is used as standard to determine the quantity of the analyte [53]. Standard and analyte contain NMR active nuclei and the purity of standard is known [54]. qNMR measures the signal of the standard and the signal of the analyte, by different approaches, and the amount of analyte is calculated using these measures, the mathematical equations for the process and the characteristics of the standard. Different mentioned approaches can be:

- Internal calibration, including in the sample a known amount of standard
- External calibration, with different options like concentric or coaxial NMR tubes
- External calibration of the internal solvent signal, which implies the calibration of the protonated solvent signal using a standard and the use of this signal as internal standard of the sample [55]

NMR has several advantages since it is not destructive, needs relative short measuring times, and several analytes can be determined simultaneously in the same sample. Its main disadvantage is the high cost of the equipment needed, in particular, high-field NMR spectrometers with high sensitivity and modern software to process and evaluate data in an accurate manner [56].

Validation protocol for ^{1}H qNMR has been described for the determination of molar ratios and amount fractions of various compounds in mixtures, obtaining results with a low measurement uncertainty associated [56]. Validation of ^{1}H qNMR (with internal standard) showed that it can provide the same quantification performance and accuracy than traditional primary methods [57]. Also, purity determination was described by ^{1}H and ^{13}C NMR techniques, achieving results consistent with gravimetry. ^{13}C NMR shows lower sensitivity than ^{1}H NMR, but it could be an alternative for analytes with no adequate ^{1}H signals for quantification [58].

Development of new quantitative applications based on nuclei other than ^{1}H and ^{13}C and characterization of mixtures by DOSY experiments are current issues, extending the scope of NMR to the quantification and quality evaluation of drugs, plant extracts, or polymers [59]. New certified RMs for qNMR, with ^{31}P and ^{19}F are being developed [54]. These atoms are present in molecules in a reduced number, so very few signals will appear in the NMR spectrum. These new materials are interesting because interferences in the measured signals will be lower since signals are better separated due to their larger spectral width and interferences with residual solvent are not usual (common deuterated NMR solvents do not contain these atoms). Due to this, they will be useful for the quantification of several pesticides or antibiotics.

Other methods to determine the amount of substance with the potential of being primary methods are optical absorption spectrometry (based on the Lambert–Beer's law) [41] and methods based on colligative properties (freezing-point depression or boiling-point elevation).

Researchers have described novel potential primary methods for certain applications, which need further investigation to be internationally accepted. Some examples are:

- High-performance liquid chromatography-circular dichroism (CD) for the quantification of chiral compounds and proteins, with mathematical equations that describe the method and uncertainties contributions completely defined [60].
- Calibration-free concentration analysis (CFCA) based on surface plasmon resonance for the differentiation between active proteins and total pure proteins, with measurement principles expressed in mathematical formulae and results traceable to the SI [61].
- A method based on flow cytometry for microbiological quantification [62], which shows good agreement with the plate-based method, can be completely described by mathematical equations and is traceable to the SI, with relative low expanded uncertainty (low for this specific field).

In summary, primary methods are an essential part of the metrology system, providing traceability through a direct link to SI units, and show several applications, like the determination of purity of substances or the certification of RMs. Although initially the number of primary methods was small, current research is showing several methods with the potential of being primary methods for certain novel applications, which fulfill the characteristics listed by CCQM. The field of primary methods will be always in permanent further development since new methods will be needed to meet new requirements.

3.5 Reference materials: characteristics, applications, and producers

As stated earlier in this chapter, RMs and certified RMs (CRMs) are used in qualitative and quantitative methods. An RM is a "material sufficiently homogeneous and stable with respect to one or more specified properties, which has been established to be fit for its intended use in a measurement process" [63]. Four important properties characterize all RMs (Figure 3.13):

- A property, qualitative or quantitative, defined by an element or a compound dissolved in an appropriate solvent or being part of a matrix.
- The homogeneity between all the units of the material, an essential characteristic to guarantee the equivalence between measurements performed in several laboratories with different units of the material.
- The stability of the material in different temperature conditions along the time. The period of validity of the material and the appropriate transport conditions are defined from this characteristic.
- The intended use of the material, necessary for its initial design, and final use.

Different forms of RMs can be available [64–66]:
- Pure substances, characterized by their purity or by the presence of trace impurities.
- Standard solutions and gas mixtures, mainly used for calibration purposes and usually prepared from pure substances using gravimetric steps.
- Matrix RMs, prepared from natural matrices with the appropriate components or from fortified synthetic matrices and characterized by the composition of certain constituents.
- Physico-chemical RMs, defined by certain properties as melting point, viscosity, pH, flash point, hardness, etc.
- Methodologically defined RMs, characterized by an analytical protocol for certain parameters as bio-availability of an element, soluble fraction, etc.
- Reference objects or artifacts, defined by functional properties as taste, odor, etc.
- RMs with an appropriate matrix composition for the calibration of certain measuring instruments.

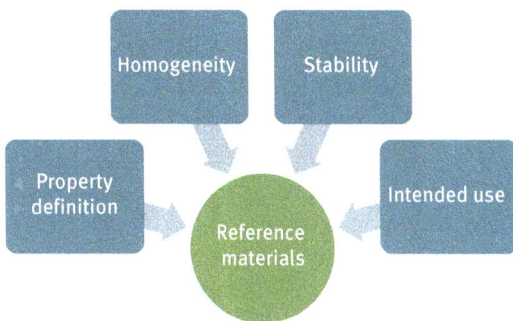

Figure 3.13: Properties of reference materials.

Also RMs can be classified according to the method used for their characterization. Primary RMs are those materials established (and/or certified) using a primary method of measurement [67], which enables its traceability. Secondary RMs are defined according to primary RMs and are used to characterize working RMs. Traceability across different categories of RMs can be represented using a pyramid from the International System of Units (SI) to working materials and routine methods used in testing and calibration laboratories (Figure 3.14).

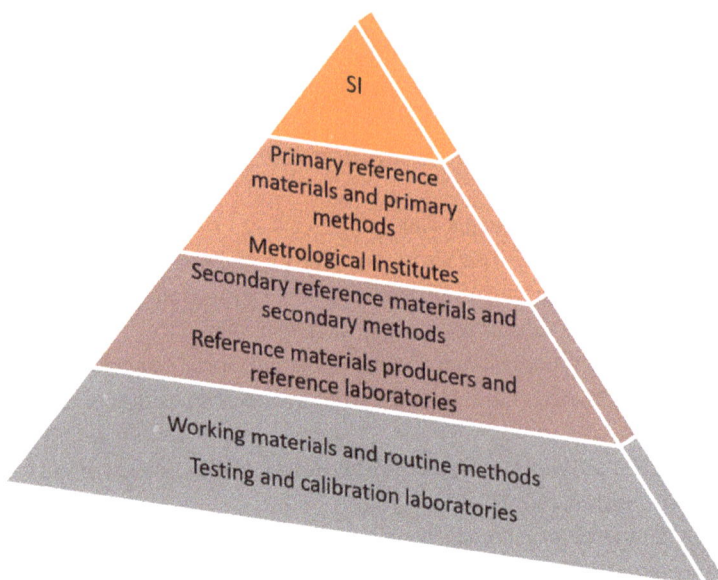

Figure 3.14: Traceability chain from the SI to working materials and routine methods.

Considering their composition (Figure 3.15), RMs can be single substance RMs or matrix RMs [68, 69]. The first category is formed by pure chemicals or solutions of pure chemicals, usually with a low uncertainty associated, which enables them to be used for calibration. In this case they are also called calibration standards or calibrators. The second category includes materials which contain the analytes in their natural form present in the environment, naturally collected or fortified, although spiked materials sometimes are not representative of real samples. The matrix is chosen to resemble the matrix of the samples that are going to be tested and the reference values of the analytes must be clearly established and correspond to the values of the problem samples. These RMs show higher uncertainties and are used to compare analytical methods and determine their characteristics, considering all the steps (Figure 3.16): extraction, clean-up, concentration, and measurement [69]. This is particularly important for analytes strongly bound to the matrix, for example, for certain contaminants present in food because this can affect the extraction efficiency [70], and for clinical RMs, which must show properties similar to those of human samples, that is, commutability [71].

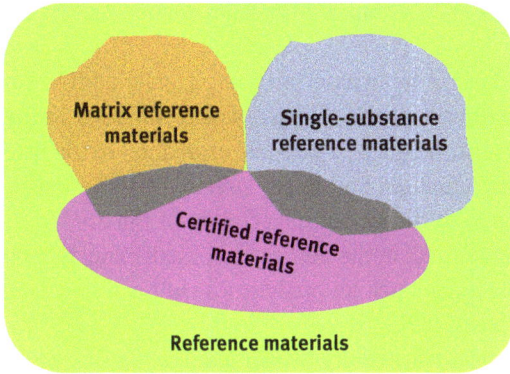

Figure 3.15: Types of reference materials.

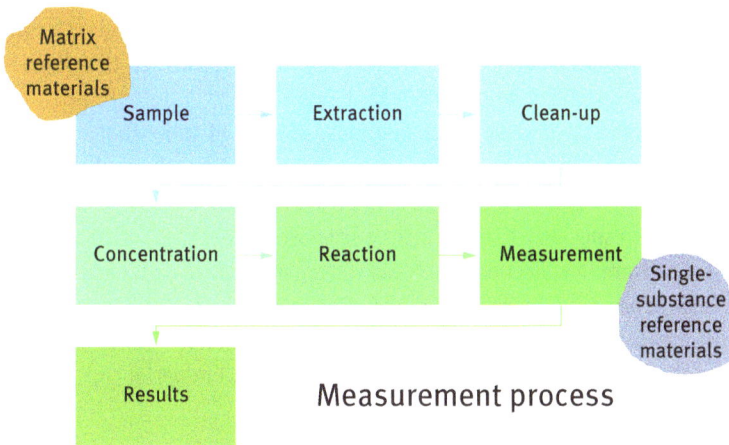

Figure 3.16: Use of single-substance reference materials and matrix reference materials in a measurement process.

The suitability of a matrix RM will depend on its correspondence with the matrix of the routine samples. Due to this sometimes they are "fit for purpose" [72]. There are a lot of specific needs of matrix RMs in all the fields of analytical science and if the need is not common the end user must develop its own laboratory RM, stable and homogeneous to carry out quality control checks. In this sense, European commission has supported the development of RMs through consecutive Framework Programmes, obtaining a high number and variety of materials [73, 74].

A CRM is a "reference material characterized by a metrologically valid procedure for one or more specified properties, accompanied by an RM certificate that provides the value of the specified property, its associated uncertainty, and a statement of metrological traceability" [63]. Hence, CRMs are a group of RMs characterized by two additional properties (Figure 3.17):

– A property value formed by a number with an associated uncertainty and the indication of its most relevant contributions, and the coverage factor used in the calculations. The reference value must be the best possible estimation of the true value of the material, being largely unbiased and highly precise [75]. The uncertainty must be small in comparison to the global uncertainty of the analytical method in which the material will be used, in the best cases is small enough to be ignored [76]. This is an important issue since an analytical result can never be better than the RM on which it is based [69].

– The establishment of the metrological traceability, which assures that the property value can be related to the appropriate units or references (preferable SI units) through an unbroken chain of comparisons. This property guarantees that all the measurements of the same magnitude are comparable because they share the same reference point, and that the uncertainty of the measurement can be correctly estimated [77]. Metrological traceability is indicated through the characterization method, the property, and the reference used.

– Appropriate references can be reference methods, RMs or SI units that must be clearly related to the property of the CRM produced. The unbroken chain of comparisons must be established to assure that there are no loss of information during the analytical procedure, especially if it requires successive experimental steps, long times of storage, or sample collection. The uncertainty associated with each step of the analytical procedure must be estimated and used to obtain the final uncertainty of the RM; this means that the smaller the chain of comparisons the better the final uncertainty and the more extensive the applications of the final material [78].

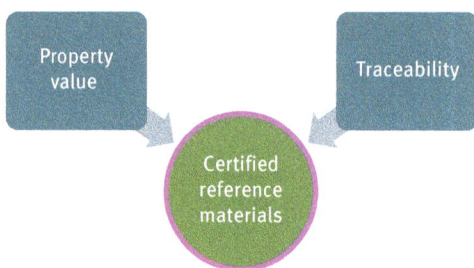

Figure 3.17: Additional properties of certified reference materials.

Sometimes RMs which are not accompanied by a certificate are defined by their use, for example, proficiency testing materials, laboratory RMs, or quality control materials [68, 79]. Other names from the literature are in-house RMs or laboratory control materials, although these terms are associated with materials produced by an analytical laboratory for its own use rather than by an external provider [80].

To avoid economical costs, final users must know the characteristics needed for their application and select the cheapest material that fulfills their requirements. For example, for analytical quality control an important characteristic will be the minimum sample intake since this material is going to be used every day or, sometimes, in an even more frequent interval. Nevertheless, for calibration purposes this parameter will be not critical since the frequency of use will be lesser [68]. In this last application a much more important characteristic is the uncertainty associated with the material; sometimes CRMs cannot be used for calibration purposes due to their too large uncertainty [79]. To select an appropriate CRM for a specific measurement method, some characteristics must be checked: its concentration must be in the same level as the measurement process, its matrix must resemble as close as possible the matrix of the measured samples, its form should be the same as the samples, its quantity must be enough to perform a complete series of experiments, its properties should be stable during the experiments, its uncertainty must be compatible with the requirements of precision and trueness of the method, and the commutability of the material and the measured samples must be stated where applicable.

Some users suggest that prices associated with RMs are very high and that production of in-house RMs will be cheaper. This approach must be considered because the production system needed to obtain an RM is complicated, with a lot of stages (Figure 3.18) performed by competent staff in calibrated equipment, with several quality controls that guarantee the reliability of data associated with the material, necessary for its use. If an independent organization decides to produce its own in-house RMs must consider the cost associated with the work needed to obtain the materials and to the time expended in these activities [68]. Also, the economic impact of using unsuitable materials must be considered, obtaining incorrect results and the loss on good image. Despite all this, as previously indicated, in some cases a laboratory needs to produce its own RMs due to various reasons; for example, the lack of appropriate commercially available ones, a very usual situation in R&D laboratories, or the need of a material that matches with a specific type of routine samples.

RMs must be sold accompanied by a document, sometimes called RM document or product information sheet [81], containing some mandatory items: title of the document, exact identification of the RM (e.g., indicating code and batch), name of the RM (which should provide a detailed description of the material), intended use, minimum sample size to be used (if it is applicable), period of validity, commutability (if it is required), and instructions for storage, handling, and use. The version of this document must be indicated, together with the total number of pages, to guarantee the traceability of the included data. On the other hand, CRMs must be accompanied by a certificate of analysis, containing all the mandatory information previously listed and the following additional items: description of the material, property of interest, property value and associated uncertainty, a statement of metrological traceability, measurement methods (for method-dependent measurands), and name and function of the approving personnel. Other information can be available in these documents, as

Figure 3.18: Steps for production of reference materials.

the measurement methods used for method independent measurands, health and safety information, use of subcontractors, indicative values, legal notice, or reference to other documents. Users of RMs (certified or noncertified) must check that documents provided contain all the mandatory information since its absence can be an indication that their quality and properties have not been correctly established.

Applications of RMs [65] are related to the previously commented characteristics and are different for certified and noncertified materials. The most important are:
– Calibration of methods and equipment
– Method validation
– Assignment of values to materials
– Quality control

Method validation includes determination of bias and precision. Bias is essential to guarantee the traceability of the results obtained whereas precision checks the laboratory standard deviation usually under repeatability conditions. These two characteristics can also be used in quality control during the routine use of the method. Assignment of values to materials is a widely used application, for example, pure materials are used to prepare solutions which are diluted to be used as calibrators, and the final concentration can be calculated using initial data and preparation done.

If the application needs an establishment of traceability only CRMs can be used, for example, the calibration of equipment or the bias control. However, if the traceability is not necessary, for example, in precision control, noncertified RMs can be used (Figure 3.19).

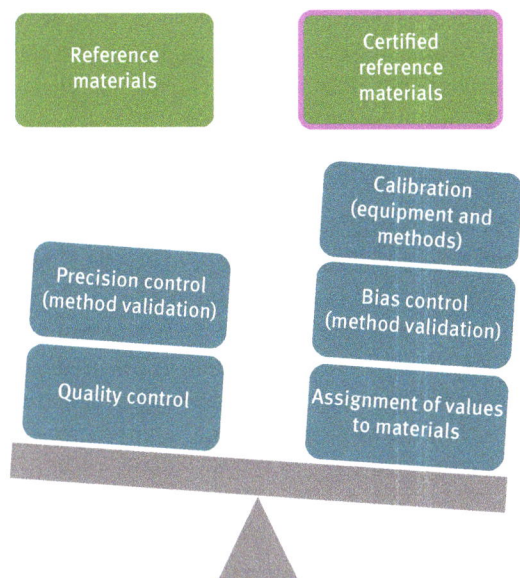

Figure 3.19: Applications of reference materials.

RMs have been used for a long time, maybe the first use was issued by the Association of Official Agricultural Chemists in 1885, when they distributed six fertilizer samples with the aim of improving interlaboratory agreement [76]. Nowadays their use has achieved every field of analytical measurements, with their production and applications internationally recognized. In food industry, for example, RMs are essential to determine the composition of food [82–84], to ensure the safety of products [85], or to assure their healthy properties [86]. Other field with several applications is clinical diagnosis; some examples are matrix RMs which fulfill the requirements of commutability with human samples [87], CRMs for the expanding field of clinical mass spectrometry [88], SARS-CoV-2 CRMs to compare sensitivity and specificity of fast kits for COVID-19 control [89] or CRMs for the measurement of antibodies related to autoimmune diseases [90]. Finally, environmental control can be cited as other field where the need of new RMs appears continuously, for example, different matrices, like fish and shellfish, or ocean, coastal, and estuarine waters, with different certified analytes, like trace elements, nutrients, and organic halogenated and nonhalogenated compounds, to study the evolution of marine ecosystems [91].

Interest in RMs has grown due to last edition of ISO 17025 [92] for testing and calibration laboratories. In previous edition of this standard [93] the use of RMs was considered as optative and other types of materials were mentioned as appropriate, without clear indications of the needed properties. Also, there were references to competent materials providers without indication of their characteristics or the needed standards to comply. In the last edition of ISO 17025 [92], the establishment of traceability using RMs is only possible if these materials are provided by competent producers with metrological traceability to SI. Standard considers explicitly that RMs producers which comply with requirements of ISO 17034 standard [94] are competent. It also states the use of RMs and CRMs from competent providers for different activities related to equipment, such as calibration or quality control, for method calibration and for quality assurance.

Requirements that RMs producers must fulfill were initially included in ISO Guide 34 [95], which is now obsolete and has been substituted by ISO 17034 [94]. Both documents are similar, but the new standard has been elaborated due to the increased interest in the production and use of RMs and the need of an international standard to be used by certain accreditation bodies [96]. A definition of RM producer appears in both documents, which includes their main functions: project planning and management, assignment and authorization of property values and related uncertainties and issue of certificates and other relevant information which accompanies the materials.

Requirements for RMs producers are classified in five categories [94]: general, structural, resource, technical, and production and management system (Figure 3.20). This classification adopts the common structure of other international standards about conformity assessment. RMs producers are conformity assessment organisms, which guarantee that their products meet the relevant requirements, like calibration and testing laboratories, certification and inspection bodies, or proficiency testing programs providers.

Figure 3.20: Requirements for reference materials producers according to ISO 17034.

General, structural, and management requirements are related to internal organization of producers, document control, customer service, internal structure, corrective actions, internal audits, and management reviews. In these categories are included requirements related to confidentiality and impartiality and the actions to address risks and opportunities, according to the new edition of ISO 9001 [97]. Management requirements can be demonstrated by a certificate according to ISO 9001 provided by an accredited certification body.

Resource, technical, and production requirements include several points related with personnel, use of subcontractors, equipment, procurement of services and supplies, facilities, production planning and control, material handling and storage, characterization, homogeneity and stability, assignment of values and uncertainties, traceability, RMs documents and labels, distribution services, quality and technical records, management of nonconforming work, and complaints.

Several activities related to these requirements can be subcontracted to other organizations. In this case producer must demonstrate that requirements are achieved, and that essential information is provided to guarantee that activities are performed in an appropriate way. There are only four activities that should be always developed by the producer: production planning, selection of subcontractors, assignment and authorization of property values and uncertainties, and authorization of certificates or other documents which accompany RMs.

One important activity is the characterization of the materials, which can be done by different approaches, guaranteeing the traceability of the characterization. Some of the possible schemes are [94]:
- A single reference measurement procedure in a single laboratory
- Two or more methods of demonstrable accuracy in one or several competent laboratories for nonoperationally defined measurands
- A network of competent laboratories for an operationally defined measurand

Extended explanation about used schemes to characterize RMs and approaches to evaluate their homogeneity, stability, purity, and qualitative properties can be checked in ISO Guide 35 [98].

The capability of an RM producer is demonstrated by an accreditation, issued by an external body according to ISO 17034, completed with a technical annex with the extent of the accreditation and the list of included RMs. Accreditation is an independent declaration of the competence of an organization, the suitability of its methods, the appropriateness of its equipment and facilities, and the assurance provided by its internal quality control. It also guarantees its technical competence and the independence of the personnel and the organization. In accreditations granted to RMs producers with technical and quality management historical capabilities, an open scope can appear in the technical annex, completed on demand by a document with the list of included RMs (Figure 3.21).

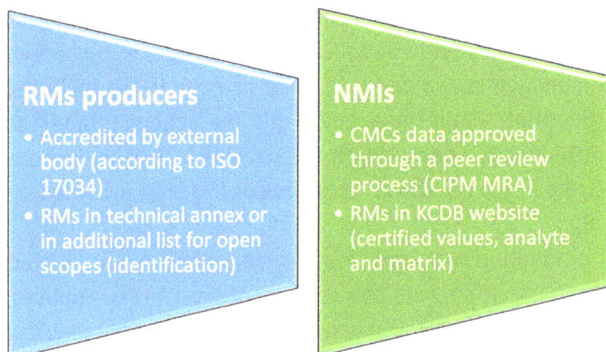

Figure 3.21: Competent reference material providers: conditions to comply.

National Metrology Institutes are an option to provide CRMs. In this case capability of the institute is demonstrated by data collected in CIPM Mutual Recognition Arrangement (MRA) database (KCDB, key comparison and calibration database). CIPM promotes the worldwide uniformity in measurements, with the aim of comparing results obtained in different countries [99]. KCDB [100] is the resource where calibration and measurement capabilities (CMCs) of participating institutes, approved by a peer-review process within the CIPM MRA, can be checked [36]. In KCDB database range of certified values in RMs for each analyte and matrix can be checked in order to demonstrate the capability of each institute as an RM producer (Figure 3.21).

RMs can be found in catalogues and websites of producers. Moreover, some international databases have been developed to compare their characteristics and guarantee their confidence. COMAR (Code d'Indexation des Matériaux de Référence) database has been proposed in late 1970s and currently is free of charge for users and available through its website [101]. It contains materials from several fields of application with complete information about producer, code, name, properties, intended use, and certificates and reports. Search in database can be done according to keywords and can be limited to CRMs, materials under an ISO 17034 accreditation, and/or materials produced by a National Metrology or Designated Institute. Although initially this database contained all types of RMs, actual recent edition only refers to matrix RMs. Also, the Technical Division on Reference Materials (TDRM) of AOAC International has developed its own RMs database [102]. It is an online searchable database [103] where RMs can be searched by analyte, by analyte and matrix, and by their use in AOAC Official Methods of Analysis. This database provides useful information to compare different RMs and the information needed to obtain their complete certificates of analysis.

In summary, RMs are a main topic in analytical research [104, 105], with several producers and international metrology organizations improving their characterization, increasing their fields of utilization and trying to cover customer needs. It is supposed that their use will increase in laboratories worldwide due to their interesting characteristics and their broad range of applications.

Keywords: Quantitative, qualitative, primary method, reference material, quality control, performance characteristics, validation

References

[1]   Commission Decision 2002/657/EC of 12 August 2002 implementing Council Directive 96/23/EC concerning the performance of analytical methods and the interpretation of results (Text with EEA relevance) (notified under document number C(2002) 3044). European Commission, 2002.

[2]   Trullols E, Ruisánchez I, Rius FX. Validation of qualitative analytical methods. TrAC Trends Analyt Chem. 2004;23(2):137–45.

[3]   Creydt M, Fischer M. Food phenotyping: Recording and processing of non-targeted liquid chromatography mass spectrometry data for verifying food authenticity. Molecules. 2020; 25(17):3972.

[4]   Eysseric E, et al. Non-targeted screening of trace organic contaminants in surface waters by a multi-tool approach based on combinatorial analysis of tandem mass spectra and open access databases. Talanta. 2021;230:122293.

[5]   Wright EJ, Beach DG, McCarron P. Non-target analysis and stability assessment of reference materials using liquid chromatography–high-resolution mass spectrometry. Anal Chim Acta. 2022;1201:339622.

[6]   Kato LS, Conte-Junior CA. Safety of plastic food packaging: The challenges about non-intentionally added substances (NIAS) discovery, identification and risk assessment. Polymers (Basel). 2021;13:13.

[7]   Tkalec Ž, et al. Suspect and non-targeted screening-based human biomonitoring identified 74 biomarkers of exposure in urine of Slovenian children. Environ Pollut. 2022;313:120091.

[8]   Petrozzi S. Practical instrumental analysis. Methods, quality assurance and laboratory management. WILEY-VCH Verlag GmbH & Co. KGaA., 2013. Weinheim, Germany.

[9]   Recommended methods of analysis and sampling. CXS 234-1999. Codex Alimentarius. FAO. WHO, 1999.

[10]   Official methods of analysis of AOAC INTERNATIONAL, AOAC International. Oxford University Press, 2023.

[11]   Commission Regulation (EU) No 519/2014 of 16 May 2014 amending Regulation (EC) No 401/2006 as regards methods of sampling of large lots, spices and food supplements, performance criteria for T-2, HT-2 toxin and citrinin and screening methods of analysis. European Union, 2014.

[12]   Commission Regulation (EU) 2017/644 of 5 April 2017 laying down methods of sampling and analysis for the control of levels of dioxins, dioxin-like PCBs and non-dioxin-like PCBs in certain foodstuffs and repealing Regulation (EU) No 589/2014. European Union, 2017.

[13]   Valcárcel M, Cárdenas S, Gallego M. Sample screening systems in analytical chemistry. TrAC Trends Analyt Chem. 1999;18(11):685–94.

[14]   Cárdenas S, Valcárcel M. Analytical features in qualitative analysis. TrAC Trends Analyt Chem. 2005;24(6):477–87.

[15]   Ríos A, et al. Quality assurance of qualitative analysis in the framework of the European project 'MEQUALAN'. Accredit Qual Assur. 2003;8(2):68–77.

[16]   Eurachem Guide: Terminology in analytical measurement – Introduction to VIM 3. 2nd ed. 2023. V.J. Barwick (Ed). Available from www.eurachem.org.

[17] Ellison SLR, Fearn T. Characterising the performance of qualitative analytical methods: Statistics and terminology. TrAC Trends Analyt Chem. 2005;24(6):468–76.

[18] Macarthur R, von Holst C. A protocol for the validation of qualitative methods of detection. Anal Methods. 2012;4(9):2744–54.

[19] Pulido A, et al. Uncertainty of results in routine qualitative analysis. TrAC Trends Analyt Chem. 2003;22(9):647–54.

[20] Ellison SLR. Uncertainties in qualitative testing and analysis. Accredit Qual Assur. 2000;5(8):346–48.

[21] Eurachem / CITAC Guide: Assessment of performance and uncertainty in qualitative chemical analysis. 1st ed. 2021. R. Bettencourt da Silva and S.L.R. Ellison (Eds). Available from www.eurachem.org.

[22] Valcárcel M. Qualitative aspects of analytical chemistry. In: Principles of analytical chemistry: A textbook. M. Valcárcel. Springer, 2000;201–45. Berlin, Germany.

[23] Simonet BM. Quality control in qualitative analysis. TrAC Trends Analyt Chem. 2005;24(6):525–31.

[24] Eurachem Guide: The fitness for purpose of analytical methods – A laboratory guide to method validation and related topics. 2nd ed. 2014. B. Magnusson and U. Örnemark (Eds). Available from www.eurachem.org.

[25] Evans EH, Foulkes ME. Analytical chemistry a practical approach. OXFORD University Press, 2019. Oxford, United Kingdom.

[26] Guidance Document on Pesticide Analytical Methods for Risk Assessment and Post-approval Control and Monitoring Purposes, SANTE/2020/12830, Rev. 2, 14 February 2023. 2023.

[27] Flanagan RJ, et al. Basic laboratory operations. In: Fundamentals of analytical toxicology. Clinical and forensic. R.J. Flanagan et al. 2nd ed. John Wiley & Sons, Ltd, 2020;52–93. West Sussex, United Kingdom.

[28] Eurachem/CITAC Guide: Use of uncertainty information in compliance assessment. 2nd ed. 2021. A. Williams and B. Magnusson (Eds). Available from www.eurachem.org.

[29] Commission Regulation (EU) 2015/705 of 30 April 2015 laying down methods of sampling and performance criteria for the methods of analysis fo the official control of the levels of erucic acid in foodstuffs and repealing Commission Directive 80/891/EEC. European Union, 2015.

[30] Eurachem/CITAC Guide: Metrological traceability in chemical measurement. A guide to achieving comparable results in chemical measurement. 2nd ed. 2019. S.L.R. Ellison and A. Williams (Eds). Available from www.eurachem.org.

[31] Corbisier P, Emons H. Towards metrologically traceable and comparable results in GM quantification. Anal Bioanal Chem. 2019;411(1):7–11.

[32] Hovind H, et al. Internal quality control – Handbook for chemical laboratories. NT Technical Report 569. 2011:1–46. Nordic Innovation Center. Oslo, Norway. Available from www.nordtest.info.

[33] ISO/IEC 17043 Conformity assessment – General requirements for the competence of proficiency testing providers. ISO/CASCO, 2023.

[34] Eurachem Guide: Selection, use and interpretation of proficiency testing schemes. 3rd ed. 2021. B. Brookman and I. Mann (Eds). Available from www.eurachem.org.

[35] de la Guardia M, Garrigues S. The concept of green analytical chemistry. In: de la Guardia M, Garrigues S, eds. Handbook of green analytical chemistry. John Wiley & Sons, Ltd., 2012;3–16. West Sussex, United Kingdom.

[36] Wielgosz RI. International comparability of chemical measurement results. Anal Bioanal Chem. 2002;374(5):767–71.

[37] Kaarls R. The Consultative Committee for Metrology in chemistry and biology – CCQM. J Chem Metrol. 2018;12(1):1–16.

[38] Milton MJT, Quinn TJ. Primary methods for the measurement of amount of substance. Metrologia. 2001;38(4):289.

[39] Duewer DL, et al. An approach to the metrologically sound traceable assessment of the chemical purity of organic reference materials. NIST Special Publication 1012, 2004.

[40] Valcárcel M. Quantitative aspects of analytical chemistry. In: Principles of analytical chemistry: A textbook. M. Valcárcel. Springer, 2000;247–82. Berlin, Germany.

[41] Richter W. Primary methods of measurement in chemical analysis. Accredit Qual Assur. 1997; 2(8):354–59.

[42] Tshilongo J, et al. Preparation of helium isotope-certified reference materials by gravimetry. Bull Korean Chem Soc. 2015;36(2):591–96.

[43] Canciani G, et al. Controlled Potential Coulometry for the accurate determination of plutonium in the presence of uranium: The role of sulfate complexation. Talanta. 2021;222:121490.

[44] Asakai T, Murayama M. Scheme and studies of reference materials for volumetric analysis in Japan. Accredit Qual Assur. 2008;13:351–60.

[45] Ari B, Can SZ, Bakirdee S. Traceable and accurate quantification of iron in seawater using isotope dilution calibration strategies by triple quadrupole ICP-MS/MS: Characterization measurements of iron in a candidate seawater CRM. Talanta. 2020;209:120503.

[46] Arı B, Bakırdere S. A primary reference method for the characterization of Cd, Cr, Cu, Ni, Pb and Zn in a candidate certified reference seawater material: TEA/Mg(OH). Anal Chim Acta. 2020;1140:178–89.

[47] Bulska E, et al. On the use of certified reference materials for assuring the quality of results for the determination of mercury in environmental samples. Environ Sci Pollut Res Int. 2017;24(9):7889–97.

[48] do Rego ECP, et al. Challenges on production of a certified reference material of ochratoxin A in roasted coffee: A Brazilian experience. J AOAC Int. 2019;102(6):1725–31.

[49] Tran TTH, et al. Certification and stability assessment of recombinant human growth hormone as a certified reference material for protein quantification. J Chromatogr B Analyt Technol Biomed Life Sci. 2019;1126–1127:121732.

[50] Bode P, Greenberg RR, De Nadai Fernandes EA. Neutron activation analysis: A primary (ratio) method to determine SI-traceable values of element content in complex samples. Chimia. 2009;63 (10):678–80.

[51] Aregbe Y, Taylor P. CCQM-K24 key comparison Cadmium amount content in rice. Final Report. GE/R/IM/37/02. IRMM Isotope Measurements Unit, 2002.

[52] Noack S, Matschat R. Final report. Key comparison CCQM-K42. Determination of chromium, copper, iron, manganese and zinc in aluminium alloy. BAM Federal Institute for Materials Research and Testing, 2008.

[53] Schoenberger T. Guideline for qNMR analysis. DWG-NMR-001, 2019.

[54] Rigger R, et al. Certified reference material for use in ^{1}H, ^{31}P, and ^{19}F quantitative NMR, ensuring traceability to the International System of Units. J AOAC Int. 2017;100(5):1365–75.

[55] Pauli GF, et al. Importance of purity evaluation and the potential of quantitative ^{1}H NMR as a purity assay. J Med Chem. 2014;57(22):9220–31.

[56] Malz F, Jancke H. Validation of quantitative NMR. J Pharm Biomed Anal. 2005;38(5):813–23.

[57] Miura T, et al. Collaborative study to validate purity determination by ^{1}H quantitative NMR spectroscopy by using internal calibration methodology. Chem Pharm Bull (Tokyo). 2020; 68(9):868–78.

[58] Violante FGM, et al. Use of quantitative ^{1}H and ^{13}C NMR to determine the purity of organic compound reference materials: A case study of standards for nitrofuran metabolites. Anal Bioanal Chem. 2021;413(6):1701–14.

[59] Diehl B, et al. Quo Vadis qNMR? J Pharm Biomed Anal. 2020;177:112847.

[60] Luo Y, et al. A novel potential primary method for quantification of enantiomers by high performance liquid chromatography-circular dichroism. Sci Rep. 2018;8(1):7390.

[61]   Su P, et al. SI-traceable calibration-free analysis for the active concentration of G2-EPSPS protein using surface plasmon resonance. Talanta. 2018;178:78–84.

[62]   Liu S, et al. Evaluation of volume-based flow cytometry as a potential primary method for quantification of bacterial reference material. Talanta. 2023;255:124197.

[63]   ISO Guide 30 reference materials – Selected terms and definitions. ISO, 2015.

[64]   The selection and use of reference materials, EEE-RM working group. European cooperation for Accreditation, 2003.

[65]   ISO Guide 33 reference materials – Good practice in using reference materials. ISO, 2015.

[66]   Maier EA. Certified reference materials for the quality control of measurements of industrial effluents and waste. TrAC Trends Analyt Chem. 1996;15:341–48.

[67]   De Bièvre P, Taylor PDP. Traceability to the SI of amount-of-substance measurements: From ignoring to realizing, a chemist's view. Metrologia. 1997;34(1):67.

[68]   Emons H, Linsinger TPJ, Gawlik BM. Reference materials: Terminology and use. Can't one see the forest for the trees? TrAC Trends Analyt Chem. 2004;23(6):442–49.

[69]   Walker R, Lumley I. Pitfalls in terminology and use of reference materials. TrAC Trends Analyt Chem. 1999;18(9):594–616.

[70]   Lauwaars M, Anklam E. Method validation and reference materials. Accredit Qual Assur. 2004;9 (4):253–58.

[71]   Panteghini M. Traceability, reference systems and result comparability. Clin Biochem Rev. 2007;28 (3):97–104.

[72]   Amigo JM, et al. Emerging needs for sustained production of laboratory reference materials. TrAC Trends Analyt Chem. 2004;23(1):80–85.

[73]   Maier EA, Boenke A, Mériguet P. Importance of the certified reference materials programmes for the European Union. TrAC Trends Analyt Chem. 1997;16(9):496–503.

[74]   Quevauviller P. Reference materials: An inquiry into their use and prospects in Europe. TrAC Trends Analyt Chem. 1999;18(2):76–85.

[75]   Kane JS. Fitness-for-purpose of reference material reference values in relation to traceability of measurement, as illustrated by USGS BCR-1, NIST SRM 610 and IAEA NBS28. Geostand Newsl. 2002;26(1):7–29.

[76]   Kane JS. The use of reference materials: A tutorial. Geostand Newsl. 2001;25(1):7–22.

[77]   Thompson M. Comparability and traceability in analytical measurements and reference materials. Analyst. 1997;122(11):1201–06.

[78]   Quevauviller P. Traceability of environmental chemical measurements. TrAC Trends Analyt Chem. 2004;23(3):171–77.

[79]   Emons H. The 'RM family' – Identification of all of its members. Accredit Qual Assur. 2006;10 (12):690–91.

[80]   Venelinov T, Sahuquillo A. Optimizing the uses and the costs of reference materials in analytical laboratories. TrAC Trends Analyt Chem. 2006;25(5):528–33.

[81]   ISO Guide 31 reference materials – Contents of certificates, labels and accompanying documentation. ISO, 2015.

[82]   Phillips MM, Sharpless KE, Wise SA. Standard reference materials for food analysis. Anal Bioanal Chem. 2013;405(13):4325–35.

[83]   Sharpless KE, et al. Filling the AOAC triangle with food-matrix standard reference materials. Anal Bioanal Chem. 2004;378(5):1161–67.

[84]   Sharpless KE, et al. Standard reference materials for foods and dietary supplements. Anal Bioanal Chem. 2007;389(1):171–78.

[85]   Wargo WF. Reference materials: Critical importance to the infant formula industry. J AOAC Int. 2017;100(5):1376–78.

[86] Chen W, et al. Development of certified reference materials for four polyunsaturated fatty acid esters. Food Chem. 2022;389:133006.

[87] Schimmel H, Zegers I. Performance criteria for reference measurement procedures and reference materials. Clin Chem Lab Med. 2015;53(6):899–904.

[88] Lingxiao S, et al. The application of certified reference materials for clinical mass spectrometry. J Clin Lab Anal. 2022;36(4):e24301.

[89] Wang D, et al. Validation of the analytical performance of nine commercial RT-qPCR kits for SARS-CoV-2 detection using certified reference material. J Virol Methods. 2021;298:114285.

[90] Monogioudi E, Zegers I. Certified reference materials and their need for the diagnosis of autoimmune diseases. Mediterr J Rheumatol. 2019;30(1):26–32.

[91] Quevauviller P. Marine chemical monitoring. Policies, techniques and metrological principles. In: Mariotti A, eds. Earth systems – Environmental engineering. ISTE Ltd, 2016. London, United Kingdom.

[92] ISO/IEC 17025 General requirements for the competence of testing and calibration laboratories. ISO/IEC, 2017.

[93] ISO/IEC 17025 General requirements for the competence of testing and calibration laboratories. ISO/IEC, 2005.

[94] ISO 17034 General requirements for the competence of reference material producers. ISO, 2016.

[95] ISO Guide 34 General requirements for the competence of reference materials producers. ISO, 2009.

[96] Trapmann S, et al. The new International Standard ISO 17034: General requirements for the competence of reference material producers. Accredit Qual Assur. 2017;22(6):381–87.

[97] ISO 9001 Quality management systems – Requirements. ISO, 2015.

[98] ISO Guide 35 reference materials – Guidance for characterization and assessment of homogeneity and stability. ISO, 2017.

[99] https://www.bipm.org/en/committees/ci/cipm. Access 2023.

[100] https://www.bipm.org/kcdb/. Access 2023.

[101] https://www.comar.org/index.htm. Access 2023.

[102] Zink D. AOAC INTERNATIONAL's Technical Division on Reference Materials (TDRM) reference materials database. J AOAC Int. 2016;99(5):1146–50.

[103] http://tdrmdb.aoac.org/. Access 2023.

[104] Phillips M, Emteborg H. Who needs reference materials? J AOAC Int. 2017;100(5):1355.

[105] Wise SA, Emons H. Reference materials for chemical analysis. Anal Bioanal Chem. 2015; 407(11):2941–43.

M. Carmen Louzao and Celia Costas

4 Toxicological studies with animals

4.1 History of toxicity studies with animals and legislation

The history of toxicity studies begins with Paracelsus (1493–1541), physician, alchemist, and astrologer regarded as the father of toxicology, when he demonstrated the harmless and beneficial effects of toxins and proved dose–response relationships for the effects of drugs. Mateo Orfila (1787–1853), a Spanish physician father of modern toxicology, determined the relationship between poisons and their biological properties and demonstrated specific organ damage caused by toxins [1].

In 1831, British physiologist Marshall Hall proposed five principles that should govern animal experimentation. First, an experiment should never be performed if the necessary information could be obtained by observations; second, no experiment should be performed without a clearly defined objective; third, scientists should be well-informed about the work of their predecessors in order to avoid unnecessary repetition of an experiment; fourth, justifiable experiments should be carried out with the least possible infliction of suffering; and finally, every experiment should be performed under circumstances that would provide the clearest possible results, thereby diminishing the need for repetition of experiments. Many of his recommendations formally instituted over a century later in the British Animals (Scientific Procedures) Act and the U.S. Animal Welfare Act.

Toxicological screening methods and toxicological research on individual substances developed in the mid-1900s. The twentieth century brought the conception that health can be impaired because of the ingestion of a single dose but also of exposure to small chemical doses over an extended period of time. There was a shift in focus from acute to chronic toxicity [2]. The use of animals in toxicity studies began in 1920, when J. W. Trevan proposed the use of the 50% lethal dose (LD_{50}) test to determine the lethal dose of individual chemicals [3]. This test became accepted as the standard metric of toxicity. After the introduction of LD_{50}, an FDA (Food and Drug Administration) scientist John Draize developed a method for testing eye and skin irritation using rabbits, which was widely accepted to evaluate the effects of chemicals and pharmaceuticals on these organs [4]. Later, the US National Cancer Institute developed a test to identify carcinogenic chemicals through the daily dosing of rats and mice for 2 years.

In the early 1960s, thousands of babies were born with birth defects caused by thalidomide. After this, the regulatory agencies concentrated on determining the toxicity profiles of all pharmaceutical substances available for regular patient use and made mandatory the submission of toxicity profiles of investigational new drugs. Toxicity testing with animals is necessary to prove that new drugs are safe before clinical

https://doi.org/10.1515/9783111014449-004

trials and first administration to humans, also to know side effects of drugs or to evaluate the harmful effects of toxins. Therefore, animal testing plays a large role in research and drug development, and it occurs regularly throughout the European Union (EU). Rats and mice are the most used animals in laboratories. Recent years have seen a surge in the use of zebrafish and nonhuman primates. However, while ethical concerns regarding zebrafish are low, those involving nonhuman primates are high. Therefore, the protection and welfare of animals used for scientific purposes is an area covered by EU laws.

The publication of "The Principles of Humane Experimental Technique" by Russell and Burch in 1959 marks the birth of the principle of the "Three Rs" the replacement, reduction, and refinement of animal testing [5]. Directive 2010/63/EU [6], updated and replaced the 1986 Directive 86/609/EEC, is an indispensable tool at the EU level to protect experimental animals. The aim of the new Directive is to strengthen legislation and improve the welfare of those animals. In addition, the Directive implements the Three Rs in Europe that took full effect on 1 January 2013 on the protection of animals used for scientific purposes [6].

4.1.1 The Three Rs

Animals are widely used for toxicological research and testing despite efforts to eliminate the use of animals and the availability of alternative methods [7]. The welfare of animals in toxicology research is very important and has been incorporated into international legislation regulating the use of animals in scientific procedures [8]. There are many benefits and good ethical, scientific, legal, and economic reasons for making sure that animals are looked after properly and used in minimum numbers [9]. If an animal is suffering stress or pain, it could affect the results of the research; therefore good animal welfare can improve the quality of science.

The guiding principles supporting the use of animals in scientific research are the Three Rs: Replace, Reduce, Refine.

- **Replace** the use of animals with alternative techniques or avoid the use of animals altogether if that is possible.
- **Reduce** the number of animals used to a minimum to obtain information from fewer animals or more information from the same number of animals.
- **Refine** the way experiments are carried out to make sure animals suffer as little as possible. This includes better housing and improvements to procedures which minimize pain and suffering and/or improve animal welfare.

The principles of Replacement, Reduction, and Refinement must be considered systematically when animals are used for the purposes of basic, translation and applied research, regulatory testing, and production as well as for the purposes of education and training in the EU. The facility, personnel involved, and experiments with animals should comply with the legislation on experimentation with animals. In some cases, it is possible to develop a whole new way of conducting a test involving fewer animals. For example, the acute toxicity test with lots of animals was used for many years to find out how toxic chemicals are. But nowadays scientists develop better tests to do the same job but using fewer animals. The European Union Reference Laboratory for alternatives to animal testing (EURL-ECVAM) is located at the Joint Research Centre of the European Commission. The EURL-ECVAM is actively involved in search for test methods for the Three Rs in animal testing and cover areas such as acute mammalian systemic toxicity, aquatic toxicity and bioaccumulation, genotoxicity, skin sensitization, and toxicokinetics (TK).

In the late 1980s, the Organisation for Economic Co-operation and Development (OECD) and the International Conference on Harmonization (ICH) brought out the guidelines for toxicity testing of pharmaceutical substances. Test Guidelines Programme of the OECD has developed standardized methods of testing that are accepted by all OECD member countries through an agreement on the mutual acceptance of data [10]. The OECD has substantially reduced the total number of animals used for certain standard tests. It also provides a focus for the introduction of new methods according to the Three Rs. Besides regulatory agencies such as OECD, Environmental Protection Agency, and FDA administered the Good Laboratory Practices that are regulations to ensure the integrity of data from nonclinical studies [11].

In 2007, the National Research Council issued a report on toxicity testing that recommended move away from the use of animals in laboratory experiments [12]. However, improved in vitro, ex vivo, in silico, or biomarker assays are needed to assess systemic interactions, downstream, and adverse long-term effects without using animals before more reliance can be placed on those assays.

Experimental animals are the biological systems to perform the toxicological testing since they can serve as accurate predictive models of toxicity in humans. Toxicity studies with animals are also needed to predict the therapeutic index of drugs and calculate the benefit–risk ratio becoming the "gold standard" for assessing human risk [13]. The way animals are handled, housed, and treated can have a very strong effect on the physiology and immunology of the animal having impact on the results of testing itself. Thus, adhering to Three Rs guidelines can benefit the outcomes and quality of research. Over time different meanings have been adscribed to the letter "R" [14]. These include the importance of Respect in the treatment of laboratory animals, Responsibility of the investigator and research team toward proper care of the animals or Relevance of a study justifying the use of animals in research, all in the framework of animal welfare, scientific integrity, and social values.

4.2 Experimental animals

The importance of animal models is unquestionable in terms of in vivo study for the implementation of any biomedical research to humans. In the process of selection of an experimental animal many factors should be taken into consideration such as the compound to test or the resemblance between animal species and humans in terms of physiological and/or pathophysiological aspects [15]. Many animal species such as *Drosophila* (insects), *Danio rerio*, or zebrafish (fish), *Caenorhabditis elegans* (nematodes), *Xenopus* (frogs), and specially mammals such as mice, rabbits, rats, cats, dogs, pigs, and monkeys have been accepted worldwide. The species selected should have the lowest welfare cost as a result of experimentation including transport to the laboratory, captivity, handling, and experimental procedures. The target receptor should be present in the test species, and it should demonstrate appropriate pharmacodynamic response. In the last two decades, researchers also produce custom-made transgenic animal models by incorporating genetic information directly into the embryo either by injecting foreign deoxyribonucleic acid (DNA) or through retroviral vectors [16]. The species chosen should be practical in terms of availability, husbandry, ability to perform procedures and assess adverse effects, and meet compound requirements. In agreement with this, the experimental animals suitable for toxicity studies are both male and female, young and healthy, and nulliparous and nonpregnant in the case of females. According to guidance documents, some tests should be conducted in two laboratory animal species, one rodent (rat, mouse) and one no-rodent (dog, pig, monkey, etc.) [17]. The choice of no-rodent species may be only justifiable if results with rodents are no conclusive or if the special studies require them; in this case a smaller number of animals is used [6].

The animals most used in experimentation are rodents due to low cost of maintenance, short lifespan, and the specific knowledge of their physiology [18]. Rat is preferable when abundant biological samples are required (i.e., blood samples). The use of animals of large size is reserved for specific toxicity studies because they need a bigger comfortable resting area, more food, and a greater amount of test compounds increasing the cost of the study. For instance, rabbits are suitable for dermal (skin and eye) and reproductive toxicity testing and guinea pigs showed basic physiology more like humans than rodents. Dogs, commonly beagles, are often chosen for chronic toxicity while nowadays nonhuman primates are rarely used.

Moreover, the suitability of the experimental animal depends on the specific methods of housing, feeding, or handling. Qualified personnel must handle the animals to minimize anxiety before, during, and after the study. Strict control of the environmental conditions as well as the use of appropriate animal care techniques specific to each species is fundamental in toxicology studies. The following parameters should be taken into account (Table 4.1) [19]

Table 4.1: Specific conditions for housing different animal species and legal minimum standards applicable from 1 January 2017 [19].

	Mouse	Rat	Rabbit	Guinea pigs	Dog	Cat	Nonhuman primates
Light	12 h/day	12 h/day	12 h/day	12 h/day	10–12 h/day	10–12 h/day	12 h/day
Temperature	20–24 °C	20–24 °C	15–21 °C	15–21 °C	15–24 °C	15–21 °C	Not restricted
Humidity	45–65%	45–65%	>45%	45–75%	Not restricted	Not restricted	40–70%
Minimum floor cage area[a] (cm²)	330	800 (<200–600 g) 1,500 (>600 g)	3,500 (<2 kg) 4,500–5,400 (2–6 kg) 6,600 (>6 kg)	1,800 (<150–450 g) 2,500 (450–>700 g)	45,000 (<10–20 kg) 80,000 (>20 kg)	15,000	Species-dependent
Minimum cage heigh (cm)	12	18 (<200–250 g) 20 (250–>600 g)	45 (<2–5 kg) 60 (5–>6 kg)	23	200	200	Species-dependent

[a]Minimum floor area for one or more animals, except in the case of rabbit and cats (one animal) or dogs (one or two animals).

1. Temperature, humidity, and ventilation.
2. Light must be artificial and satisfy the biological requirements of each animal species.
3. Feeding: Conventional laboratory diets along with water shall be provided ad libitum. The diet should cover all the nutritional requirements of the animal.
4. Housing options depend on the type of study and shall be adapted to the animal species, providing a comfortable resting area of adequate size, clean, and dry.
5. Environmental enrichment should be regarded equally important as nutrition and veterinary care [20]. The types of enrichment are typically categorized as social and physical. Some animals need social interaction and perform a specific activity. Depending on the study, the animals will be housed individually or in collective cages of no more than five animals per cage, all of the same sex. Physical enrichment includes complex enclosures (i.e., nesting material) and both sensory and nutritional stimuli (Figure 4.1).

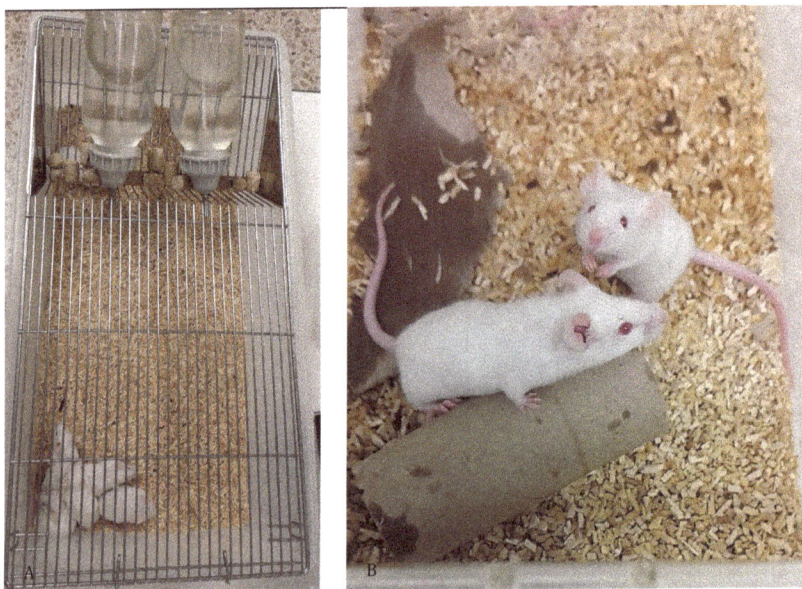

Figure 4.1: Mice housed in large cages (A). Example of physical environment enrichment from mice (B).

Transportation and introduction of animals into unfamiliar surroundings are potentially stressful events. An acclimation period allows animals time to stabilize in a new environment and promotes both animal welfare and reproducible experimental results. The acclimation period is 48 h for rodents and nonmammalian vertebrates (including birds, amphibians, and reptiles) and from 72 up to 5 days for larger mammals (including rabbits, cats, swine, sheep, and goats) [21]. During this time animals randomly selected and identified are caged in groups of the same age and sex. Weight variation should not exceed ±20% of the mean weight of all the animals for each sex. If a fasting period prior to treatment is required, it should be appropriate to the animal species used.

4.3 Administration routes

Toxicology tests usually start with the administration of a **single dose** of compound to each animal. Administration of substances to animals is often a critical component of the test and requires careful consideration and planning to optimize delivery of the agent to the animal while minimizing unintentional adverse effects from the procedure. The route of administration of the test compound should mimic the normal exposure in humans as long as toxicity is route-dependent. The characterization of the test substance (physical and chemical properties, bioavailability, mode of action, etc.) also determines the choice of the administration route.

For all species, many different routes are available for the administration of substances [22, 23]. Once the route is selected, issues such as volume of administration, site of delivery, and pH of the substance must be considered to refine the technique [24]. The maximum volume of test substance that can be administered depends on the species, animal size, and administration route (Table 4.2) [22].

Table 4.2: Administration volumes according to animal species and route of administration considered good practice [22].

Animal	Route and volumes (mL/100 g)				
	Oral	Subcutaneous	Intraperitoneal	Intramuscular	Intravenous
Mouse	1	1	2	0.005	0.5
Rat	1	0.5	1	0.01	0.5
Rabbit	1	0.1	0.5	0.025	0.2
Minipig	1	0.1	0.1	0.025	2.5
Dog	0.5	0.1	0.1	0.025	0.25
Macaque	0.5	0.2	Not available	0.025	0.2

The main routes of administration are described further.

4.3.1 Oral

Administration of test substance through the mouth via the diet, drinking water, gavage, or encapsulation is common in laboratory animals. The oral route is economical, convenient, and relatively safe, and some animals can be trained to cooperate voluntarily to the administration. Voluntary ingestion can improve animal well-being and represents an opportunity for refinement (Figure 4.2). In this administration the compound of interest is mixed with an attractive vehicle. Among vehicles, jam, gelatin, cookie dough, nut paste, or bread have been successfully used in rodents [25]. Capsules administration would be more appropriate when taste and odor of the test substance could hamper the oral administration in diet or drinking water. Capsules are useful in large animals, such as dogs. The number and size of capsules administered should be proportional to the size of the animal to minimize regurgitation. Some limitations of oral dosage include a slower onset of action compared with parenteral delivery, a potential first-pass effect by the liver for those substances metabolized through this route, lack of absorption of substances due to chemical polarity or degradation, poor compliance with voluntary consumption, and difficulty in determining the exact dose ingested and inability to use this route in animals that are unconscious or have clinically significant emesis [26]. **Gavage** (oral intubation) is often used in research settings to ensure precise and accurate dosing of animals [27]. Although this method could induce stress in a daily handling of the animals, selection of appropriate tubing size for gavage is important to minimize discomfort while optimizing delivery of substances [23]. In oral gavage administration, animals should be fasted to some extend prior to dosing the test substance.

Figure 4.2: Mouse eating bread as the vector to oral treatment.

4.3.2 Intranasal, intratracheal, and inhalation

Intranasal techmique may be used for either local or systemic delivery of substances. Volumes administered intranasally are small to minimize the potential for suffocation and death. The nasal mucosa lines the nasal cavity and is richly supplied with blood vessels, potentially resulting in a rapid absorption. In research settings, animals generally are sedated or anesthetized for intranasal and intratracheal routes of delivery.

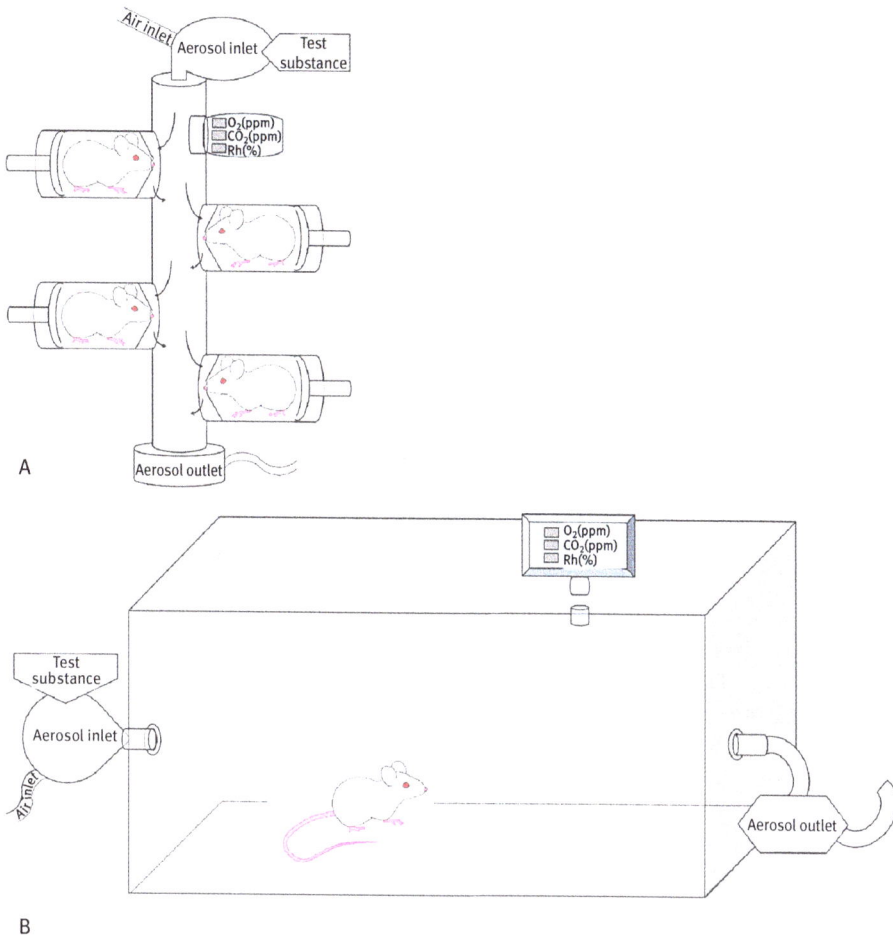

Figure 4.3: Schemes of inhalation chambers: (A) scheme of the nose-only inhalation exposure tower configured with multiple chambers. Mice were immobilized inside individual compartments and exposed to the aerosolized/nebulized test substance, whose flow (arrows) only comes into contact the mice nose incorporating the excess to the internal flow of the camera. Levels of oxygen (ppm), carbon dioxide (ppm), and relative humidity are controlled; (B) Scheme of whole-body inhalation chamber. The test substance is aerosolized/nebulized prior to entry into the chamber. Single mouse exposure is recommended and atmosphere conditions (oxygen (ppm), carbon dioxide (ppm), and relative humidity (%)) must be controlled.

Intratracheal instillation is an easy delivery method for intrapulmonary delivery, involves injecting small volumes of solutions directly into the trachea of animals, and results in rapid inhomogeneous distribution over a relatively small volume of the lung [28]. In inhalation delivery, animals may be exposed to the test compound as a gas, vapor, aerosol, or a mixture thereof and mimic human exposure well [29]. Administration is determined by the goal of the study and may be nose-only (head-only, nose-only, or snout-only) or whole body [6]. For testing compounds such as liquid, solid aerosols, and vapors preferred the nose-only exposition (Figure 4.3A). Whole-body exposition implies single-caged individuals, and the test animal should not exceed 5% of the total volume of the chamber to ensure atmosphere stability. Concentration of oxygen (at least 19%) and carbon dioxide (not exceed 1%) as well as the relative humidity in the breathing zone should be controlled (Figure 4.3B). Inhalation time is adjusted for each animal species, 4–6 h in rodents.

4.3.3 Dermal

The skin is a route of exposure that needs to be considered when conducting a toxicity study. It is necessary to identify the potential for dermal penetration by a test compound [30]. The substance is administered directly to the skin surface and must be extended to 10% of the body surface area. The absorption through the skin into the systemic circulation depends on factors such as the concentration of the substance administered, their lipid solubility, the skin thickness at the site of application, the length of time that the compound is in contact with the skin surface, and the degree of skin hydration. This site is typically used for the assessment of immune, inflammatory, or sensitization response [22]. Material may be formulated with an adjuvant and volumes of 0.05–0.1 mL can be used. Transdermal of percutaneous delivery represents a similar route of administration except that materials are applied to the skin surface usually by means of a patch, for absorption across the epithelial barrier into the systemic circulation. This method produces very constant blood levels of the substance being administered.

4.3.4 Intravenous

The intravenous route is the most efficient means of delivering substances to animals because it bypasses the need for solute absorption. In treatments requiring single or repeated intravenous administration, technique refinements that may enhance animal comfort should be considered. For instance, the use of topical anesthetic creams prior to the injection, the smallest needle size possible, catheters, and vascular access ports for animal comfort and locomotion freedom. Substances administered intravenously must be delivered aseptically and should be sterile and free of particulates that may induce foreign body emboli. The use of chronically implanted catheters and vascular access ports has simplified the administration of drugs and sampling of blood. It requires regular cleaning and maintenance to ensure patency and prevent infection [31]. For intravenous route, distinctions are made between bolus injection, slow intravenous injection, and intravenous infusion [22].

Bolus injection. In most studies using the intravenous route the test substance is given over a short period of approximately 1 min. Such relatively rapid injections require the test substance to be compatible with blood and not too viscous. When large volumes are required to be given, the injection material should be warmed to body temperature. It is suggested that, for rodents, the rate of injection should not exceed 3 mL/min.

Slow intravenous injection. It may be necessary to consider administering substances by slow intravenous injection over the course of 5–10 min because of the expected clinical application of the compound or because of limiting factors such as solubility or irritancy. In this case the possibility of extravascular injection of material should be minimized.

Continuous infusion. This intravenous administration could be chosen for solubility reasons or clinical indication. The volume and rate of administration will depend on the substance being given and the fluid therapy. As a guide, the volume administered on a single occasion will be <10% of the circulating blood volume over 2 h. Precision electronic infusion pumps equipped with microdrop infusion sets could be used to ensure accurate intravenous delivery.

4.3.5 Intraperitoneal

Injection of substances into the peritoneal cavity is a common technique in laboratory rodents but rarely is used in larger mammals and humans. In this route drug absorption from the peritoneal cavity is dependent on the properties of the drug particles and the vehicle, and the drug may be absorbed into both systemic and portal circulations. Intraperitoneal administration of compounds is a justifiable route for pharmacological or toxicological studies where the goal is to evaluate the effects of target engagement rather than properties of drug formulation and or its pharmacokinetics for clinical translation [32].

4.3.6 Intramuscular

Intramuscular administration of substances is a common parenteral route in large animals but often is avoided in smaller species because of the reduced muscle mass. Generally, intramuscular injections result in uniform and rapid absorption of substances because of the rich vascular supply. Intramuscular injections may be painful because muscle fibers are necessarily placed under tension by the injected material. Sites need to be chosen to minimize the possibility of nerve damage. Repeated injections may result in muscle inflammation and histopathological lesions [33].

4.3.7 Subcutaneous

This route is frequently used since it is a rapid, inexpensive, and simple method of parenteral substance administration. Substances administered subcutaneously often are absorbed at a slower rate compared with other parenteral routes, providing a sustained effect. The rate and extent of absorption depend on the formulation (aqueous or oily fluids). Subcutaneous is large and can be an excellent site for large volume fluid delivery in small or dehydrated animals.

4.3.8 Epidural and intrathecal

For rapid effects of substances on cerebrospinal tissues or meninges, substances can be administered into the epidural or subarachnoid (intrathecal) space of the spinal cord. Epidural injection volumes range from 10 to 50 μL in mice and 10–300 μL in rats. Intrathecal injection volumes range from 5 to 10 μL in mice and 10–50 μL in rats [34]. The technique often requires animals to be sedated and given a local anesthetic over the spinal needle insertion site.

4.4 Types of toxicity studies with animals

The basic principles guiding toxicity test in animals are to check the effect of the substances on laboratory animals evaluating also exposition to high doses and long periods of time. Toxicity studies could be divided into:
– Acute toxicity studies: Examination of adverse effects that may occur on first exposure to a single dose of a substance.
– Repeated dose, subchronic, or chronic toxicity studies: Identify whether toxicity occurs after continuous exposure to a substance.

The presence and severity of the toxic effect depends on the level (dose and length) along with route of exposure and animal species susceptibility. The minimum time for each type of study may vary according to regulations of different regions (the United States, Europe, and Japan). The **dose** is the amount of test substance to which an individual is exposed. The test substances are administered in a **vehicle** whose selection is an important consideration in animal investigations. Vehicles themselves should offer optimal exposure but should not influence the results obtained for the compound under investigation, and they should be biologically inert, have no effect on the biophysical properties of the substance, and have no toxic effects on the animals [22]. Simple vehicles used to administer compounds include aqueous isotonic solutions, buffered solutions, cosolvent systems, suspensions, and oils. Therefore, toxicity studies require several groups of animals (controls and treated groups):

- Treatment groups of animals are administered with different doses of the compound tested.
- Control groups of animals receive the vehicle to assure that it does not influence the results obtained for the compound under investigation.

Animals for each group of the study must be randomly selected to ensure a representative population. All animals, in control and treatment groups, should be handled in the same way. Clinical signs of toxicity, as well as their frequency, severity, and duration shall be recorded. In some toxicity studies, surviving animals must be humanly euthanized to avoid suffering [35]. Research animals should undergo complete necropsy quickly and immediately after dead to yield valuable information about the effects in target organs and examine morphopathological changes (Figure 4.4). Proper and thorough fluids and tissue collection, storage, and assessment are paramount in attaining the most efficient use of animals in research [36].

Figure 4.4: Gross necropsy of mouse: abdominal and thoracic organs.

Results from animal tests are used in combination with data on the efficacy of drugs to the risk-benefit assessment. For instance, they help to decide whether the beneficial effects of a treatment would outweigh the risks of adverse side effects. About toxins, data of toxicological response with information on human exposure are integrated to produce a risk assessment and to identify control measures necessary to manage and reduce any identified risk. Besides, tests on species such as fish and amphibians are used to assess the potential environmental effects of chemicals.

There are different groups of toxicity studies regarding to duration and number of doses administered.

4.4.1 Acute toxicity testing

Acute toxicity is carried out to determine the effects of a single dose of compound on a particular animal species administered preferably by oral administration in a short period exposition. However other routes of exposure (inhalation, dermal, etc.) could also be used.

It could be recommended that acute toxicity testing be carried out in two different animal species [1]. The aims of acute toxicity testing are to define the nature and duration of any toxicity of the test substance, determine susceptible species, identify target organs, and provide information for risk assessment. Mortality during the exposure and observation period is recorded. Dead animals or animals sacrificed at the end of the study are examined for morphological, histological, and pathological changes.

Those tests also determine the lethal doses and provide preliminary information relevant to single exposure or overdosage in humans. Dose–response curves are obtained from results of the acute toxicity studies and allow determination of the indicator of acute toxicity (Figure 4.5).

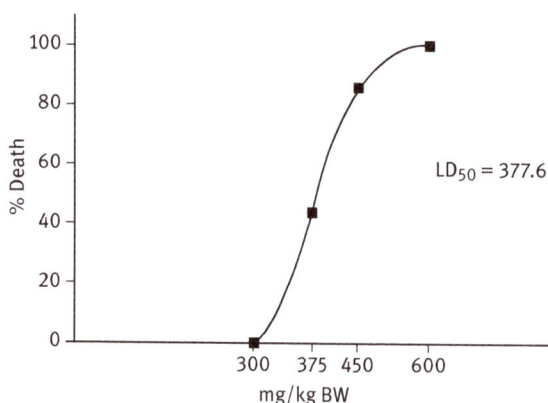

Figure 4.5: Dose–response curve of mortality and the LD_{50} generated.

LD_{50}: 50% lethal dose is a statistically derived single dose of a substance that can be expected to cause death in 50% of treated animals when administered by the given route of exposure. The LD_{50} value is expressed in terms of quantity of test substance per unit weight of test animal (mg/kg).

Lethal dose tests are still used to assess the safety of certain foods (shellfish) for the presence of some toxins in EU [37], food ingredients, cosmetics, industrial products, chemicals, or pharmaceuticals prior their approval for human uses [38]. Several methods have been developed to perform acute toxicity testing with the goal of use fewer animals and in some cases replace death as the endpoint with signs of significant toxicity [1].

4.4.1.1 Fixed dose procedure

The fixed dose procedure (FDP) is frequently used to assess the nonlethal toxicity rather than the lethal dose [39]. The first dose to produce evident toxicity is selected by a sighting study from the fixed dose levels of 5, 50, 300, and 2,000 mg/kg body weight (one animal per dose). When there is no previous information of the test substance, the starting dose will be 300 mg/kg in the main study with five animals group-caged compose each dose level. Higher or lower fixed doses will be administered depending on the presence or absence of signs of toxicity or mortality. An observation period for a minimum of 14 days is recommended. The final point of the study is reached when toxicity is evident or no more than one death occurs. The information obtained can allow the test chemical to be classified in the Globally Harmonized System.

4.4.1.2 Acute toxic class method

The acute toxic class method is a sequential procedure in which three animals of the same sex are used for each step [40]. Same preidentified starting doses as in the FDP may be used and the test dose. When 2–3 animals die, the next level dose will be lower, but if 0–1 animals die the dose level is tested again and under the same result a higher dose level will be next. The standard test for acute toxicity would use a mean of 11 animals [41] and requires death of animals as an endpoint. This test will enable the classification of the test substance according to the Globally Harmonised System.

4.4.1.3 Up-and-down procedure

Test Guidelines Programme of the Organization for Economic Cooperation and Development (OECD) has developed standardized methods following the Three Rs. One of these is the up-and-down procedure (UDP) that substantially reduced the total number of animals used. This is the method developed and statistically evaluated for acute toxicity testing most recommended by various regulatory agencies [42].

This method provides a way to determine the toxicity of chemicals while achieving significant reductions in animal use by performing sequential dosing steps [43]. The four level up-and-down procedures will require only three to nine animals in each level of dosification (Figure 4.6). Starting with a dose below LD_{50} and a factor of 3.2 between levels are recommended. Testing in females is recommended as they were generally more sensitive in the toxicological studies though males can be used when previous reports indicate that they are more sensitive than females to the test substance. In the up-and-down procedure, a small number of animals are dosed one at a time. If most of the animals survive, the dose for the next group of animals is increased; if most of them die, the dose is decreased. The animals (usually mouse or rats) are observed for 24 or 48 h before dosing the next group. The procedure for estimating the LD_{50} takes into account all deaths and may be performed using widely available computer program packages. As a disadvantage this method should not be used for testing compounds where deaths could be beyond 2 days postdosing.

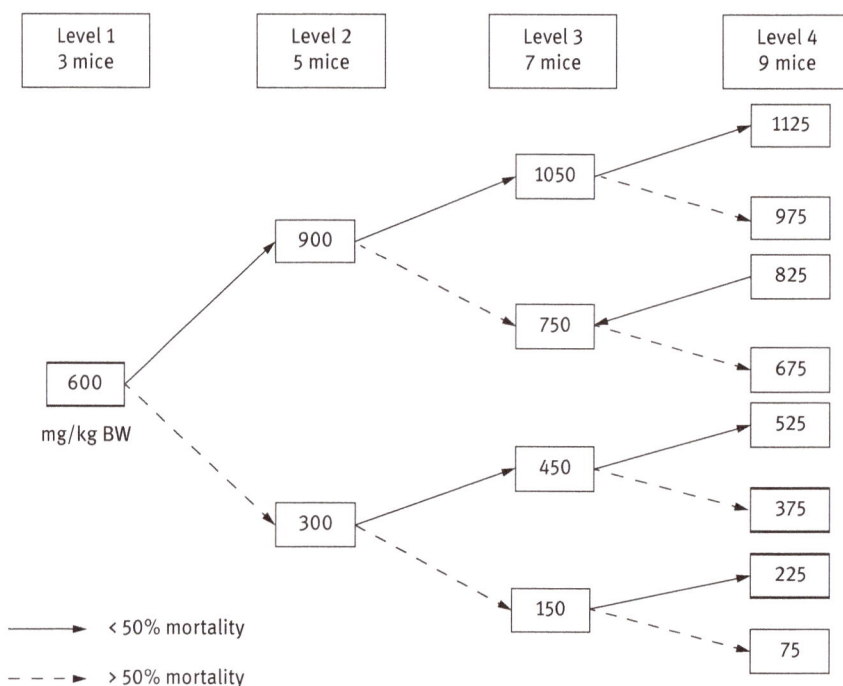

Figure 4.6: Representative four-level up-and-down procedure for an acute toxicity testing. At each dose level the number of animals increased up to 9. If most of the mice survive after the treatment, the dose for the next level is increased (continuous arrows); if most of them died the dose was decreased (discontinuous arrow). The animals are observed for 24 or 48 h before dosing the next group.

4.4.1.4 Acute toxicity testing for inhalation

Acute inhalation toxicity testing is performed for aerosol-like preparations with two possible approaches: a traditional study or a concentration-time assay [44]. In the traditional study, five animals per sex and concentration group are exposed for a fixed time typically of 4 h. Three concentrations of the test substance are required. In the concentration-time assay, treatment should be administered by nose-only exposure. Two animals per sex, preferably rats, or four animals of the susceptible sex for each concentration and time point. A total of four concentrations and five periods of exposure are recommended. For both tests, animals are monitored for 14 days. Clinical signs, body weight, and mortality during the exposure and observation period are noted. Dead animals or animals sacrificed at the end of the study are examined for histological and pathological changes.

4.4.1.5 Acute toxicity testing for topical preparations

The eye irritation test and skin irritation test are very important for topical preparations. The eye irritancy test and the skin irritancy test are used to measure the harmfulness of chemicals and pharmaceutical substances in rabbits and guinea pigs. At the end of the study, the animals are sacrificed and pathological changes are evaluated.

4.4.1.6 Acute eye irritation/corrosion

In the test the substance is administered to an animal's eyes to evaluate the irritation based on lesions' scoring [45]. Albino rabbit is the preferably animal. Eyes should be checked the day prior to treatment in order to ensure no lesion is present. The test substance should be placed in the conjunctival sac of one eye of the animal, thus the other eye serves as a control. An initial study with one animal determines if the test substance is corrosive or a severe irritant. In case of negative response, a confirmatory test must be performed using two animals simultaneously; if the response is positive, the two animals' exposition will be sequential. Dose for liquid and solids testing is 0.1 mL and a maximum weight of 100 mg, respectively. For pressurized aerosol, the test substance is administered to the eye in a simple burst from 10 cm of the eye. The eye may be rinsed with saline or distilled water at 24 h after the treatment or 1 h if the solid test substance remains in contact with the eye. The administration of general analgesics and topical anesthetics is needed before and after application of the test substance to minimize pain. They are observed for 21 days or shorter time when reversibility occurs.

4.4.1.7 Acute dermal irritation/corrosion

Skin irritancy is normally assessed by applying a single dose of the test substance to shaved areas of the backs of albino rabbits. Albino rabbits are normally exposed for 4 h to test substance, which is then removed. Doses of 0.5 mL for liquid or 0.5 g for solid test substance are applied to the surface of the skin of animals individually housed. Generally this study requires an initial test conducted in one animal by three sequentially exposures to know whether the test substance is corrosive. If a negative result is obtained, a confirmatory test with two animals treated with different patches can be performed. In case the initial test is not performed, two or three animals may be treated with a single patch [46]. Animals should be observed for signs of stress and reversibility of the test substance up to 14 days.

4.4.1.8 Acute dermal toxicity

Acute dermal toxicity test allows the classification of the test substance into the Globally Harmonized System. This is based on the outcome (death/survival) to sequential predefined fixed doses. These are 50, 200, 1,000, and 2,000 mg/kg. The starting dose may be chosen taking into consideration the previous toxicity data or by performing a range-finding study. The animals used more are adult rabbits, rats, mice, and guinea pigs. Female rats of 8–10 weeks old are preferably the experimental animals. The test chemical is applied to the naked skin (10% of the total body surface area) with a porous gauze for 24 h. Each level dose is composed by two animals individually caged until the end of the exposure [47]. Afterward, the test is removed and animals are caged in groups for up to 14 days of observation.

The local lymph node assay is widely accepted that meets regulatory requirements as a skin sensitization method. In this test the substance is applied on the surface of the ears of a mouse for 3 consecutive days. At the end the mouse is euthanized and the early stages of sensitization are detected by measuring the proliferation of lymphocytes in the draining lymph node [48].

4.4.2 Repeated dose toxicity testing

Dose level selection is an important consideration when designing repeated-dose toxicity testing to ensure that exposure levels that lead to relevant hazards are identified [49]. Repeated-dose toxicity testing is conducted to determine the existence of effects derived from an exposure of extended duration. Administration of the test substance is performed daily in the medium term between 14 and 90 days depending on the purpose of the study and no lasting more than the 10% of the lifespan of the animal. Rats and mice are generally used but test may also be conducted in nonrodent animals such as dogs, pigs, or macaques. At least 5 or 10 animals per sex should be used at each dose level. In the case of using nonrodent animals, only four per sex are needed. Repeated-dose studies must be composed by at least three dose levels, which higher concentration should result in toxic effects without causing lingering signs or lethality and lower concentration should produce little or no evidence of toxicity; therefore these studies are also termed subacute toxicity testing [50]. There is an extensive evaluation of toxic effects (body weights, clinical signs of toxicity, food consumption, clinical pathology, biochemical parameters).

The interpretation of human safety details is essential in repeated dose toxicity studies. The test data allow an assessment of the parameter NOAEL:

NOAEL is the abbreviation for "no observed adverse effect level" and is the highest dose without significant adverse effects.

NOEL is used in risk assessment [51]. For instance, in the case of food toxins, these studies are used to assign a reference dose to which safety factors are applied to give acceptable daily intake (ADI).

ADI is typically a hundredfold less than the observed NOAEL. ADI can be defined as the dose level to which humans may be exposed through residues of foodstuffs and drinking water with the practical certainty that no adverse health effects will ensue.

Another important concept is maximum tolerated dose (MTD):

MTD is the highest dose that produces toxicity but no death.

MTD studies often replace acute studies, especially in the case of larger species such as dog and primates. They involve steadily increasing the dose given to an animal until adverse effects indicate that an MTD has been reached. This is normally determined by careful observation of the animals. Effects such as vomiting and convulsions are sometimes evaluated as signs of the MTD.

These kinds of repeated dose toxicity studies are used as preliminary to a long-term chronic toxicity study and preferably route is oral [52, 53], but the test substance may be administered by dermal [11], inhalation [54], or other parenteral route.

4.4.2.1 Oral

Rodent species, preferably the rat, are used, although may also be conducted in pigs and dogs. Animals are treated daily 7 days each week. Gavage is recommended though administration of the test substance in drinking water or in the diet is also contemplated. Three dose levels are set by two to fourfold intervals. Generally, a 28-day study [55] can serve as a preliminary evaluation to perform the 90-day oral toxicity [52].

4.4.2.2 Dermal

The adult rat, rabbit, or guinea pig may be used [56]. It is recommended that at least three doses are tested. Treatments are applied for at least 6 h per day, 7 days per week. Signs of toxicity, changes in body weight, and food consumption should be monitored. The lowest dose level should not produce any evidence of toxicity, whereas the highest dose should induce clear toxicity signs without causing death.

4.4.2.3 Inhalation

The preferable species is the rat. Nose-only administration is highly recommended, though whole body may be required depending on the chemical and the aim of the study. Animals are exposed 6 h per day, 5 days per week. At least three doses could be tested with five animals per sex and dose. Changes in body temperature and respiration are characteristic.

In the repeated dose toxicity testing, blood removal could be a procedure performed on the animals. Table 4.3 indicates the maximum blood sample volumes that can be removed without significant disturbance to the animal's normal physiology. These values do not include a terminal sample, which can be taken when the animal is euthanized. Recommended routes for bleeding are the lateral tail vein, the sublingual vein, and the lateral tarsal vein for all rodents and the marginal ear vein, central ear artery, and the jugular vein for rabbits while sampling by cardiac routes is only carried out as a terminal procedure under general anesthesia.

Table 4.3: Total blood volume (mL) of the animal species more commonly used in experimentation and the maximum blood sample volume recommended for extraction (mL).

Animal (weight)	Blood volume (mL)	Maximum blood sample extraction (mL)
Mouse (25 g)	1.8	0.4
Rat (250 g)	16	3.2
Rabbit (4 kg)	224	45
Minipig (15 kg)	975	195
Dog (10 kg)	850	170
Macaque rhesus (5 kg)	280	56
Macaque cynomolgus (50 kg)	325	65

At the end of the repeated dose toxicity testing, tissues from most of the organs are removed, and histological changes are recorded (Figure 4.7). If possible, immunotoxicity (adverse effects on the immune system) studies are performed on the same animals. Immunotoxicological analysis is not feasible beyond the period of 14 days. Parameters such as delayed-type hypersensitivity (DTH), mitogen- or antigen-stimulated lymphocyte

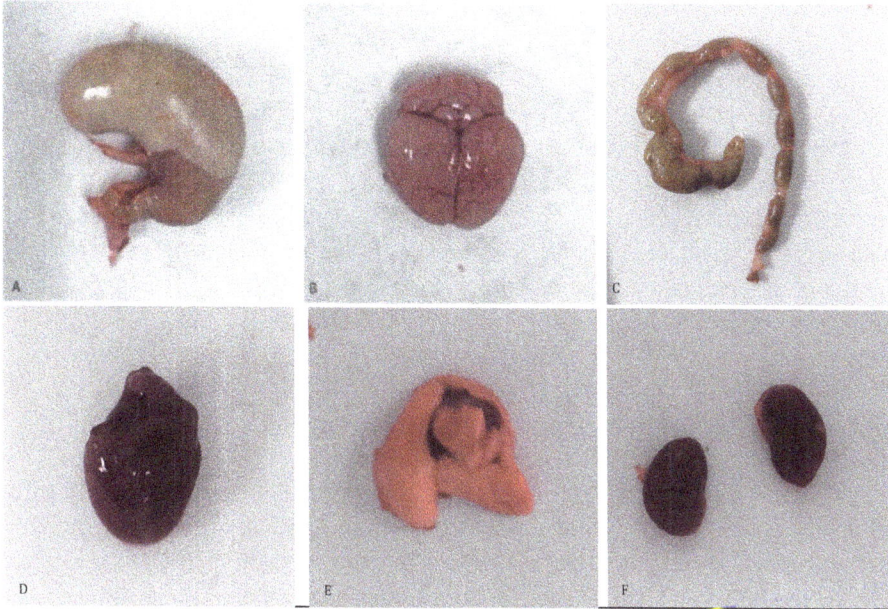

Figure 4.7: Mouse organs: (A) stomach, (B) brain, (C) large intestine, (D) heart, (E) lungs, and (F) kidneys.

proliferative responses, macrophage function, and primary antibody response to T-cell-dependent antigen are assessed in immunotoxicological studies.

4.4.2.4 Genotoxicity testing

Genotoxicity refers to the induction of damage of the genetic material (DNA and chromosomes) and regulatory cellular components for the potential to cause heritable mutations [57]. In animal studies rodents are preferred, usually mice or rats. The purpose is to demonstrate that the chemical can or cannot reach a sensitive tissue and cause genetic changes in the intact animal. When toxicity occurs, the dose levels should range from dose producing little or no toxicity to the MTD. If toxicity was previously not observed, the highest dose should be 1,000 or 2,000 mg/kg/body weight/day for administration periods of more than or less than 14 days, respectively. The test substances may be administered as a single dose or separated doses. Single dose can be administered as a split dose within a maximum range of 2–3 h (split dose) to facilitate administration. Separated doses or repeated doses should be given 24 h apart. Besides, a negative control group and a positive control group should be taken into consideration. Route of administration should mimic the way humans are exposed to the test substance. In order to accomplish the principles of the Three Rs, it is highly encouraged to perform some of the genotoxicity assays simultaneously [58–60].

4.4.2.5 Mammalian erythrocyte micronucleus test

In this assay, erythrocytes from bone marrow or peripheral blood are analyzed in search of damage in the chromosomes or the mitotic apparatus of erythroblasts indicated by an increase formation of micronuclei [58]. Animals, usually rodents (rat or mice), are exposed to the test substance by a single or few daily doses. Control and each treated animal group must be composed by at least five of one sex or five animals/sex. The collection of samples will depend on the routine administration of the substance. After extraction, samples are stained using DNA staining dyes and studied by microscopy or flow cytometer.

4.4.2.6 Mammalian bone marrow chromosomal aberration test

This in vivo test is an early predictor of carcinogenic activity [59]. Aberrations in chromosomes may be detected by this test in bone marrow cells. A single dose of compound is administrated to rats or mice. Bone marrow samples should be taken after 12–18 h and 24 h after the first sample. When repeated doses are administrated, sampling should be taken 12–18 h following the last treatment. Between treatment and the end of the assay, a metaphase-arresting agent is injected. From bone marrow cells immediately obtained after euthanasia fixed chromosome preparations are stained and analyzed for chromosomal changes. It is expected that the highest dose level used will show evidence of adverse effects if the substance is genotoxic, and the MTD is normally used to set this dose level.

4.4.2.7 In vivo mammalian alkaline comet assay

Single cells/nuclei suspensions are prepared from different tissues dissected for analyzing DNA strand breaks in animals, most often rodents (6–10 weeks), although studies are carried out in other mammalian and nonmammalian species too [58]. Treatments are daily given for 2 or more days and samples collected once between 2 and 6 h after the last treatment. Tissues disaggregated into cell suspension are lyzed in a highly alkaline buffer and then undergo electrophoresis and DNA staining. Only preparations that allow (semi)quantification are analysis. Dissected tissues should also be analyzed by histopathology analysis.

4.4.2.8 Neurotoxicity studies

The effects of a test substance on the nervous system can be studied through neurotoxicity studies [61]. Structurally the nervous system has two components: the central nervous system and the peripheral nervous system. The central nervous system is made upon of the brain, spinal cord, and nerves. The peripheral nervous system is further divided into somatic and autonomic nervous systems. Neurotoxic studies may be employed to evaluate the specific histopathological and behavioral neurotoxicity of a chemical and are used to characterize neurotoxic responses such as neuropathological lesions and neurological dysfunctions (loss of memory, sensory defects, and learning and memory dysfunctions) (Figure 4.8). Usually neurotoxicological studies are carried out in adult rodents, preferably rats. The test substance may be administered orally in a stand-alone study or daily for 14 days or even more than 90 days, being in this case a subchronic toxicity testing and neurological changes are evaluated [62]. Routine observation of clinical signs and behavior alteration as well as possible moribund animals should be carried out. In-depth histopathology examination of nervous tissues along with biochemical blood analysis is performed at the end of the test.

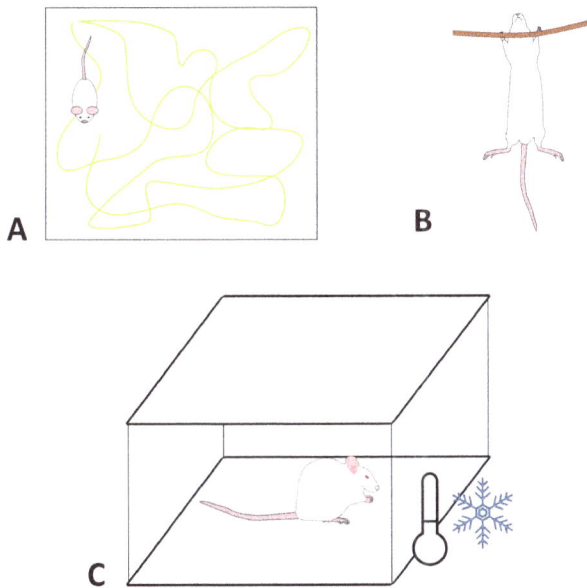

Figure 4.8: Tests to evaluate neurological dysfunctions: (A) The open field test is broadly used to evaluate motor activity by tracking animal movement for several minutes; (B) The hanging wire or hanging grid test allows to measure loss in limbs strength by recording latency to fall; (C) Test measuring sensitivity is strongly recommended, an example is the cold plate test, which allows to evaluate changes in response to cold.

4.4.3 Subchronic toxicity testing

The major difference between repeated dose and subchronic toxicity studies is the duration: repeated dose toxicity studies are conducted between 14 and 90 days, and subchronic toxicity studies are carried out over 90 days [63]. Rodents and nonrodents are used to study the subchronic toxicity of a substance. The test substance is administered preferably by oral route for 90 days but other routes could also be used [64, 65]. Weekly body weight variations, monthly biochemical and cardiovascular parameters changes, and behavioral changes are observed. At the end of the study, the experimental animals are sacrificed. Gross pathological changes are observed, and all the tissues are subjected to histopathological analyses. In the study protocol a control group and a high-dose group may be included.

4.4.4 Chronic toxicity testing

Chronic toxicity tests determine the effects of long-term exposure to a test substance [66]. Administration of the test substance is normally oral, although dermal and inhalation are used too. Test substance is daily administered to animals for 12 months. For dermal and inhalation administration treatments are given for at least 6 h per day [67]. The duration of exposure has to be long enough to allow manifestation of effects due to cumulative toxicity.

Chronic toxicity studies are conducted mainly in rodents though nonrodent species are also considered. The studies must be composed of at least three levels doses, each one of 20 animals per sex for rodents or 4 per sex in nonrodents studies. Spacing of the dose levels is frequently of two to fourfold intervals, and the highest dose level should evidence toxicity but avoiding suffering, morbidity, or death. During the study period, the animals are observed periodically for normal physiological functions, behavioral variations, and alterations in biochemical parameters. At the end of the study, the animals are sacrificed, and gross pathological changes are noted and histopathological studies are carried out. The report on chronic oral toxicity is essential for new drugs.

4.4.4.1 Carcinogenicity testing

Rodents are preferably used in carcinogenicity testing, though nonrodent animal species could also be used with adequate modifications [48]. The preferred via of administration may be preferably oral, though dermal or inhalation administration may be required in some cases, trying to mimic the route of human exposure. At least 50 animals should be used at each dose level. The tests are carried out over the greater portion of an animal's lifespan 24 months for rodents. Depending on species and strain of the animal, the duration may vary. During and after exposure to test substances, the experimental animals are observed for signs of toxicity and development of tumors. At the end of the study, a screening of tests is performed to assess the health of the animals. This may include hematological analysis, macroscopic, and histopathology evaluation. It is also recommended to perform simultaneously the chronic toxicity testing and the carcinogenicity testing [68].

4.4.4.2 Reproduction/developmental toxicity testing

These studies evaluate the effects on reproductive function and the ability to produce birth defects. The test compound is administered to both male and female animals. Oral administration is for the duration of at least one complete spermatogenic cycle in male animals and for two complete estrous cycles for female animals. Rat is the animal used, but in some cases testing is carried out on other rodents under justified reasons. Adult animals and offspring are used in reproductive and development studies. The test substance should be administered before, during, and after mating. Moreover, administration should also last for the duration of pregnancy and 13 days postpartum in females. These tests provide information on fertility, mating behavior, parenteral behavior, and development of the neonate adulthood, including endocrine disruption and altered growth [69].

4.4.5 Toxicokinetic studies

TK studies that are an extension of pharmacokinetics of a substance are conducted to obtain information on its absorption, distribution, biotransformation, and excretion to aid in relating concentration to the observed toxicity [70]. Animal TK data helps to understand the toxicology studies by demonstrating that the animals are systematically exposed to the test substance and by revealing which are the circulating moieties (parent substance/metabolites). In some situations, TK data can be collected as part of the evaluation in toxicological studies [71]. Basic TK parameters determined from these studies will also provide information on the potential accumulation of the test substance in tissues and/or organs and the potential for induction of biotransformation as a result of exposure to the test substance. Toxicokinetic can contribute to the assessment of the relevance of animal toxicity data for extrapolation to human hazard.

TK studies are preferably carried out in rodents but other rodents, rabbits, dogs, nonhuman primates, and swine are also used. Oral administration is preferable; however other routes of administration (dermal, inhalation, intraperitoneal, and intravenous) may be applicable for certain chemicals [72]. Single dose may be adequate; however in some circumstances repeated dose administration may be needed. These studies may be conducted at one dose or, more likely, at two or more doses. Each animal is placed in a separate metabolic unit for the collection of urine and feces (Figure 4.9).

Figure 4.9: Mouse in a metabolic cage.

The purpose of these studies is to obtain estimates of basic TK parameters for the test substance:

C_{max}: Maximal concentration in blood after administration or maximal excretion (in urine or feces) after administration

T_{max}: Time to reach C_{max}

Half-life ($t_{1/2}$): The time taken for the concentration of the test substance to decrease by one-half in a compartment. It typically refers to plasma concentration or the amount of the test substance in the whole body.

AUC: Area under the curve in a plot of concentration of substance in plasma over time. It represents the total amount of substance absorbed by the body within a predetermined period of time.

Following the administration of the test substance blood samples, urine, and feces samples should be collected from each animal at suitable time points using appropriate sampling methodology. Identification and quantitation of unchanged test substance and metabolites could be done by analytical systems such as high-performance liquid chromatography, liquid chromatography/mass spectrometry (LC-MS), or nuclear magnetic resonance (NMR) spectrometry.

Knowledge of tissue distribution of a test substance and its metabolites is important for the identification of target tissues, understanding of the mechanisms of toxicity, and the potential for test substance and metabolite accumulation and persistence. The percent of the total dose in tissues as well as residual carcass should at a minimum be measured at the termination of the excretion experiment (e.g., typically up to 7 days postdose or less depending on test substance specific behavior). Tissues that should be collected include liver, fat, gastrointestinal tract, kidney, spleen, whole blood, residual carcass, target organ tissues, and any other tissues (e.g., thyroid, erythrocytes, reproductive organs, skin, eye, and brain) of potential significance in the toxicological evaluation of the test substance [73]. Analysis of additional tissues at the same time points should be considered to maximize the utilization of animals.

Toxicokinetic models may have utility for various aspects of hazard and risk assessment as for example in the prediction of systemic exposure and internal tissue dose. Furthermore specific questions on mode of action may be addressed, and these models can provide a basis for extrapolation across species, routes of exposure, dosing patterns, and for human risk assessment.

4.5 Future of toxicity studies with animals

This chapter presents the toxicity studies with animals as a relatively standardized and guideline-driven process of testing compounds that relied on various in vivo studies in rodent and nonrodent animal species. Animal research faces the regulatory challenge to simultaneously improve the welfare of animal and keep the standard of science. The complexity of animal experiments can lead to the implementation of alternatives for these experiments. Currently several in vitro, ex vivo, and in silico tools as well as computational models and informatics approaches to toxicity testing are in stages of development and being incorporated into safety guidances for different sectors [74]. The information gleaned from these approaches will contribute both to the winnowing and prioritizing of chemicals that need to be tested in animals as well as to provide a basis for hypothesis-driven whole animal testing. Besides this will assist in reducing the number of animals needed and will potentially shorten the overall time necessary to complete safety assessments [75]. The cosmetic industry is pioneering the integration of new approach methodologies into safety assessments, and the OECDs. In food safety evaluations, global regulatory bodies such as EFSA and FDA are embracing newer tools in risk assessment but at slow pace. Although some progress in the field has already been achieved and several methods have been validated and are also used for regulatory toxicity testing of chemicals, a complete replacement of animals in toxicology studies is not justified [76]. This is due to the lack of validated models for systemic toxicity, which are a prerequisite for establishing safety reference values for food. Considerable efforts are being made to integrate in vitro data with other alternative tools such as computational models and use data generated for hazard assessment [74]. However, the ultimate adjudication of a chemical's toxic potential must reside in well-designed and conducted experiments using experimental animals.

Acknowledgments: The research leading to these results has received funding from the following grants. From Campus Terra (USC), BreveRiesgo (2022-PU011), and CLIMIGAL (2022-PU016). From Conselleria de Cultura, Educacion e Ordenación Universitaria, Xunta de Galicia, GRC (ED431C 2021/01). From Ministerio de Ciencia e Innovación Grant CPP2021-008447 is funded by MCIN/AEI/10.13039/501100011033 and by The European Union Next-GenerationEU/PRT, IISCIII/PI19/001248, and PID 2020-11262RB-C21. From European Union, Interreg EAPA-0032/2022 – BEAP-MAR, HORIZON-MSCA-2022-DN-01-MSCA Doctoral Networks 2022 101119901-BIOTOXDoc and HORIZON-CL6-2023-CIRCBIO-01 COMBO-101135438.

Keywords: Pharmacokinetics, animal test species, administration route, acute testing, subchronic testing, chronic testing, toxicokinetics

Abbreviations: ADI, acceptable daily intake; C_{max}, maximal concentration; DNA, deoxyribonucleic acid; EU, European Union; EURL ECVAM, European Union Reference Laboratory for Alternatives to Animal Testing; FDA, Food and Drug Administration; FDP, fixed dose procedure; ; LD_{50}, iethal dose 50%; LC-MS, iiquid chromatography-mass spectrometry; MTD, maximum tolerated dose; NMR, nuclear magnetic ressonance; NOAEL, no observed adverse effect level; OECD, Organisation for Economic Cooperation and Development; T_{max}, time to reach C_{max}; TK, toxicokinetics; UDP, up-and-down procedure.

References

[1]   Parasuraman S. Toxicological screening. J Pharmacol Pharmacother. 2011;2(2):74–79.

[2]   Jacobs AC, Hatfield KP. History of chronic toxicity and animal carcinogenicity studies for pharmaceuticals. Vet Pathol. 2013;50(2):324–33.

[3]   Rowan A. Shortcomings of LD50-values and acute toxicity testing in animals. Acta Pharmacol Toxicol (Copenh). 1983;52(Suppl 2:):52–64.

[4]   Wilhelmus KR. The Draize eye test. Surv Ophthalmol. 2001;45(6):493–515.

[5]   Russell WS, Burch RL. The principles of human experimental technique. London: Methuen, 1959.

[6]   Union E.P.C.o.t.E. Directive 2010/63/EU of the European Parliament and of the Council of 22 September 2010 on the protection of animals used for scientific purposes, C.o.t.E.U. European Parliament, Editor, 2010.

[7]   Stokes WS. Animals and the 3Rs in toxicology research and testing: The way forward. Hum Exp Toxicol. 2015;34(12):1297–303.

[8]   Maestri E. The 3Rs principle in animal experimentation: A legal review of the state of the art in Europe and the case in Italy. BioTech. 2021;10:1–11.

[9]   Lewis DI. Animal experimentation: Implementation and application of the 3Rs. Emerg Top Life Sci. 2019;3(6):675–79.

[10]   OECD. Guidelines for the testing of chemicals, Section 4: Health effects. Paris: Organisation for Economic Cooperation and Development Publishing, 1981.

[11]   OECD. Principles of good laboratory practice. Paris: Organisation for Economic Cooperation and Development Publishing, 1981.

[12]   Krewski D, et al. Toxicity testing in the 21st century: A vision and a strategy. J Toxicol Environ Health B Crit Rev. 2010;13(2–4):51–138.

[13]   Everitt JI. The future of preclinical animal models in pharmaceutical discovery and development: A need to bring in cerebro to the in vivo discussions. Toxicol Pathol. 2015;43(1):70–77.

[14]   Brink CB, Lewis DI. The 12 Rs framework as a comprehensive, unifying construct for principles guiding animal research ethics. Animals. 2023;13(1128):1–15.

[15]   Mukherjee P, et al. Role of animal models in biomedical research: A review. Lab Anim Res. 2022;38 (18):1–17.

[16]   Boverhof DR, et al. Transgenic animal models in toxicology: Historical perspectives and future outlook. Toxicol Sci. 2011;121(2):207–33.

[17]   Agency EM. ICH guideline M3(R2) on non-clinical safety studies for the conduct of human clinical trials and marketing authorization for pharmaceutical (CPMP/ICH/286/95). London: EMEA/CPMP/ ICH/286/1995, December 2009.

[18]   Prior H, et al. Justification for species selection for pharmaceutical toxicity studies. Toxicol Res (Camb). 2020;9(6):758–70.

[19]   Europe C. European Convention for the protection of vertebrate animals used for experimental and other scientific purposes. E. Union, Editor, 2006.

[20]   Baumans V. Environmental enrichment for laboratory rodents and rabbits: Requirements of rodents, rabbits, and research. ILAR J. 2005;46(2):162–70.

[21]   Obernier JA, Baldwin RL. Establishing an appropriate period of acclimatization following transportation of laboratory animals. ILAR J. 2006;47(4):364–69.

[22]   Diehl KH, et al. A good practice guide to the administration of substances and removal of blood, including routes and volumes. J Appl Toxicol. 2001;21(1):15–23.

[23]   Turner PV, et al. Administration of substances to laboratory animals: Routes of administration and factors to consider. J Am Assoc Lab Anim Sci. 2011;50(5):600–13.

[24]   Morton D, et al. Refining procedures for the administration of substances. Report of the BVAAWF/ FRAME/RSPCA/UFAW Joint Working Group on Refinement. British Veterinary Association Animal Welfare Foundation/Fund for the Replacement of Animals in Medical Experiments/Royal Society for the Prevention of Cruelty to Animals/Universities Federation for Animal Welfare. Lab Anim. 2001;35(1):1–41.

[25]   Ruvira R, et al. Evaluation of parameters which influence voluntary ingestion of supplements in rats. Animals. 2023;13:1–13.

[26]   Wang Y, et al. Relationship between lethal toxicity in oral administration and injection to mice: Effect of exposure routes. Regul Toxicol Pharmacol. 2015;71(2):205–12.

[27]   Sekihashi K, et al. A comparison of intraperitoneal and oral gavage administration in comet assay in mouse eight organs. Mutat Res. 2001;493(1–2):39–54.

[28]   Wu L, et al. Quantitative comparison of three widely-used pulmonary administration methods in vivo with radiolabeled inhalable nanoparticles. Eur J Pharm Biopharm. 2020;152:108–15.

[29]   Pauluhn J, Mohr U. Inhalation studies in laboratory animals – current concepts and alternatives. Toxicol Pathol. 2000;28(5):734–53.

[30]   Mattie DR, Grabau JH, McDougal JN. Significance of the dermal route of exposure to risk assessment. Risk Anal. 1994;14(3):277–84.

[31]   de Wit M, et al. Implantable device for intravenous drug delivery in the rat. Lab Anim. 2001;35 (4):321–24.

[32] Al Shoyaib A, Archie SR, Karamyan VT. Intraperitoneal route of drug administration: Should it be used in experimental animal studies? Pharm Res. 2019;37(1):12.

[33] Thuilliez C, et al. Histopathological lesions following intramuscular administration of saline in laboratory rodents and rabbits. Exp Toxicol Pathol. 2009;61(1):13–21.

[34] Rahman MM, et al. Epidural and intrathecal drug delivery in rats and mice for experimental research: Fundamental concepts, techniques, precaution, and application. Biomedicines. 2023;11: (5)1413.

[35] OECD. Guidance document on the recognition, assessment, and use of clinical signs as human endpoints for experimental animals used in safety evaluation. Paris: OECD Publishing, 2000.

[36] Hampshire V, Rippy M. Optimizing research animal necropsy and histology practices. Lab Anim (NY). 2015;44(5):170–72.

[37] EURLMB. EURLMB Standard operating procedure for PSP toxins by Mouse Bioassay, version 1 March 2014, 2009.

[38] Erhirhie EO, Ihekwereme CP, Ilodigwe EE. Advances in acute toxicity testing: Strengths, weaknesses and regulatory acceptance. Interdiscip Toxicol. 2018;11(1):5–12.

[39] OECD. Test No. 420: Acute oral toxicity – Fixed dose procedure. OECD Publishing, 2002 Paris.

[40] OECD. Test No. 423: Acute oral toxicity-acute toxic class method. OECD Publishing, 2001;1–14 Paris.

[41] Schlede E, et al. Oral acute toxic class method: A successful alternative to the oral LD50 test. Regul Toxicol Pharmacol. 2005;42(1):15–23.

[42] OECD. Test No. 425. Acute oral toxicity-up and down procedure. Organisation for Economic Co-operation and Development. Paris: OECD Publishing, 2022;1–27.

[43] Abal P, et al. Characterization of the dinophysistoxin-2 acute oral toxicity in mice to define the Toxicity Equivalency Factor. Food Chem Toxicol. 2017;102:166–75.

[44] OECD. Test No. 403: Acute inhalation toxicity. Organisation for Economic Co-operation and Development. Paris: OECD Publishing, 2009;1–14.

[45] OECD. Test No. 405: Acute eye irritation/corrosion. OECD Publishing, 2023 Paris.

[46] OECD. Test No. 404: Acute dermal irritation/corrosion. OECD Publishing, 2015.

[47] OECD. Test No. 402: Acute dermal toxicity. OECD Publishing, 2017 Paris.

[48] OECD. Test No. 442B: Skin sensitization: Local lymph node assay: BrdU-ELISA or –FCM. Paris: OECD Publishing, 2018.

[49] Sewell F, et al. Recommendations on dose level selection for repeat dose toxicity studies. Arch Toxicol. 2022;96(7):1921–34.

[50] Ferreiro SF, et al. Subacute cardiovascular toxicity of the marine phycotoxin azaspiracid-1 in rats. Toxicol Sci. 2016;151(1):104–14.

[51] Abal P, et al. Acute oral toxicity of tetrodotoxin in mice: Determination of Lethal Dose 50 (LD50) and No Observed Adverse Effect Level (NOAEL). Toxins (Basel). 2017;9(3):75.

[52] OECD. Test No. 408: Repeated dose 90-day oral toxicity study in rodents. OECD Publishing, 2018 Paris.

[53] OECD. Test No. 409: Repeated dose 90-day oral toxicity study in non-rodents. OECD Publishing, 1998 Paris.

[54] OECD. Test No. 412: Subacute inhalation toxicity: 28-day study. OECD Publishing, 2017 Paris.

[55] OECD. Test No. 407: Repeated dose 28-day oral toxicity study in rodents. OECD Publishing, 2008 Paris.

[56] OECD. Test No. 410: Repeated dose dermal toxicity: 21/28-day study. OECD Publishing, 1981 Paris.

[57] Eastmond DA, et al. Mutagenicity testing for chemical risk assessment: Update of the WHO/IPCS Harmonized Scheme. Mutagenesis. 2009;24(4):341–49.

[58] OECD. Test No. 474: Mammalian erythrocyte micronucleus test. OECD Publishing, 2016 Paris.

[59] OECD. Test No. 475: Mammalian bone marrow chromosomal aberration test. OECD Publishing, 2016 Paris.

[60] OECD. Test No. 489: In vivo mammalian alkaline comet assay. OECD Publishing, 2016 Paris.

[61] OECD. Test No.424: Neurotoxicity study in rodents. OECD Publishing, 1997 Paris.

[62] Leiros M, et al. The Streptomyces metabolite anhydroexfoliamycin ameliorates hallmarks of Alzheimer's disease in vitro and in vivo. Neuroscience. 2015;305:26–35.

[63] Products C.f.H.M. Note for guidance on repeated dose toxicity (CPMP/SWP/1042/99 Rev 1 Corr*). European Medicines Agency, 2010;1–9 London.

[64] OECD. Test No. 411: Subchronic dermal toxicity: 90-day study. OECD Publishing, 1981 Paris.

[65] OECD. Test No. 413: Subchronic inhalation toxicity: 90-day study. OECD Publishing, 2017 Paris.

[66] OECD. Test No. 452: Chronic toxicity studies. OECD Publishing, 2018;1–16 Paris.

[67] OECD. Test No. 451: Carcinogenicity studies. OECD Publishing, 2018 Paris.

[68] OECD. Test No. 453: Combined chronic toxicity/carcinogenicity studies. OECD Publishing, 2018 Paris.

[69] OECD. Test No. 421: Reproduction/developmental toxicity screening test. OECD Publishing, 2016 Paris.

[70] OECD. Test No. 417: Toxicokinetics. OECD guidelines for the testing of chemicals. Paris: OECD Publisher, 2010;1–20.

[71] Van Kesteren P, et al. Implementation of toxicokinetics in toxicity studies-Example of 4-mthylanisole. Toxicol Lett. 2017;280:S33–S34.

[72] Otero P, et al. Pharmacokinetic and toxicological data of spirolides after oral and intraperitoneal administration. Food Chem Toxicol. 2012;50(2):232–37.

[73] Vieira AC, et al. Brain pathology in adult rats treated with domoic acid. Vet Pathol. 2015;52 (6):1077–86.

[74] Reddy N, et al. Alternatives to animal testing in toxicity testing: Current status and future perspectives in food safety assessments. Food Chem Toxicol. 2023;179:113944.

[75] DeSesso JM. Future of developmental toxicity testing. Curr Opin Toxicol. 2017;3:1–5.

[76] Bluemel J. Toxicity testing in the 21st century: Challenges and perspectives. J Drug Metabl Toxicol. 2012;3:1–2.

Carmen Vale and Sandra Raposo-García

5 Toxicological studies with cells

5.1 Introduction

Cell culture provides a model system that provides direct access and evaluation of the effects of chemicals on tissues and constitutes a valuable tool to analyze cell toxicity mechanisms. In vitro model systems, in general, have been used to study the mechanism of action of chemical and to analyze the cellular basis for chemical-induced toxicity as well as to develop rapid and high-throughput screening systems for the evaluation of the toxicity of chemicals, which may complement in vivo toxicity testing or may replace some in vivo models if scientifically validated and accepted by regulatory agencies [1, 2].

The term "in vitro" ("in the glass") refers to the technique of performing a given experiment in a test tube or in a controlled environment outside a living organism. In vitro methods are based on the use of cells or tissues that are cultured under controlled conditions in flasks and plates. Cells/tissues are exposed to chemicals, and their toxic effect is measured. Increasingly, human cells are used since they better predict possible effects on humans. The commonly used in vitro models for assessing chemical toxicity include perfused organ preparations, isolated tissue preparations, single-cell suspensions, and cell culture systems, such as primary cell cultures and mammalian cell lines. Of these in vitro models, cell culture systems have been widely used because they are reliable, reproducible, and relatively inexpensive experimental systems to assess chemical toxicity at the cellular level [3]. Nevertheless, it should be pointed out that whether in vitro tests are based on primary cells, immortalized (e.g., SV40 transformation) and cancer-derived cell lines, stem cells, or reconstituted tissue cultures, it is important to have in vitro systems that adequately mimic key events of the in vivo mechanisms of action triggered in humans upon exposure to a toxic compound [4].

Safety testing of chemicals is required under several directives of the European Union (EU) and international regulatory environments. At an international level, the OECD (Organization for Economic Cooperation and Development) is developing a Guidance Document on Good In Vitro Method Practices [5] which aims to ensure that efficacy and efficiency of the process between in vitro method development and its implementation for regulatory use [6]. The first guidance report on good cell culture practice was released in 2005 [7] and a second one related with the use of stem cells in 2017 [8], and some of the recommendations exposed in these reports are summarized below.

Carmen Vale, Sandra Raposo-García, Department of Pharmacology, Facultade de Veterinaria, Universidade de Santiago de Compostela, Campus Universitario s/n, 27002 Lugo

https://doi.org/10.1515/9783111014449-005

5.2 In vitro culture conditions

Cells or tissues in culture require specific conditions that differ from in vivo systems. However, cell culture conditions vary for each cell type. The consequences of deviating from the culture conditions required for a particular cell type can range from the expression of aberrant phenotypes to a complete failure of the cell culture [7]. Specific elements of culture conditions include culture media, supplements and other additives, culture-ware, and incubation conditions.

– **Basal medium:** This refers to a complex nutritive media designed to obtain a good cell viability or cell proliferation or to maintain a desired cell differentiation. Many solid or liquid medium formulations are commercially available with subtle changes in their medium formulation (including phenol red, glutamine, and other additives). Most of the commercial media are derived from the Eagle's minimum essential medium which contains amino acids, salts, glucose, and vitamins, or its modifications such as Dulbecco's modified Eagle's medium (DMEM) which contains a fourfold higher concentration of amino acids and vitamins, as well as additional supplementary components and was initially developed for the culture of embryonic mouse cells. The Ham's nutrient mixtures were originally developed to support the growth of several clones of Chinese hamster ovary (CHO) cells, as well as clones of HeLa and mouse L-cells. The RPMI medium (developed by Moore and Hood [9] at Roswell Park Memorial Institute, hence the acronym RPMI) uses a bicarbonate buffering system and alterations in the concentrations of amino acids and vitamins and has been used for the culture of human normal and neoplastic leukocytes. However, their supplementation has demonstrated wide applicability for supporting growth of many types of cultured cells, including fresh human lymphocytes. However, even subtle changes in the media formulation can alter the characteristics of certain cells or tissues; therefore, the medium to be used should be precisely specified.
– **Serum:** This is a complex mixture essential for the maintenance and/or proliferation of many cell types. However, due to its complexity and batch-to-batch variations, it introduces unknown variables in the culture system and in addition it represents a potential source of microbiological contaminants such as mycoplasma and bovine viruses due to its animal origin [7].
– **Serum-free media**: These are commercially available in order to decrease the batch-to-batch variability problems associated with the use of serum and offer better reproducibility and the potential for selective culture and differentiation of specific cell types such as cell lines [10]. However, serum-free supplements can also include poorly defined components such as pituitary extracts, chick embryo extracts, bovine milk fractions, or bovine serum albumin (http://www.oecd.org/env/ehs/testing/OECD %20Draft%20GIVIMP_v05%20-%20clean.pdf) which can exhibit batch-to-batch variation in biological activity.
– **Antibiotics:** These agents are used in cell culture to protect against contamination and for the selection of recombinant clones that express antibiotic-resistant

genes. The use of antibiotics should be avoided if possible or minimized since those agents can interfere with normal cell biology [7].

- **Cell culture matrix**: All cell culture plastics or glasses need to be sterile and are usually commercially available disposable plastic culture vessels. Most of the culture vessels are manufactured from polystyrene, a long carbon chain polymer with benzene rings attached to every other carbon. Polystyrene was chosen because it had excellent optical clarity and could be sterilized by irradiation. However, polystyrene is very hydrophobic and this fact can difficult cell attachment to untreated surfaces. Therefore, to increase cell attachment to the plastic surface, polystyrene is usually treated to create a more hydrophilic surface and negatively charged after medium addition. However, some cells are not able to attach in polystyrene-treated surfaces and require the coating with biological materials including extracellular matrix, attachment and adhesion proteins (collagen, laminin, and fibronectin), and mucopolysaccharides, such as heparin sulfate, hyaluronidate, and chondroitin sulfate, both individually and as mixtures. Another useful coating surface is the synthetic polymer poly-D-lysine (PDL) which creates a positive charge on polystyrene, and consequently, for some cell types, enhances cell attachment, growth, and differentiation, especially in serum-free and low serum conditions [11]. However, most of the coated culture surfaces need to be washed before cell seeding to avoid the possible toxicity of the coating material [7]. Figure 5.1 shows images of commercial presentations of plastic cultures (flask and plates), commercial culture medium, and culture plates after coating with laminin and PDL and washed with sterile phosphate-buffered saline.

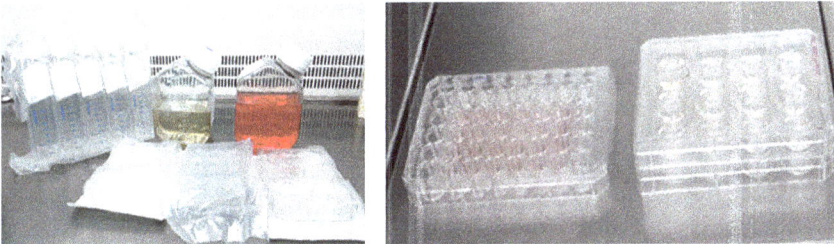

Figure 5.1: Different formats of plastic culture vessels (flasks and plates), commercial culture media (left panel), and precoated and washed culture plates ready for cell culture (right panel).

5.3 Handling and maintenance of cell cultures

Cell lines and primary tissues may contain microorganisms or pathogens able to cause human diseases or alter the in vitro results. To avoid these hazards, cell culture should be handled at biosafety (hazard) level 2. As a minimum, cell culture should be performed in a class II biological safety cabinet (Figure 5.2). Class II cabinets are designated to prevent biological exposure to personnel and the environment and to protect experimental material from being contaminated. Biological safety cabinets use high-efficiency particulate air (HEPA) filters in their exhaust and/or supply systems. These filtered cabinets are primarily designed to protect against exposure to toxics, including biological agents used in the cabinet. Air flow is drawn from the room around the operator into the front grille of the cabinet, which provides personnel protection. In addition, the downward laminar flow of HEPA-filtered air provides protection for experimental material inside the cabinet. Because cabinet air has passed through the exhaust HEPA filter, it is contaminant-free, providing environmental protection, and may be recirculated back into the laboratory (Class II Type A biological safety cabinets) or ducted out of the building (Class II Type B biological safety cabinets). A scheme of the air flow through a biological safety cabinet is shown in Figure 5.2.

In addition, cells or tissues in culture should be kept in controlled ambient conditions. The optimal culture temperature depends on the cell type; thus, insect cells have low optimal growth temperatures than mammalian cells while most mammalian cells grow normally well at a temperature of 37 °C [7]. Moreover, oxygen and carbon dioxide are vital for cell growth, and for most cell types, the appropriate atmosphere is normally 5% v/v carbon dioxide in air. Cell culture incubators maintain a constant temperature and high humidity for the growth of tissue culture cells under a CO_2 atmosphere. Typical control of temperature settings in the cell culture incubator ranges

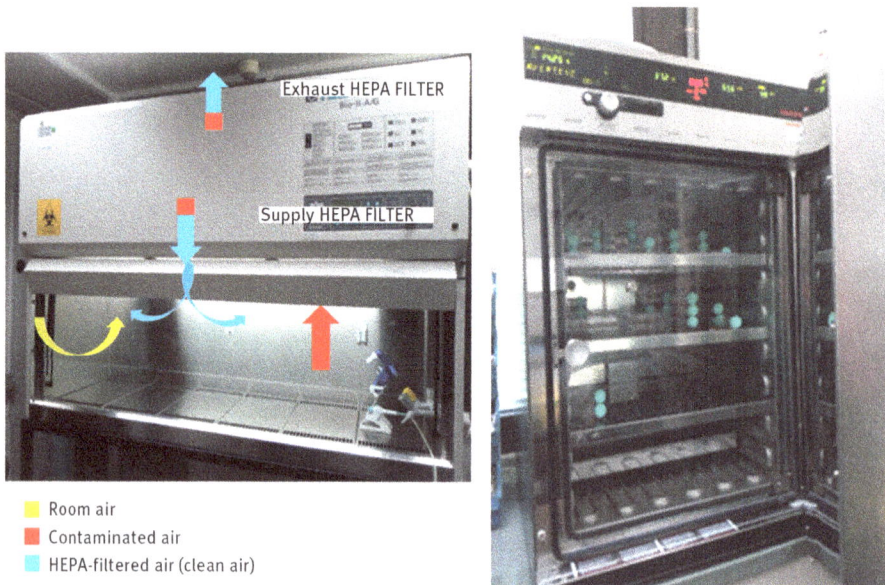

Exhaust HEPA FILTER

Supply HEPA FILTER

- Room air
- Contaminated air
- HEPA-filtered air (clean air)

Figure 5.2: Air flow through a biological safety cabinet for handling of cell cultures (left) and a culture incubator (right).

from 4 to 50 °C, and CO_2 concentrations run from 0.3% to 19.9%. Moreover, the temperature is typically controlled either by a water bath that circulates through the walls of the cabinet, or by electric coils that give off radiant heat. Some units also include refrigeration for cooling. Relative humidity is maintained between 95% and 98% by an atomizer system or a water reservoir. An image of a cell culture incubator is shown in Figure 5.2.

5.4 In vitro culture models

In vitro model systems are frequently grouped into three broad groups listed below:
1. **Isolated organs or tissues:** These cells are normally obtained directly from an animal or a donor. These in vitro systems are widely used in toxicological applications and may be used in different preparations that include:
   a. **Slices of certain tissues** (liver, lung, kidney, and brain) that, temporarily, retain some structural and functional features of the original organ.
   b. **Isolation and reaggregation of cells from different organs** given rise to two- or three-dimensional cultures that also retain some functional properties of the original organ and tissue.
   c. **Cells from blood or other body fluids** can be prepared as homogeneous preparations and kept in vitro for several days, or even used to generate stem cells (umbilical cord cells and bone marrow cells). An image showing the steps for purification of human T lymphocytes in our laboratory [12] as well as a confocal image for the expression of the voltage-dependent potassium channel KCNC1 (or $K_v3.1$) in control human T lymphocytes and in the same cells after 48 h of treatment with the lectin concanavalin A at 50 µg/mL is shown in Figure 5.3.
2. **Primary cultures and early passage cultures:** These in vitro cultures are harvested cells and tissues obtained directly from animals or humans but maintained in vitro for several times depending on the cell type. Frequently, those cultures retain key morphological and functional features of the in vivo system and are widely used for basic research and in different in vitro applications. Depending on the tissue of origin, cells in the primary culture can proliferate; however, their life span is limited and may change their initial characteristics with time in culture. With some exceptions, these systems normally represent heterogeneous cell populations that may be maintained either in suspension or in monolayers in glass or plastic surfaces. Examples of such cultures are primary cultures of cortical or cerebellar neurons widely used for the study of toxicity and mechanism of action of neurotoxic compounds [13] such as marine toxins [14–16]. Figure 5.4 shows a scheme for the isolation of primary cultures of cortical neurons and the in vitro development of such cells.

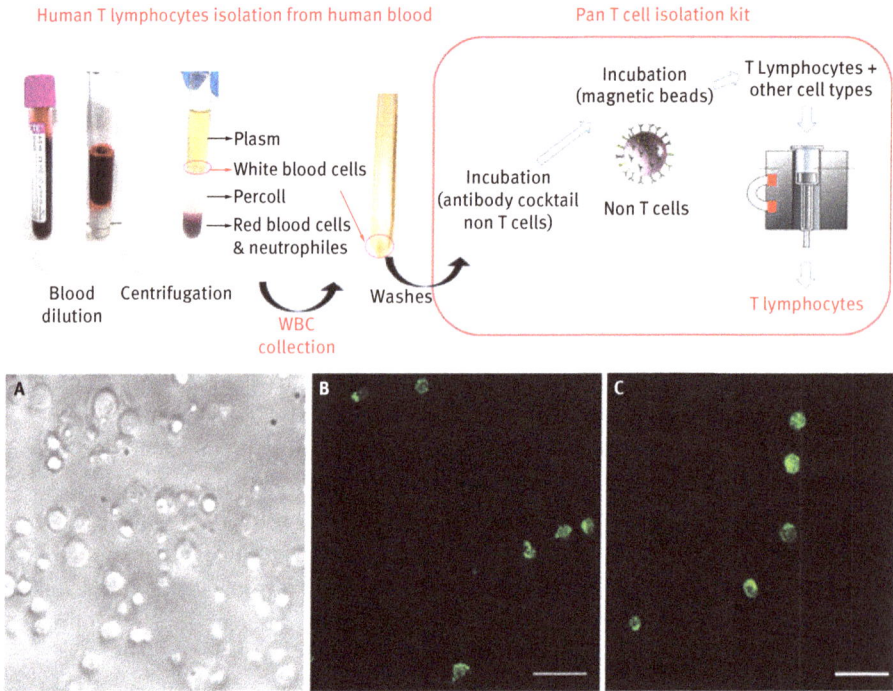

Figure 5.3: Upper panel: Scheme showing the isolation of human T lymphocytes from human blood donors. Lower panel: Phase contrast microscopic image of purified human T lymphocytes (A). (B) Confocal image showing the expression of the Kv3.1 potassium channel in human T lymphocytes cultured for 48 h, and (C) the expression of the same channel after treatment of the cells for 48 h with concanavalin A. WBC, white blood cells.

3. **Cell lines:** This term refers to cells that are able to multiply in vitro for extended periods of time and can be maintained in subculture. Cell lines can be divided into finite cell lines, continuous cell lines, and stem cell lines.

    a. **Finite cell lines:** These refer to cell lines that can be subcultured or passaged for a period of time but finally the cells stop to replicate but still maintain viability. Numerous finite cell lines have been established and they are genetically stable and remain diploid for many passages but generally reach senescence after 50–60 passages.

    b. **Continuous cell lines**: Those are cells that can be subcultured indefinitely and do not reach senescence. These are cells usually derived from tumors or normal embryonic tissues. These cell lines can either generate spontaneously or can be produced through a variety of techniques including radiation or treatment with chemical mutagens or carcinogens, isolated from cultures in-

A

B

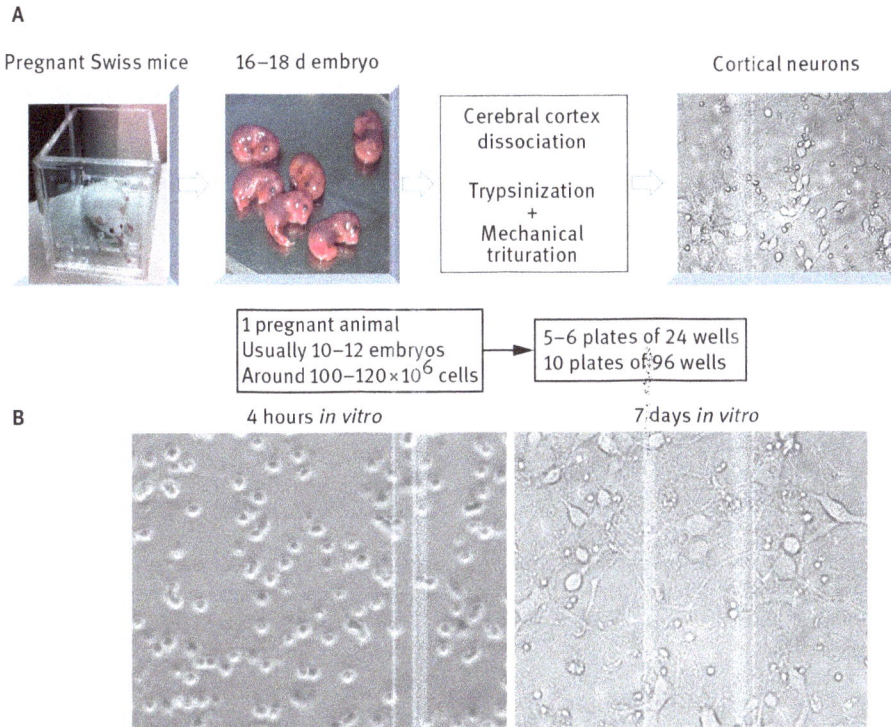

Figure 5.4: The procedure for the isolation of primary cultures of cortical neurons (A). Primary cultures of cortical neurons develop their typical morphology and functional characteristics with time in culture (B).

fected with viruses, through genetic modification by transfection or obtained from transgenic animals.

Widely used cell lines include epithelial cell lines obtained from the ovary of an adult Chinese hamster (CHO cells) or the human liver cancer cells HepG2 shown in Figure 5.5.

c. **Stem cell lines**: These are continuous cell lines that retain the characteristics of stem cells and can produce diverse differentiated cell types. They require great care in their maintenance, handling, and preservation. One example of a human stem cell line is the CTX0E16 cell line which is a human neuronal stem cells line, obtained from the cerebral cortex of a fetus of 12 weeks of gestation and immortalized by the ectopic expression of the c-mycER^{TAM} transgene and kindly provided to our laboratory by a material transfer agreement with Re-Neuron Limited (Guildford, Surrey GU2 7AF, UK). Human neural progenitor cells CTX0E16 hNPCs were cultured, following the provider instructions as previously reported [17]. Briefly, proliferating cells were maintained in reduced modified medium containing DMEM:F12 with 15 mM HEPES and sodium bicarbonate (Sigma) supplemented with 0.03% human serum albumin (Sigma), 100 µg/mL

CHO cell line   Hep G2 cell line

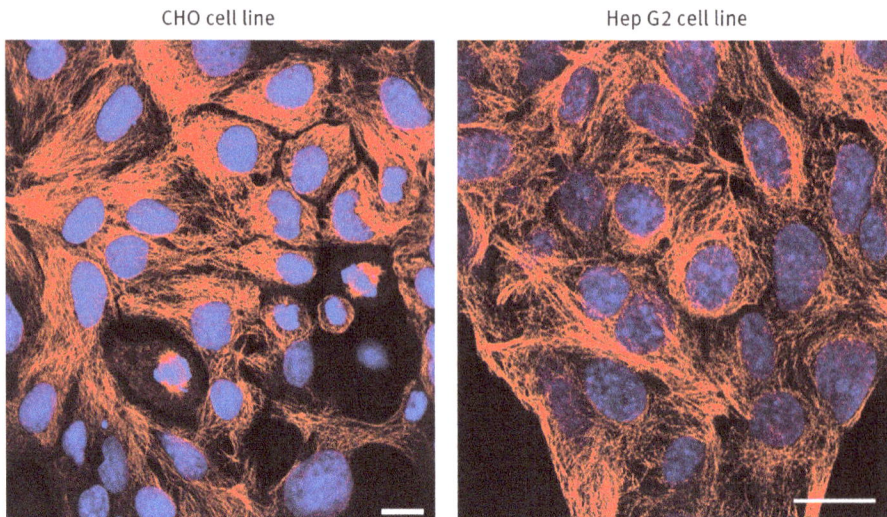

Figure 5.5: Confocal microscopy images showing adherent CHO cells and HepG2 cells after immunocytochemistry for the cytoskeleton marker tubulin (red). Nuclei were labeled with the nucleic acid stain Hoechst (blue). Images were kindly provided by Dr. J.A. Rubiolo (University of Santiago de Compostela).

apotransferrin (Scipac Ltd, Kent, UK), 16.2 μg/mL putrescine (Sigma), 5 μg/mL human insulin (Sigma), 60 ng/mL progesterone (Sigma), 2 mM L-glutamine (Sigma), and 40 ng/mL of sodium selenite (Sigma). Under proliferative conditions, cells were cultured in the presence of 10 ng/mL of human fibroblast growth factor 2, 20 ng/mL of human epidermal growth factor, both from PeproTech (Rocky Hill, NJ), and 100 nM hydroxytamoxifen (4-OHT, Sigma). CTX0E16 hNPCs were seeded onto PDL (5 μg/cm^2, Sigma) and laminin-coated (1 μg/cm^2, Sigma) tissue culture flasks, with full media changes occurring every 2–3 days. Cells were passaged once they reached 70–80% confluence using Accutase (Sigma) and maintained between 25 and 30 passages; all experiments were carried out using cells from passages 12 to 30. For differentiation, CTX0E16 cultures were washed twice with nonsupplemented DMEM:F12 medium and passaged onto PDL and laminin-coated tissue culture plates or glass coverslips at a density of 50,000 cells per mL. Cells were then washed in warm Dulbecco's phosphate-buffered saline (Thermo Fisher) and maintained in neuronal differentiation media (Neurobasal Medium, Thermo Fisher) supplemented with 0.03% human serum albumin, 100 μg/mL apotransferrin (Scipac Ltd, UK), 16.2 μg/mL putrescine (Sigma), 5 μg/mL human insulin (Sigma), 60 ng/mL progesterone (Sigma), 2 mM L-glutamine (Sigma), 40 ng/mL sodium selenite (Sigma), and 1 × B27 serum-free supplement (Thermo Fisher). Half medium changes were performed every 2–3 days, and cultures were differentiated for up to 60 days (days differentiated).

Undifferentiated and differentiated phase contrast microscopy images of these human neuronal stem cells are shown in Figure 5.6.

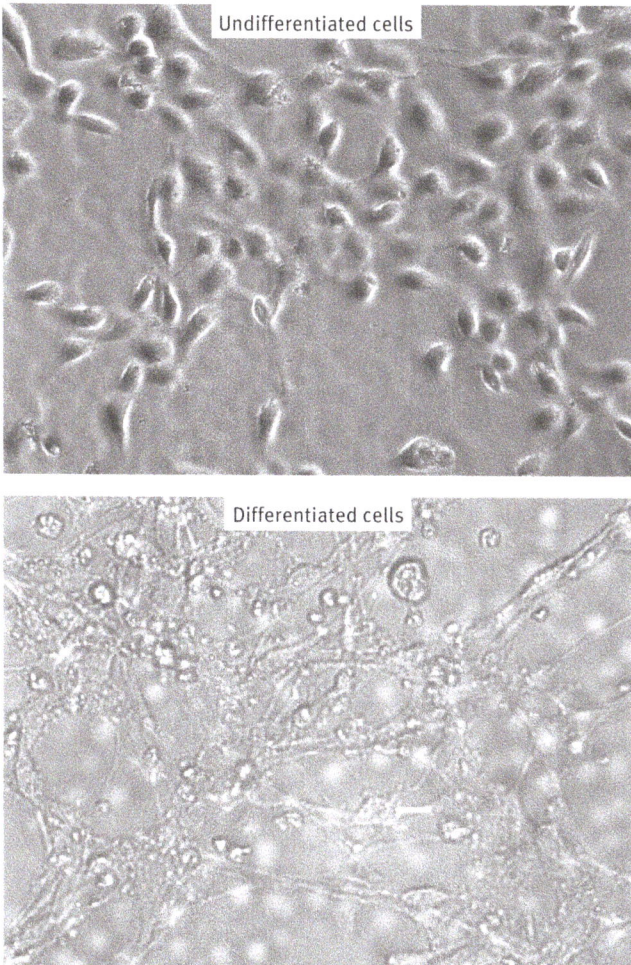

Undifferentiated cells

Differentiated cells

Figure 5.6: Phase contrast image showing undifferentiated (left) and differentiated human neuronal stem cells (right). Arrow heads indicate neuronal cell bodies.

5.5 In vitro cytotoxicity tests and methods to evaluate cellular function

In vitro tests may measure cellular function and cell death. Cell function is normally determined by the evaluation of the effect of chemicals on events or cellular signaling cascades that are related to cell injury or toxicity although several cell death tests are commonly used to evaluate cellular toxicity [3]. In vitro methods are common and widely used for screening and ranking chemicals, and have also been taken into account sporadically for risk assessment purposes in the case of food additives; however, a major promise of in vitro systems is to obtain information on the mechanism of action of chemical that is considered pivotal for adequate risk assessment [18].

In Europe, the European Union Reference Laboratory for alternatives to animal testing (EURL-ECVAM) has been established in 2011 and promotes the development and validation of methods alternative to animal testing. This laboratory has promoted the creation of the EURL ECVAM DataBase Service on ALternative Methods to animal experimentation (DB-ALM), a public and freely available service that provides evaluated information on the development and applications of advanced and alternative methods to animal experimentation in biomedical sciences and toxicology, both in research and for regulatory purposes (https://ecvam-dbalm.jrc.ec.europa.eu/). Actually, this database has evaluated more than 300 methods and almost 200 protocols, following the OECD guidelines. In total, detailed DB-ALM method descriptions, presented as comprehensive method summaries or individual protocols, are available for 26 different topic areas. Since the recompilation of all the available in vitro methods is out of the scope of this chapter, some of the methods are listed accordingly to the proposed endpoint following the DB-ALM classification, and other methods are listed in more detail, focusing on those frequently used in our laboratory to analyze toxicity of marine toxins.

Usually, cytotoxicity is considered primarily as the potential of a compound to induce cell death and is most frequently related to necrosis, a term used to define a class of irreversible cell death that most often results from acute cellular injury that causes metabolic failure of the cell that coincides with rapid depletion of ATP [19]. Most in vitro cytotoxicity tests measure necrosis. However, an equally important mechanism of cell death is apoptosis, also known as "programmed cell death," which is a type of cell death that is mediated by a genetically controlled, energy-requiring program that requires different methods for its evaluation [19]. The inhibition of apoptosis is also of toxicological importance. Furthermore, detailed studies on dose and time dependence on toxic effects to cells, together with the observation of effects on the cell cycle and their reversibility, can provide valuable information about mechanisms and type of toxicity, including necrosis, apoptosis, or other events [18]. It is widely accepted that in vitro cytotoxicity tests are useful and necessary to define basal cytotoxicity, for example, the intrinsic ability of a compound to cause cell death as a consequence of damage to basic cellular functions. Cytotoxicity tests are also necessary to define the concentration range for further and more detailed in vitro testing to provide meaningful information on parameters such as genotoxicity, induction of mutations, or programmed cell death. By establishing the dose at which 50% of the cells are affected (i.e., TC_{50}, TC represents toxic concentration), it is possible to compare quantitatively responses of single compounds in different systems or of several compounds in individual systems [18].

5.6 In vitro methods to evaluate the effects of compounds on reproduction

As mentioned above, the in vitro methods included in the DB-ALM database are listed in function of the endpoint analyzed, and each endpoint may or may not include several subgroups. For example, the in vitro methods to evaluate the effects of compounds on reproduction are divided into three categories analyzing male and female fertility and developmental toxicity (Figure 5.7); however, as stated in the review document of these methods, none of these ex vivo/in vitro/in silico methods have yet gained regulatory acceptance to date [20]. For this reason, these alternative methods do not allow for full replacement of animal testing. Therefore, these methods are mostly used for screening purposes and are intended to be applied in test batteries and as part of integrated testing strategies. However, with reference to "Reduction and Refinement" of animal use, the Extended One-Generation Reproductive Toxicity Study (EOGRTS) has been adopted as OECD Test Guideline 443 [21]. For male and female fertility, the list of cell-based protocols is summarized in Figure 5.7, while for developmental toxicity only the protocols are listed.

5.7 In vitro methods in cancer research

Cancer research and carcinogenicity and tumor promotion is one of the priority topics established in the DB-ALM database. For the field of cancer research, the methods are divided into three categories: drug discovery and activity testing (six methods), carcinogenicity (four methods), and tumor promotion (one method). The methods included in this section are listed in Figure 5.8.

5.8 In vitro methods for environmental toxicity

For testing the environmental toxicity of chemicals and biological agents, the DB-ALM provides several tests divided into different sections: aquatic short-term toxicity (6 methods), genotoxicity/mutagenicity (11 methods and 8 protocols), hematotoxicity (1 method and 1 protocol), hepatotoxicity/metabolism-mediated toxicity (31 methods and 9 protocols, however, under this epigraph several methods for cell culture are in-

Female fertility (total: 10 methods and protocols; 8 with cells)	• Follicle Culture Bioassay (FBA) • Granulosa and Theca Cell Culture Systems • *In vitro* Fertilization Assay • Ishikawa cell Test • Oocyte *In Vitro* Maturation Assay • Ovarian Culture • Permanent Embryonic Germ Cell Line Assay • Transactivation Assays to Detect Estrogen Receptor Agonists and Antagonists *In Vitro* with Stably Transfected Human Cell Lines
Male fertility (11 methods and protocols; 10 with cells)	• AR-CALUX: Androgen-Responsive Transactivation Bioassay • Computer-Assisted Semen Analysis (CASA) • *In vitro* Fertilization Assay • *In vitro* Sperm Chromatin Structure Assay (SCSA) • Leydig Cell-enriched Cultures • ReProComet Assay • Seminiferous Tubules Cultures • Sertoli Cell-Enriched Cultures • Sertoli-Germ Cells Cocultures • Testicular Organ and Tissue Culture Systems
Developmental toxicity (23 methods and 9 protocols)	• Embryonic Stem Cell Test (EST) • Embryotoxicity Testing in Post-Implantation Whole-Embryo Culture (WEC) - Method of Piersma • Embryotoxicity testing using a whole-embryo culture (WEC) procedure • *In Vitro* Micromass Teratogen Assay • *In Vitro* Syrian Hamster Embryo Cell Transformation Assay (SHE CTA) • Lung Cell Assay • Rabbit articular chondrocyte functional toxicity test • Rat Whole-Embryo Culture (WEC) • The Micromass Test - Method of Brown

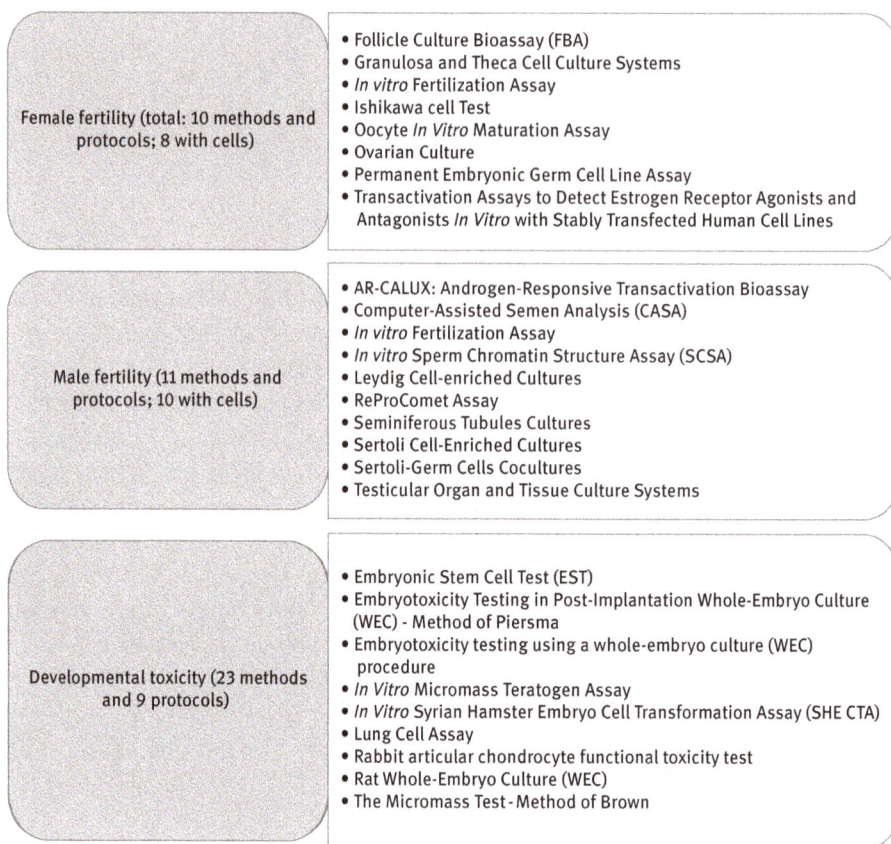

Figure 5.7: List of the reviewed in vitro methods to assess the effects of compounds on reproduction.

cluded), and 1 method and 1 protocol for immunotoxicity. A summary of the DB-ALM methods to test for environmental toxicity is provided in Figure 5.9.

5.9 In vitro methods for basal toxicity and cytotoxicity

Under this section, about 17 methods were reviewed and approved/recommended by the experts of the DB-ALM to test for the basal toxicity and cytotoxicity of compounds and biological agents, including marine toxins. These methods are detailed below, summarizing the cell type employed and the cellular function or activity evaluated. Since these group of methods are usually employed as starting points, they are examined in detail later.

Drug discovery and activity testing (6 methods)	• Annexin V assay • Colorimetric Cytotoxicity Assays for Anchorage-Dependent Cells(MTT based) • DNA fragmentation stains usung the bisbenzimide dyes (Hosecht 33258 and Hoescht 3334) • Determination of DNA fragmentation with 4'-6-diamidine-2-phenylindole (DAPI) staining • Diphenylamine assay • Terminal deoxynucleotidy transferase -mediated dUTP Nick End Labeling (TUNELL) assay
Carciogenicity (4 methods)	• Alkaline Unwinding Gentoxicty Test • Bhas 42 Cell Transformation Assay in 6-and 96-well plates • Cell transformation assay with BALB/c 3T3 cells (BALB/c 3T3 CTA) • *In vitro* Syrian Hamster Embryo Cell Transformation Assay(SHE CTA)
Tumor promotion (3 methods)	• Lucifer Yellow Intercellular Exchange Assay for Tumor Promotes • Screening System of promoters using RAS Transfected BALA 3T3 Clone(Bhas 42) • Serum-Free Liver mitogen test

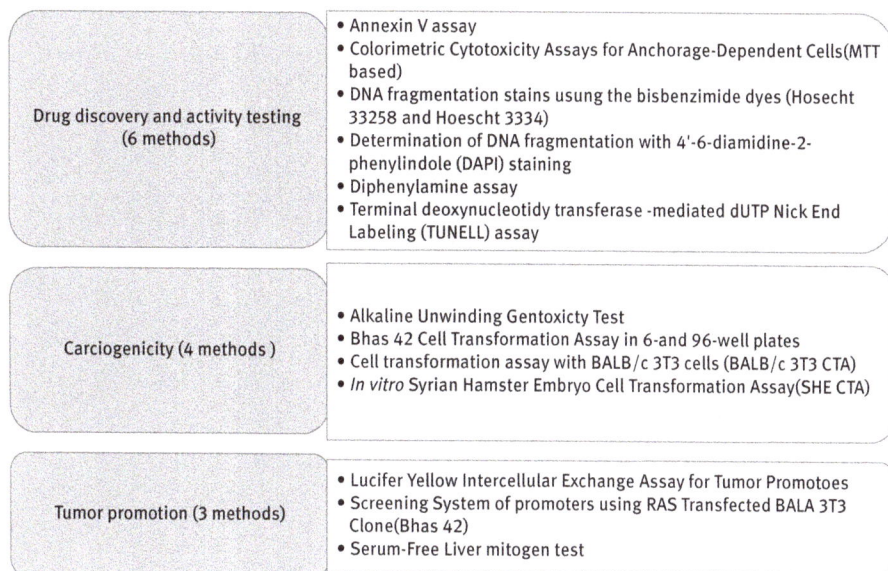

Figure 5.8: List of the reviewed in vitro methods for cancer research.

CHO cell proliferation test and CHO cell Na$^+$/K$^+$/ATPase activity test: These are two different methods using the same cellular model. CHO cells constitute a system useful for ecotoxicological studies. The proliferation rate of CHO cells correlates with physiological membrane functions, such as adenosine uptake and the activity of Na$^+$/K$^+$/ATPase. In the proliferation assay, CHO cells are cultured for 1 week in Petri dishes with various concentrations of test substance and the number of cells is counted twice daily, and the proliferation rate is determined from the logarithmic growth phase. The CHO cell Na$^+$/K$^+$/ATPase test measures the activity of the enzyme in the membrane of CHO cells.

Cytoskeletal alterations as a parameter for assessment of toxicity: The method is based on the determination of changes of cytoskeletal proteins (α- and β-tubulin and vimentin) after exposure to compounds by indirect immunofluorescence microscopy or quantitative biochemical methods (extraction of tubulin from the cells and measuring the tubulin content of the extracts using a colchicine binding assay). Immunocytochemistry and immunofluorescence microscopy are useful methods to evaluate the effects of marine toxins on cell integrity [22] and a protocol for a such methods is shown in Figure 5.10.

The main advantage of this method is that the microscopic visualization of cytoskeletal proteins provides a detailed view of changes in morphology; however, the technique is not adequate to quantify for protein expression changes. An example of the results obtained after the treatment of neuronal cells with chemicals and the analysis of their

Aquatic Short-Term Toxicity (6 methods)

- Allium Test
- Fish Embryo Acute Toxicity Test with Zebrafish
- Tetrahymena Assay for membrane-Stabilising Activity
- *Tetrahymena* Proliferation Rate and Maximal Density
- Tetrahymena thermophyla Chemosensory Response
- The Dunaliella tertiolecta Test, a marine Angal assay

Genotoxicity/Mutagenicity (11 methods, 8 protocols)

- Prostaglandin H Synthase (PHS) - Mediated Genotoxicity of Xenobiotics
- Bacterial Mutation Assay (with S.typhimurium and E.coli)
- Bacterial reverse mutation test (Ames test)
- Cytotoxicity and Genotoxicity In primary Cultures of Human Hepatocytes
- DNA Binding Studies for Alkylating Compounds Using Isolated Perfused Rat Liver
- DNA Binding In Bacteria
- GreenScreen HCtm Genotoxicity Test
- Permanent Embryonic Germ Cell line Assay
- Unscheduled DNA Synthesis in Hepatocyte Cultures Assessed by the Nuclei procedure

Hematotoxicity (1 method)

- Colony Forming Unit-Granuloyte/Macrophage (CFU-GM)Assay

Hepatotoxicity/metabolism-mediated Toxicity (31 Methods)

- Alginate entrapped primary hepatocytes (LiverBeadsTM)
- CYP1A1-inducing potency and Cytotoxicity Test in the Hepa-1 mouse Hepatoma Cell Line
- H-4-II-E Rat Hepatoma Cell Bioassay
- Liver Slice Hepatotoxicity Screening System
- Rat Hepatocyte Flow Cytometric cytotoxicity Test
- Reactive Metabolite Formation by fortified Liver microsomes
- The use of avian embryonic tissue for studies on metabollsm and metabollsm-mediated toxicity of chemicals
- The use of stem or progenitor cell-derved hepatocyte-like cells
- The use of subcellular fractions of liver tissue and liver homogenate
- Use of stable cell line Expressing Cytochromes CYP cDNA in Toxicity Testing

Immunotoxicity

- Polymerphonuclear Leucocytes Locomotion

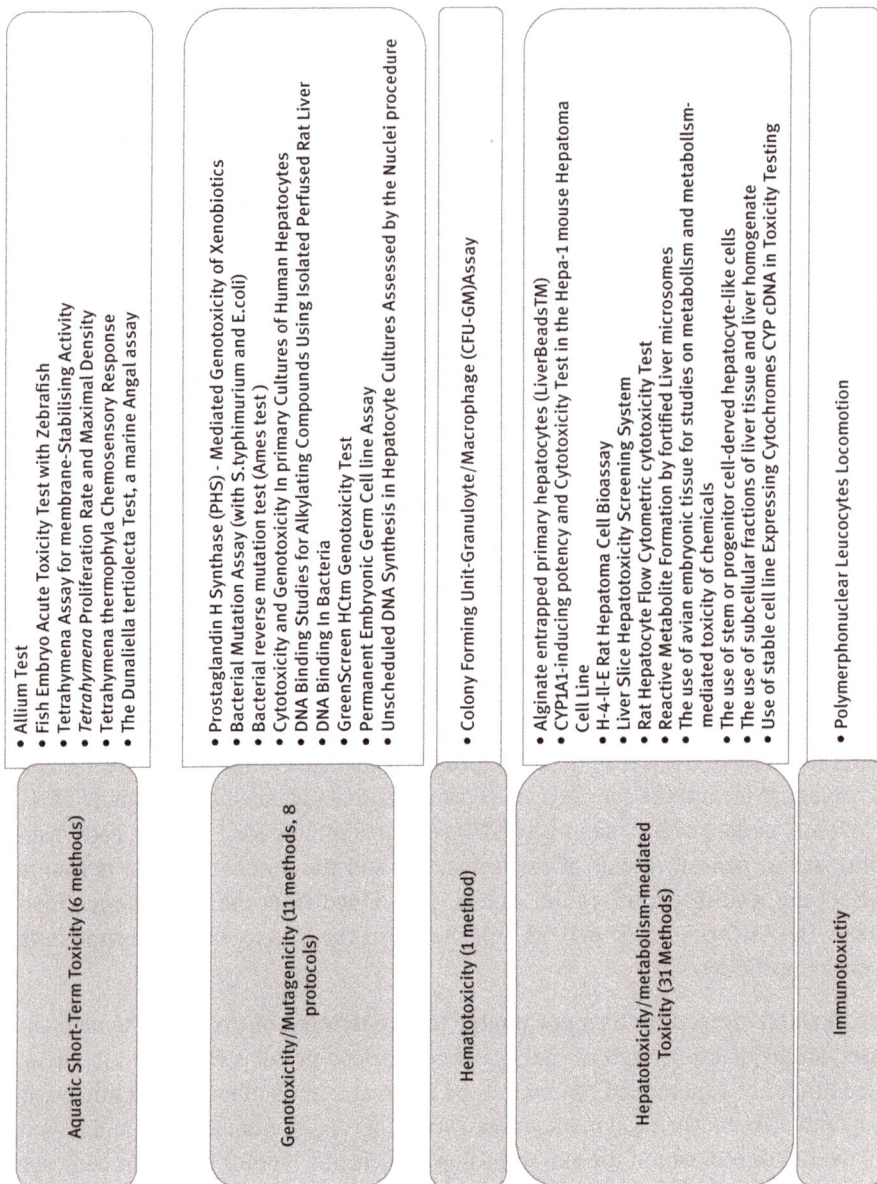

Figure 5.9: List of the reviewed in vitro methods to assess the environmental toxicity of compounds.

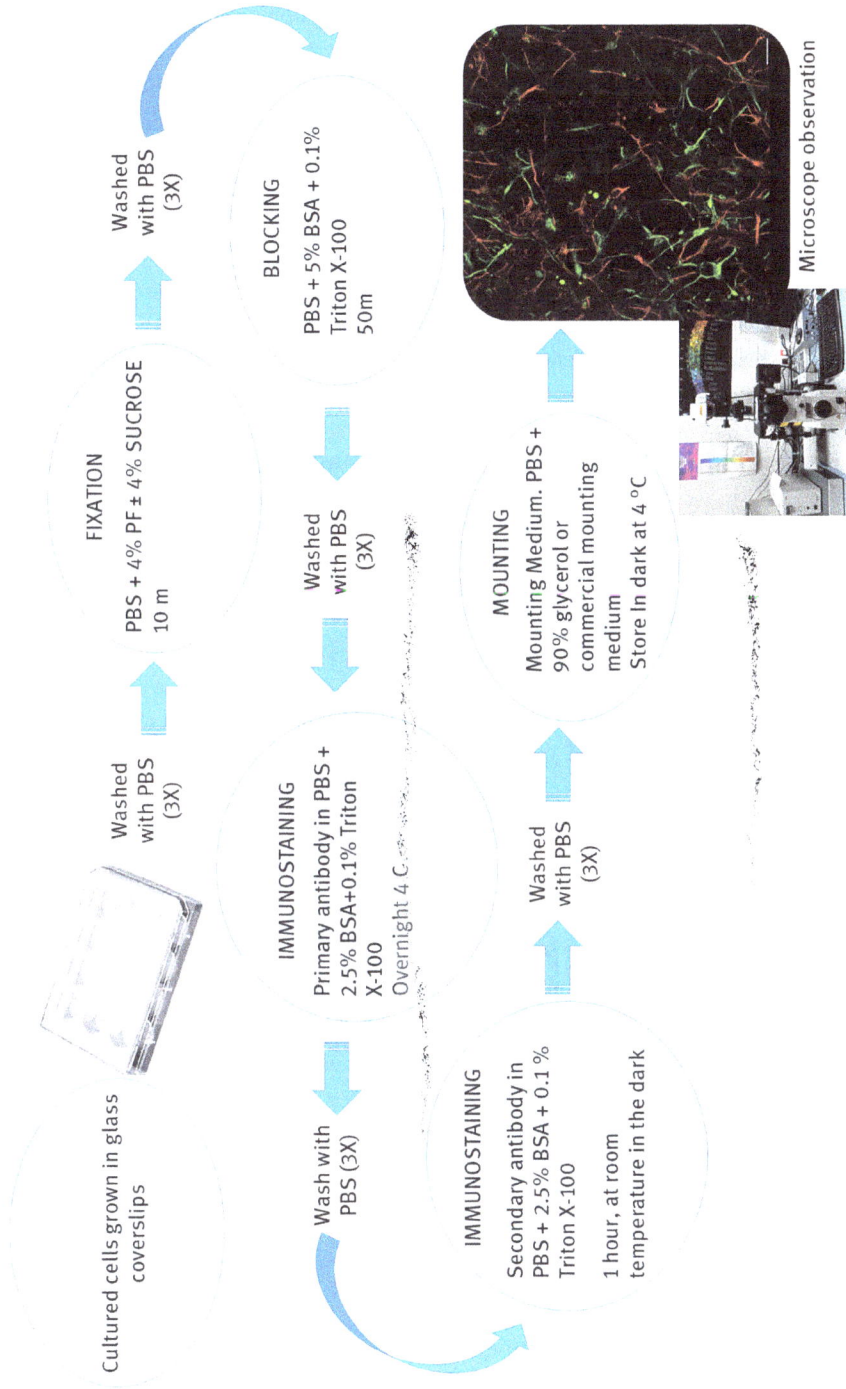

Figure 5.10: Scheme showing the protocol for immunostaining of adherent cells in culture.

effect on the cytoskeleton is shown in Figure 5.11. In the upper panel, control cortical neurons of 7 days in culture and age-matched neurons were treated with the marine ciguatoxin CTX3C for 24 h, and a double staining was performed with antibodies against the neuronal microtubule-associated protein MAP2 and against the GluR2/3 subunit of glutamate receptors. No apparent changes in protein expression or cell morphology were observed. In contrast, in the lower panel, human neuronal stem cells were treated with vehicle or with active derivative of the neurotoxins 6-hydroxydopamine or 1-methyl-4-phenyl-1,2,3,6-tetrahydropyridine known as 1-methyl-4-phenylpyridinium ion (MPP+), which is capable of inducing degeneration in human neuronal stem cells [23], and then they were stained with the neuronal cytoskeletal marker b3-tubulin and the neuronal nuclear marker NeuN. As shown in Figure 5.11, morphological changes were evident in cells treated with the neurotoxin MPP+.

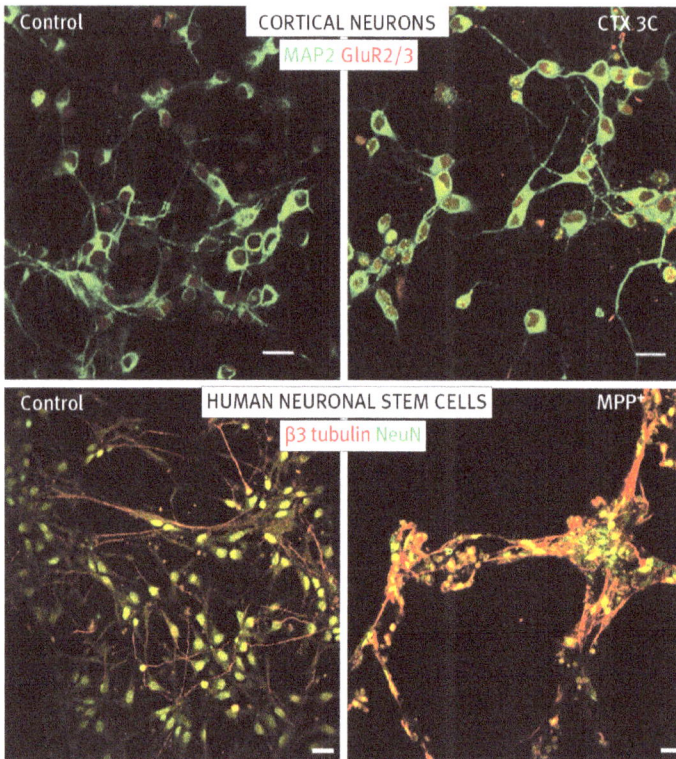

Figure 5.11: Confocal microscope images showing the staining of cytoskeletal proteins in primary cultures of cortical neurons (upper panel) and human neuronal stem cells (lower panel). Cultured cortical neurons nontreated (control) and treated with the marine toxin CTX3C were stained with MAP2 (green) and the glutamate receptor subunit GluR2/3 (red), showing that the toxin did not alter neuronal morphology. In the lower panel, human neuronal stem cells were exposed to the neurotoxin MPP+ and stained with the neuronal cytoskeletal marker β3-tubulin (red) or the nuclear protein NeuN (green). MPP+ treatment caused a dramatic change in cell morphology.

As mentioned above, one of the main disadvantages of immunocytochemistry is its difficulty to objectively quantify changes in protein expression levels. The data shown in the upper panel of Figure 5.11 does not indicate any change in the level of expression of the glutamate receptor subunit GluR2/3 after ciguatoxin treatment in cortical neurons. However, quantitative analysis of the expression levels of GluR2/3 by western blotting showed a significant decrease in GluR2/3 expression in cortical neurons after ciguatoxin treatment [15].

Western blotting is an important technique used in cell and molecular biology to identify specific proteins from a complex mixture of proteins extracted from cells and it allows to quantify the protein expression as well [24]. The technique uses three elements to identify a specific protein that include:
– Separation of the protein by size or molecular weight
– Transfer the proteins to a solid support
– Identification of proteins using primary and secondary antibodies

Although several different methods can be used to quantify proteins in biological samples, sodium dodecyl sulfate (SDS) polyacrylamide gel electrophoresis is the most widely used analytical method to resolve separate components of a protein mixture. The technique is used by many laboratories to investigate or demonstrate expression changes of a given protein between control states and experimental conditions and is still the method of choice for basic research, but also a useful tool in clinical applications and may be applied to any biological sample including cell or tissue extracts and body fluids, such as plasma or serum or urine [25]. A scheme of the procedure used for western blot in lysates from adherent cell cultures is shown in Figure 5.12. In brief, cell lysates are one of the most common systems used in western blot. Protein extraction from mammalian tissues usually requires mechanical disruption and the use of reagents containing detergents. Moreover, cell lysis should be performed at low temperature and with protease and phosphatase inhibitors to prevent protein degradation. Once extracted, protein concentration is measured, usually with a spectrophotometer to quantify the amount of protein loaded into each well. Afterward, the sample is diluted into a loading buffer, which contains glycerol that increases the density of the sample relative to the surrounding running buffer, making it easier to load in the well; a tracking dye (bromophenol blue) is used to follow the run of protein sample on the gel; SDS is used to denature proteins and load the proteins with a strong negative charge that will allow each protein to migrate in the electrophoretic field in a measure proportional to its size; and thiol reagents (β-mercaptoethanol) are used to reduce disulfide bonds. It is also very important to have positive and negative controls for the sample. For a positive control, a known source of target protein such as purified protein or a control lysate is used. This helps to confirm the identity of the protein and the activity of the antibody. A negative control is a null cell line, such as β-actin, which is used to confirm that the staining is not nonspecific. Western blot uses two different types of agarose gel: stacking and separating gel. The higher stacking gel is slightly acidic (pH 6.8) and has a lower acrylamide concentration making a porous gel, which separates protein poorly but allows them to form thin, sharply defined bands. The lower gel called the separating or resolving gel is basic (pH 8.8), and has a higher polyacrylamide content, making the gel's pores narrower. Protein is thus separated by their size in this gel, as the smaller proteins travel more easily and rapidly than larger proteins. After separating the protein mixture, it is transferred to a membrane (blotting) using an electric field causing proteins to move out of the gel and onto the membrane. In this step, it is necessary to ensure a close contact between the gel and the membrane and the placement of the membrane between the gel and the positive electrode. The membrane must be placed as such so that the negatively charged proteins can migrate from the gel to the membrane. This type of transfer is called electrophoretic transfer and can be done in semi-dry or wet conditions. Wet conditions are usually more reliable as it is less likely to dry out the gel and is preferred for larger proteins while semi-dry transfer is faster. Membranes are made either of nitrocellulose or polyvinylidene difluoride (PVDF). Nitrocellulose is used for its high affinity for protein and its retention abilities but does not allow the membrane to be used for reproving. In contrast, PVDF membranes provide better mechanical support and allow the blot to be reproved and stored but have a higher background than nitrocellulose membranes. Regarding the washing, blocking, and antibody incubation steps, it should be mentioned that blocking is very important to prevent antibodies from binding to the membrane in a nonspecific manner. Blocking is usually performed in 5% bovine serum albumin or nonfat dried milk to reduce background. Frequently, the concentration of the antibody used is chosen following the instructions provided by the manufacturer, and primary and secondary antibodies are diluted in washing buffers, either phosphate-buffered saline (PBS) or tris-buffered saline containing Tween 20 (TBST). Afterward, proteins in the membrane are usually detected with a secondary antibody linked to an enzyme such as horseradish peroxidase (HRP) and visualized by chemiluminescence detection. The use of chemiluminescence detection allows multiple film exposures and enables optimization of signal to noise. Chemiluminescent detection occurs when energy from a chemical reaction is released in the form of light. The most popular chemiluminescent western blotting substrates are luminol-based. For example, in the presence of HRP and peroxide buffer, luminol oxidizes and forms an excited state product that emits light as it decays to the ground state. Light emission occurs only during the enzyme–substrate reaction; therefore, once the substrate in proximity to the enzyme is exhausted, the signal output ceases. The two most common enzyme reporters that catalyze chemiluminescent reactions are HRP and alkaline phosphatase. Enzyme-conjugated secondary antibodies are used for western blotting, and light-producing reactions are captured with X-ray film or with charge-coupled device camera-based digital imaging instruments. The detection reagents can be removed and the entire blot reproved to visualize another protein or to optimize detection of the first protein. Finally, it should be emphasized that quantification of protein levels is always semiquantitative because western blot data provides a relative comparison of protein levels, but not an absolute measure of quantity. The reason for the semiquantitative nature of western blot measurements is due to the variations in loading and transfer rates between the samples in separate lanes and to the fact that the signal generated by the detection is not linear across the concentration range of the samples [24]. However, the differences in protein loading are normally standardized, using an internal control or loading control protein which is a protein derived from

1. Preparation of lysates from cell cultures

| Adherent cell culture | Wash with PBS (phosphate buffered saline) 3x | Add lysis buffer With proteases and phosphatases inhibitor cocktail 5 m | Scrape adherent cells from the dish | Centrifugate cell suspension 4°C 20 m 12000 rpm |

2. Preparation of samples and electrophoresis: separate proteins by size

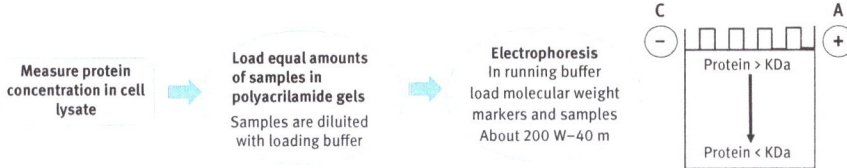

| Measure protein concentration in cell lysate | Load equal amounts of samples in polyacrilamide gels Samples are diluted with loading buffer | Electrophoresis In running buffer load molecular weight markers and samples About 200 W–40 m |

C (−)   A (+)
Protein > KDa
Protein < KDa

3. Electrotransfer (semi-dry): transfer the protein to membranes

Before transfer...
Equilibrate de gel in PBS-0.01% Tween (PBST) 30 m
Extra thick blot papers and membranes must be equilibrated in transfer buffer

Electrotransfer
About 10 W–30 m

A (+)
Extra Think
Gel
PVDF or nitrocellulose membrane
Extra Think
C (−)

4. Membrane blocking, washing, antibody incubation and chemiluminiscence detection

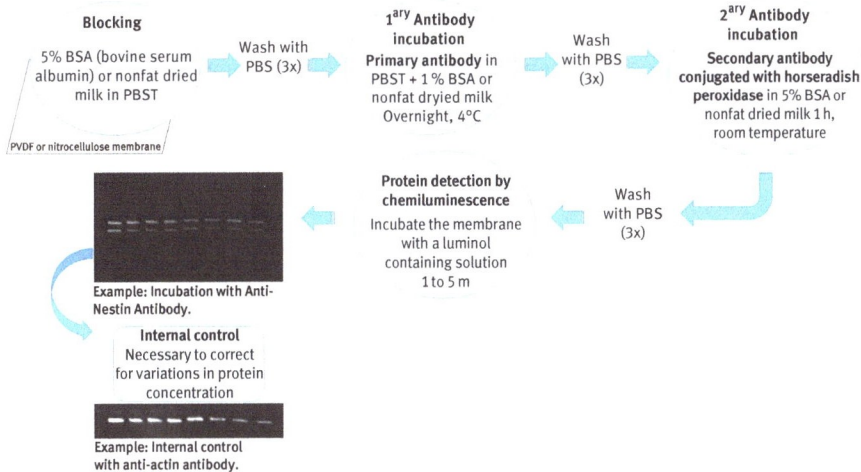

| Blocking 5% BSA (bovine serum albumin) or nonfat dried milk in PBST | Wash with PBS (3x) | 1^ary Antibody incubation Primary antibody in PBST + 1 % BSA or nonfat dryied milk Overnight, 4°C | Wash with PBS (3x) | 2^ary Antibody incubation Secondary antibody conjugated with horseradish peroxidase in 5% BSA or nonfat dried milk 1 h, room temperature |

PVDF or nitrocellulose membrane

Protein detection by chemiluminescence
Incubate the membrane with a luminol containing solution
1 to 5 m

Wash with PBS (3x)

Example: Incubation with Anti-Nestin Antibody.

Internal control
Necessary to correct for variations in protein concentration

Example: Internal control with anti-actin antibody.

Figure 5.12: Schematic representation of a western blot protocol using cell lysates obtained from adherent cells.

ubiquitously expressed "housekeeping" genes and has been widely used due to its presumed consistent level of expression across a diverse range of samples. Actin and tubulin are two of the most frequently used loading controls in biomedical research [26].

HEL-30 cytotoxicity test: This determines the anabolic competence of the cell. HEL-30 cells are incubated in the presence of radiolabeled leucine with or without test chemical for a short period of time. Uptake of the radio-labeled leucine is terminated by the addition of unlabeled leucine. Cell protein is precipitated with trichloroacetic acid and harvested onto glass fiber filters. The radioactivity of the samples is measured by liquid scintillation counting.

Human lymphocyte cytotoxicity assay: This measures the leakage of DNA and lactate dehydrogenase (LDH, EC. 1.1.1 27) from lymphocytes into the surrounding medium as an indicator of cytotoxicity and also includes a measure of mitochondrial activity through the MTT (3-(4,5-dimethylthiazol-2-yl)-2,5-diphenyltetrazolium bromide) assay. The method uses lymphocytes isolated from anticoagulated, normal, human blood samples and grown for 5 days, centrifuged and resuspended in complete medium, aliquoted into 24-well plates and exposed to the test chemicals at appropriate dilutions.

LS-L929 cytotoxicity test: This test uses L-929 fibroblasts (mouse) maintained in suspension culture and incubated in the presence of a test material in a range of concentrations for 4 h. Cell viability is then determined by uptake of the dyes ethidium bromide (EB) and fluorescein acetate. The resultant cytotoxic effect is quantified using two complementary fluorimetric assay procedures. After exposure to a test compound, viable cells may be identified by their ability to accumulate fluorescein on incubation with fluorescein diacetate (FDA). The nucleic acid of both viable and nonviable cells can be stained with EB. Nonviable cells (stained only with EB) may be distinguished from viable cells (stained with EB and with FDA) by the selective use of filters.

Laser diffraction measurement of tumor spheroids: Tumor cell lines cultured as aggregates can be utilized for in vitro radiosensitivity and/or chemosensitivity tests. Chemical effects are monitored by studying the changes in spheroid diameter measured by laser diffraction. This determines cell viability and morphology and has been used in human cervical, colorectal, and lung cancer cell lines.

Quantitative video microscopy of intracellular motion and mitochondria-specific fluorescence: The test uses either IMR-90 fibroblasts (human) or L-929 fibroblasts (mouse). IMR 90 cells are cultured on cover slips, mounted on slides, and incubated for 1–24 h in the presence or absence of test compounds. The movement of cell organelles is observed by means of video-enhanced contrast microscopy. The cells are maintained at a stable temperature and pH in an incubation chamber. At the same time, the lysosomes and mitochondria are specifically stained with fluorescent vital dyes so that their number and morphology may be assessed. The analog video signal is enhanced, digitalized, and subjected to several steps of image processing. The final images are recorded and later analyzed to provide plots of organelle velocity versus incubation time.

MTT assay: The tetrazolium salt MTT is taken up into cells and reduced in a mitochondria-dependent reaction to yield a formazan product [14, 16]. The product accumulates within the cell due to the fact that it cannot pass through the plasma membrane. On solubilization of the cells, the product is liberated and can readily be detected and quantified by a simple colorimetric method. The method can be used almost with any cell type.

The ability of cells to reduce MTT provides an indication of mitochondrial integrity and activity which, in turn, may be interpreted as a measure of viability and/or cell number. The test measures the tissue viability which is determined by a reduction in mitochondrial dehydrogenase activity and measured by formazan salt production from MTT. The result is expressed as the percentage of the negative control. The endpoint value is the tissue viability (%) calculated as the ratio $(OD_{treated})/(OD_{negative\ control}) \times 100\%$. It is useful to determine the TC_{50} or the IC_{50} which is defined as the concentration of a test substance that decreases the MTT reduction to formazan by 50%, from a dose–response curve.

The assay is useful to determine the viability of many cell types but not for cells with low mitochondrial activity. However, the number of cells initially plated, the period of exposure to chemicals, the concentration of MTT, the total duration of the experiment, and so on must be standardized for each cell line. The protocol used for the MTT assay in human neuronal CTX0E16 cells is summarized in Figure 5.13 and is detailed below.

1. Cells seeded in 96- or 48-well plates are cultured for the desired time and treated with the compound of interest (in this case, several DMSO concentrations are 0.001%, 0.01%, 0.1%, 1%, 2.5%, 5%, and 10% DMSO v/v) and control wells are treated with the corresponding vehicle used to solubilize the compound and are incubated in the culture incubator for the desired time (in this case, 24 h in the incubator).
2. Remove the culture medium and wash three times with Locke's buffer containing 154 mM NaCl, 5.6 mM KCl, 1.3 mM $CaCl_2$, 1 mM $MgCl_2$, 10 mM HEPES, and 5.6 mM glucose (pH 7.4) for 1 h at 37 °C.
3. Incubate the plates with 200 μL of 500 μg/mL MTT dissolved in Locke's buffer per well for 60 min at 37 °C. MTT solution should be prepared immediately before the experiment, kept in the dark, and discarded after using.
4. Carefully wash off excess MTT in order to avoid cell detachment and add 200 μL of 5% SDS at 5%, 200 μL per well to disaggregate the cells and keep the plate overnight in the dark.
5. Transfer disaggregated cells from each well to 96-well plates and treat with 10 μL isopropanol to eliminate air bubbles.
6. Read the absorbance of the colored formazan salt at 595 nm in a spectrophotometer plate reader.

Cell treatment with compound

	1	2	3	4	5	6	7	8	9	10	11	12
A												
B	C	TI	T2	T3	T4	T5	T6	T7	T8	T9		
C	S	C	TI	T2	T3	T4	T5	T6	T7	T8		
D	T9	S	C	TI	T2	T3	T4	T5	T6	T7		
E	T8	T9	S	C	TI	T2	T3	T4	T5	T6		
F	T7	T8	T9	S	C	TI	T2	T3	T4	T5		
G	T6	T7	T8	T9	S	S	C	C	C	C		
H												

48 well plate with MTT

00 : 60
HRS MIN

Solubilizated cells with MTT

12 : 00
HRS MIN

Live cells in purple

Absorbance reading at 595 nm

C	0,455	0,480	0,460	0,480
T1	0,463	0,447	0,469	0,460
T2	0,445	0,448	0,377	0,423
T3	0,411	0,339	0,437	0,396
T4	0,479	0,388	0,392	0,420
T5	0,417	0,385	0,385	0,396
T6	0,253	0,264	0,307	0,275
T7	0,260	0,335	0,404	0,333
T8	0,242	0,347	0,333	0,307
T9	0,111	0,142	0,124	0,126
S	0,113	0,143	0,136	0,131

Plot data and calculate IC_{50}

Mitochondrial function (% of control)

% DMSO

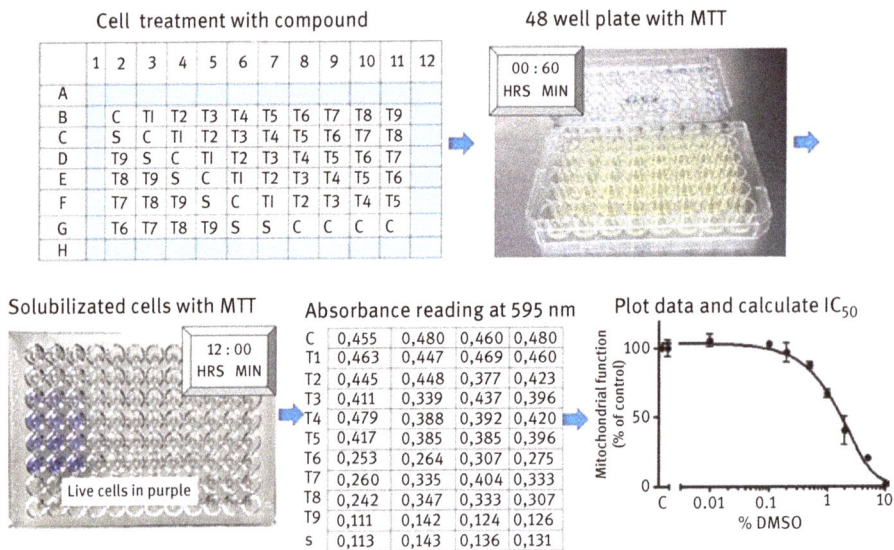

Figure 5.13: MTT test to measure cell viability in CTX0E16 human neuronal stem cells: Experimental design is shown in the left upper panel (in blue are empty wells): C, control cells; T, treated cells. Treatments: T1, DMSO 0.001%; T2, DMSO 0.01%; T3, DMSO 0.1%; T4, DMSO 0.2%; T5, DMSO 0.5%; T6, DMSO 1%; T7, DMSO 2%; T8, DMSO 5%; T9, DMSO 10%; S, saponin (cell death control). Upper right panel: addition of MTT to treated cells. Lower left panel: MTT-colored formazan salt in solubilized cells. Middle lower panel: 595 nm absorbance values reading with the corresponding treatment for one representative experiment. Right lower panel: Dose–response plot showing the means of at least three independent experiments.

Resazurin (Alamar Blue™) cytotoxicity assay: This test is also used to determine the viability of cells in culture after exposure to chemicals. Resazurin (7-hydroxy-3H-phenoxazin-3-one-10-oxide, also called Alamar Blue™ by some authors) is a nontoxic, nonfluorescent blue dye which is reduced by metabolically active cells to the fluorescent red dye resorufin. This redox indicator is a widely used nontoxic reagent that exhibits both fluorimetric and colorimetric properties in response to metabolic activity. Resorufin has a pink color and is highly fluorescent. As the number of living cell decreases, the intensity of the fluorescence also decreases. The reduction-induced color change varies proportionately with cell number and time, changing from a nonfluorescent color blue to a reduced pink fluorescent form [27]. A scheme of the procedure used to evaluate cytotoxicity using the resazurin assay is shown in Figure 5.14.

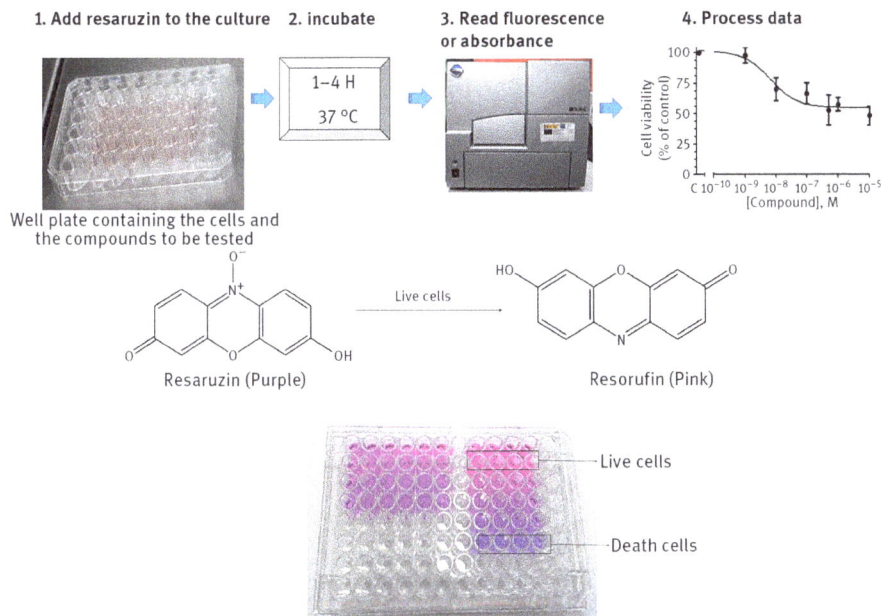

Figure 5.14: Scheme showing the protocol to evaluate cytotoxicity using the resazurin cytotoxicity assay and its mechanism of action.

LDH cytotoxicity assay: The measurement of LDH levels is a reliable tool to determine the toxicity of chemical compounds. The release of LDH to the cell culture media is an indicator of irreversible cell death resulting from the cell membrane damage [28, 29]. The LDH release can be measured through a two-step enzymatic reaction. Initially, LDH catalyzes the conversion of lactate to pyruvate by reducing NAD^+ to NADH. Then, diaphorase utilizes NADH to reduce the tetrazolium salt, iodonitrotetrazolium chloride, to produce a red formazan product. The quantity of formazan produced is directly correlated to the quantity of LDH released to the medium. The level of formazan formation can be easily detected by measuring the absorbance at 490 nm. Absorbance should also be measured at 680 nm to detect the background absorbance value. The absorbance value obtained at 680 nm should be subtracted from the absorbance value measured at 490 nm to obtain the final absorbance value and, therefore, the quantification of formazan. Currently, there are multiple kits that allow us to measure LDH release. Each assay kit brings specifications to follow, but in general, the schematic representation of LDH protocol along with the chemical reaction is represented in Figure 5.15.

This method has many advantages such as the reliability, speed, and the direct measurement of the LDH in the culture media, avoiding the need of wash the cultured cells. Therefore, LDH can provide a reliable method for measuring cell damage in con-

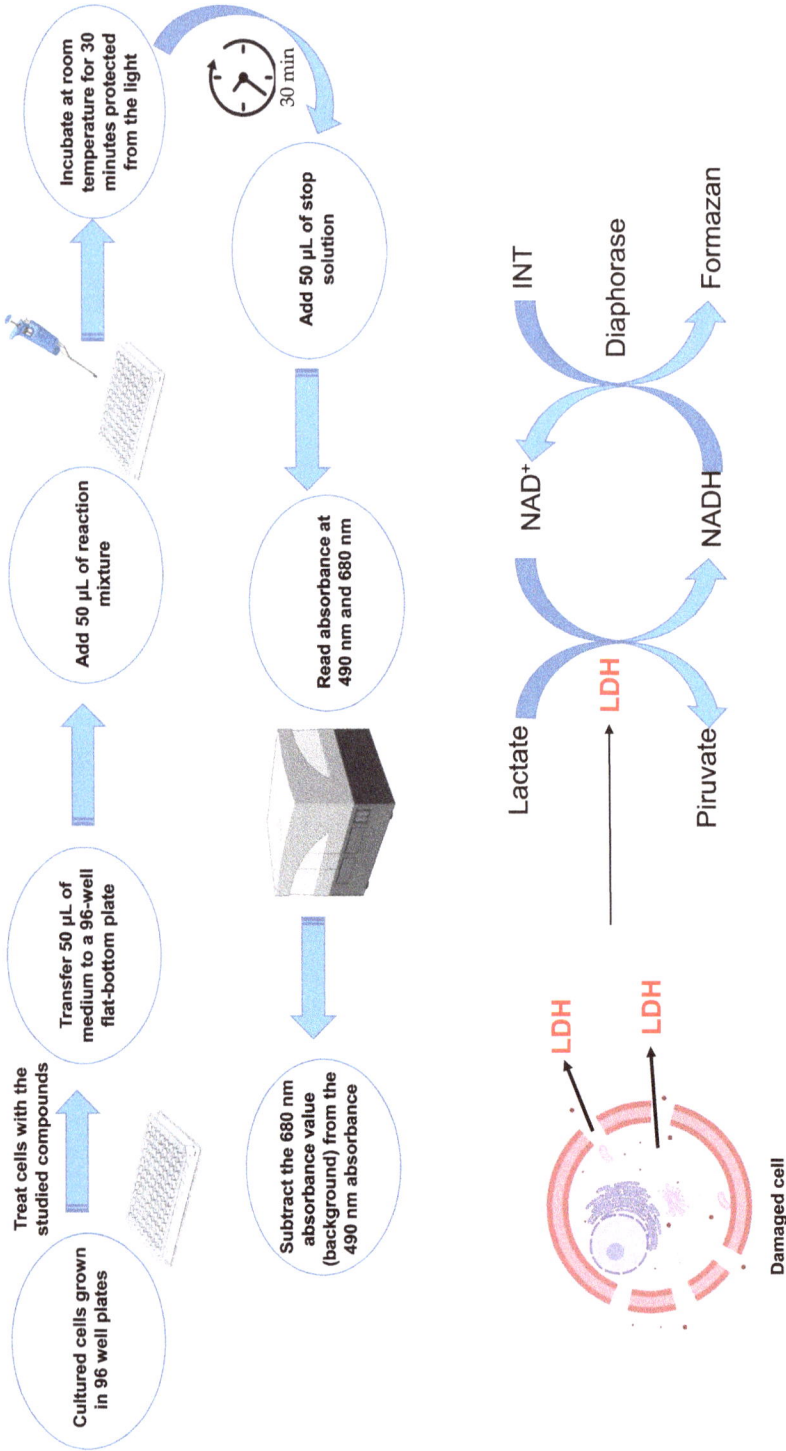

Figure 5.15: Schematic representation of the protocol to determine LDH release and the chemical reactions involved.

Table 5.1: Summary of the different endpoints proposed to evaluate in vitro cytotoxicity as well as their main advantages and disadvantages (adapted from [5]).

Endpoint	Assay	Mechanism	Advantages/disadvantages
1. Structural cell damage (noninvasive)	Evaluation of overall cell shape, cytoplasmic structure, flatness, and outline properties on a good phase contrast light microscope	Screening assay	**Advantages:** Noninvasive **Disadvantages:** Qualitative data, no exact cell death definition
	LDH-release test	Lactate dehydrogenase (LDH) enzyme is released to the culture medium when cell membranes rupture (nonviable cells), and the enzyme can then be measured in the supernatant	**Advantages:** Measurement of a definite/unambiguous cell death endpoint; can be combined with cell function assays. Allows cells to be used for other purposes, if only supernatant is sampled. **Disadvantages:** Normalization necessary (extra wells for controls). Frequently high background LDH levels are observed (e.g., from serum components).
2. Structural cell damage (invasive)	Membrane penetration by dyes to detect "cytotoxicity" (e.g., naphthalene black, trypan blue, propidium iodide, ethidium bromide, and EH-1)	Involves the use of dyes that stain nonviable cells, but do not enter viable cells with an intact cell membrane	**Advantages:** Rapid and usually easy to interpret. Gives information on the single cell level. **Disadvantages:** May overestimate viability since apoptotic cells continue to have intact membranes and may appear viable. Some dyes (e.g., trypan blue, H-33342) are cytotoxic so that the evaluation has to be performed rapidly. Usually need to be combined with dyes that stain both live and death cells.

(continued)

Table 5.1 (continued)

Endpoint	Assay	Mechanism	Advantages/disadvantages
	Retention of dyes within intact cells to detect "viability" (e.g., fluorescein diacetate or calcein-AM)	The lipid-soluble dyes are transformed by cellular enzymes (esterases) into lipid-insoluble fluorescent compounds that cannot escape from cells with intact membranes	**Advantages:** Rapid and usually easy to interpret. **Disadvantages:** Some cells leak the dyes and some dyes, and some dyes can suffer photo-bleaching.
	Evaluation of programmed cell death/apoptosis markers	Activation of caspases (enzymatic analysis or staining) Activation or endonucleases (detectable as DNA fragmentation) Chromatin condensation (detectable by DNA stains) Detection of phosphatidylserine on the outside of the plasma membrane (annexin staining)	**Advantages:** Adds mechanistic information to cytotoxicity data. **Disadvantages:** Not all types of cell death may be detected by a given endpoint. Needs to be combined with a general cytotoxicity test. Some endpoints are prone to artifacts (annexin staining) and some staining techniques (TUNEL, caspase-3) lead to an unintentional selection of subpopulations. Caspase activity measurement does not easily yield a prediction model for the extent of cell death.
3. Cell growth	Cell counting	For some cell populations impaired growth is considered as a reduction of viability	**Advantages:** growth can be a sensitive parameter of cell well-being. **Disadvantages:** growth is not necessarily linked to cytotoxicity; artifacts Needs careful control in combination with cytotoxicity assays.
	BrdU or EdU incorporation	Measures new DNA synthesis based on incorporation of the easily detectable nucleoside analogs BrdU (or EdU) into DNA	**Advantages:** Measurement on single cell level. Easy to quantify. **Disadvantages:** BrdU/EdU can be cytotoxic; high cost and effort compared to counting.

	Staining of cellular components that are proportional to overall cell mass (proteins by, e.g., sulforhodamine B or crystal violet; DNA by Hoechst H-33342)	These assays evaluate a surrogate measure of overall cell mass and assume that it correlates with total cell number	**Advantages:** Simple and cheap; lots of historical data. **Disadvantages:** Mostly not a single cell measure but only population level. Protein staining is only a surrogate endpoint of real cell number. For DNA quantification with Hoechst 33342, fluorescent probe penetration, bleaching, and cytotoxicity are issues to be considered.
4. Cellular metabolism	3-(4,5-Dimethylthiazol-2-yl)-2,5-diphenyltetrazolium (MTT) assay, or similar tetrazolium dye reduction assays	Measures the reduction of the tetrazolium dye by viable cells	**Advantages:** High throughput, easy, robust, low cost. Used in several ISO standards and OECD test guidelines. High sensitivity. Can be used for tissue constructs. **Disadvantages:** Measures amount of viable cells and needs control for contribution of proliferation. Cells with reduced mitochondrial function may appear non-viable. Measurement usually not on single cells.
	Resazurin (Alamar blue) reduction assay	Fluorescent resorufin is formed from resazurin through mitochondrial metabolism of viable cells	**Advantages:** Many tests can be performed rapidly in multiwell dishes and cells can be tested repeatedly (noninvasive measurement). High sensitivity. **Disadvantages:** Cells with reduced mitochondrial function may appear nonviable. Some test items interfere with the assay (e.g., superoxide also reduces the dye).

(continued)

Table 5.1 (continued)

Endpoint	Assay	Mechanism	Advantages/disadvantages
	Mitochondrial depolarization assays (based on fluorescent indicator dyes)	Measurement of mitochondrial membrane potential by addition of potential sensing fluorescent dyes like JC-1, TMRE, and MitoTracker	**Advantages:** Fast, cheap, high throughput; single cell information. **Disadvantages:** As for MTT (measures cell function, not cytotoxicity). Artifacts by test items that affect mitochondria specifically. Artifacts by test items that affect plasma membrane potential, bleaching, quenching and unquenching, and shape changes and clustering of mitochondria.
	Neutral red assay (ISO 10993)	Active cells accumulate the red dye in lysosomes, and the dye incorporation is measured by spectrophotometric analysis	**Advantages:** Low cost. Used in several ISO standards and OECD test guidelines. **Disadvantages:** Normalization required for quantitative measurement, e.g., with protein content or number of cells. Gives usually information only at the population level. Not suited for tissue constructs and certain cell lines.
	ATP assays	Measurement of the total ATP content. Dying cells fail to produce ATP, have an increased ATP consumption, and may lose ATP through perforations of the plasma membrane. For the test, cell lysates are prepared, and the ATP content is assessed by a luminometric assay.	**Advantages:** Fast, high throughput. **Disadvantages:** No single cell data, expensive, not a direct measure of cytotoxicity.

trast to other assays to assess cell viability, such as the previously mentioned MTT study, where the number of cells at the beginning of the assay may not be the same as at the end since viable cells, but damaged ones can be removed by multiple washes that are needed. One of the major drawbacks of LDH assay is that the serum commonly used in culture media has an intrinsic LDH activity [30, 31]. Therefore, it is important to measure the absorbance obtained from the LDH activity in medium with the amount of serum used in the study. To reduce the background signal, it is recommended to use the minimum serum percentage appropriate for each cell line without compromising cell viability.

Finally, in this section, it should be mentioned that a draft guidance document on Good In Vitro Method Practices (GIVIMP) for the development and implementation of in vitro methods for regulatory use in human safety assessment has been coordinated by the European validation body EURL ECVAM and accepted on the work plan of the OECD test guideline program since April 2015 [5] and has been reviewed in 2017 [32]. This draft provides a list of several endpoints useful to evaluate cell toxicity, and the proposed tests are summarized in Table 5.1.

5.10 In vitro methods to test cellular function

To evaluate cellular function in living cells, there are multiple techniques, and several of them including cytosolic calcium determination, depolarization membrane detection, and electrophysiological measurements are commonly used to evaluate the effects of marine compounds [12, 14–16, 33, 34].

5.10.1 Electrophysiology

Due to the complex equipment and specialized personnel required to perform electrophysiological determinations, next only the basic principles of electrophysiology are going to be summarized. In brief, electrophysiological recordings allow gaining insight on the effect of compounds on voltage-gated and ligand-gated ion channels. The technique is based on the use of a recording pipette in narrow contact with the cell membrane (seal) and apply negative pressure in order to break the cell membrane and be able to measure either the current (in the modality of voltage clamp) or the voltage (in the modality of current clamp) flowing through the cell membrane. A scheme showing some of the equipment required in an electrophysiology setup as well as the basis of the procedure to obtain access to the cell is shown in Figure 5.16.

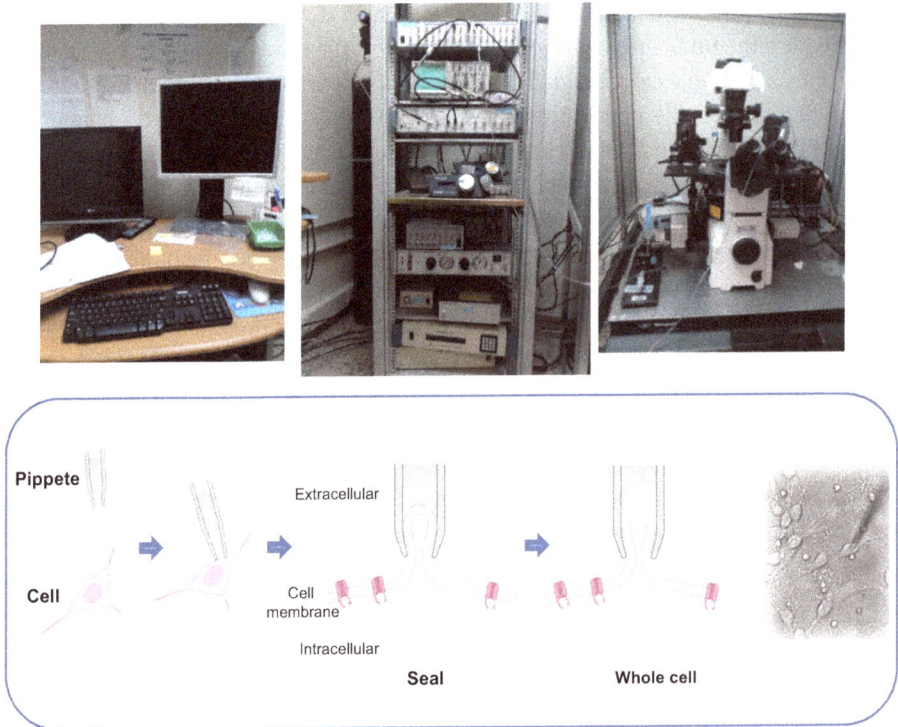

Figure 5.16: Electrophysiological equipment and schematic representation of the procedure to obtain a seal between the recording pipette and the cell.

5.10.2 Cytosolic calcium determination

Another powerful technique to assess the cell function after chemical application is cytosolic calcium imaging. The increase in cytosolic calcium concentration has been shown to play an important role in vital cellular functions such as muscle contraction, cell secretion, oocyte fertilization, nerve conduction, embryo development and apoptosis in animals, plants, and microbes, and in the invasion of mammalian cells by parasites, bacteria, and viruses. Therefore, live cell imaging of increases in cytosolic calcium concentration in cellular compartments has been investigated intensively. Multiple calcium imaging systems are now available commercially [35]. Among the first calcium indicators used for monitoring, the dynamics of cellular calcium signaling were bioluminescent calcium-binding photoproteins, such as aequorin. However, for calcium imaging in living cells, the most important achievement was the development of more sensitive and versatile fluorescent calcium indicators and buffers by Roger Tsien [36]. These indicators were the result of the hybridization of highly calcium-selective chelators like EGTA or BAPTA with a fluorescent chromophore. The first gen-

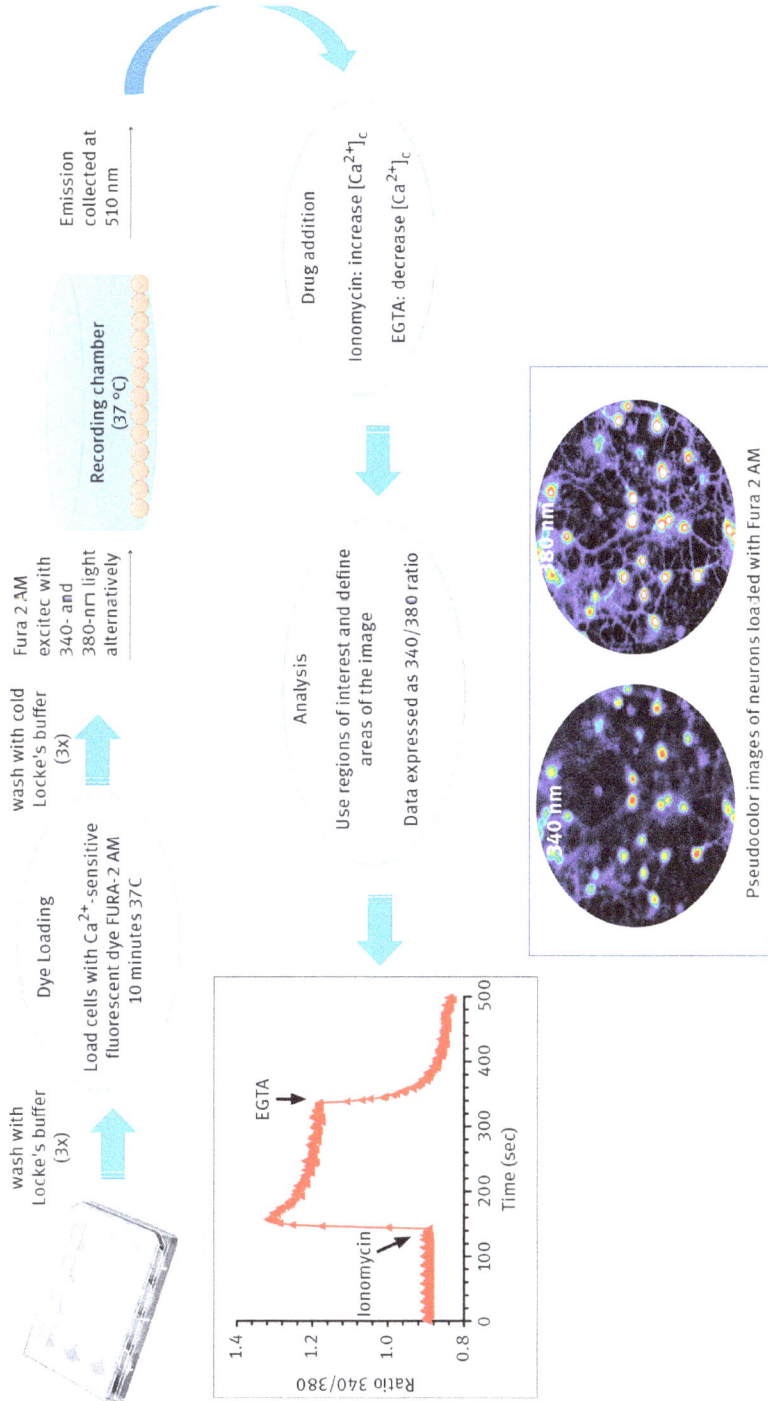

Figure 5.17: Schematic representation of the protocol for the determination of the cytosolic calcium concentration in primary cultures of cerebellar granule cells using the cell-permeable calcium-sensitive dye Fura-2 AM.

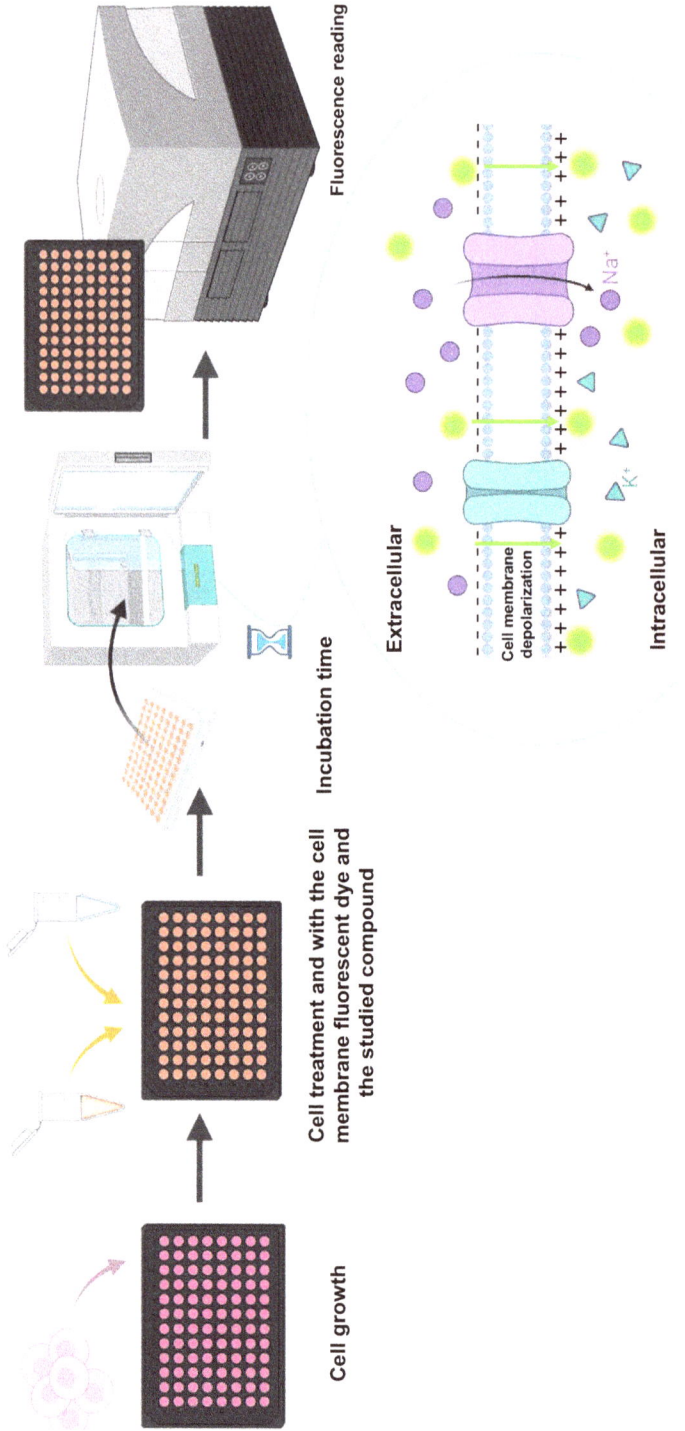

Figure 5.18: Schematic representation of the protocol to detect changes in membrane potential.

eration of fluorescent calcium indicators consisted of quin-2, fura-2, indo-1, and fluo-3. Quin-2 is excited by ultraviolet light (339 nm) and was the first dye of this group to be used in biological experiments (see [37] for review). The widely used calcium-sensitive fura-2 AM is a combination of calcium chelator and fluorophore. It is excitable by ultraviolet light (e.g., 350/380 nm), and its emission peak is between 505 and 520 nm [36]. The binding of calcium ions causes intramolecular conformational changes that lead to a change in the emitted fluorescence. A summary illustrating the protocol for cytosolic calcium imaging in primary cultures of cerebellar granule cells is shown in Figure 5.17.

5.10.3 Membrane potential indicators

Membrane potential is an important physiological parameter of live cells with important different roles, especially in membrane-signaling processes as nerve-impulse propagation or muscle contraction [38]. In addition to the direct measurement using electrophysiological techniques, there are also indirect membrane potential measurement techniques. These indirect measurements consist of ionic cellular dyes that stain the cell membrane and emit fluorescence. When membrane potential changes, the florescence emitted changes as well, increasing with depolarization and decreasing with cell membrane hyperpolarization [39]. Therefore, by measuring the fluorescence emitted by cells after the exposure to a compound, it is possible to determine whether the compound has caused a change in membrane potential. Membrane potential indicators are classified into two groups: slow and fast responding dyes. The more conventional dye used is bis(1,3-dibutylbarbituric acid) trimethine oxonol or $DiBAC_4(3)$. These dyes enter depolarized cells and bind to intracellular proteins exhibiting an increase in fluorescence emitted. An increase in cell membrane depolarization results in an increase in the influx of the dye. Therefore, the fluorescence signal gets more intense. Contrary to that, hyperpolarization is detected by a decrease in fluorescence emission [40, 41]. There are a large number of commercial kits with the different membrane potential indicators that allow to detect changes in this parameter. Based on the objective of the study, the most appropriate one will be chosen in each case. The detailed protocol for each dye varies. But in general, it consists in the measurement of the emitted fluorescence at the appropriate wavelength (excitation/emission) after cell exposure to the dye and the studied compound with the aim to detect either an increase or a decrease in fluorescence emission. The general protocol and cellular basis are represented in Figure 5.18.

References

[1] Almeida A, Sarmento B, Rodrigues F. Insights on in vitro models for safety and toxicity assessment of cosmetic ingredients. Int J Pharm. 2017;519(1–2):178–85.

[2] Braakhuis HM, et al. Simple in vitro models can predict pulmonary toxicity of silver nanoparticles. Nanotoxicology. 2016;10(6):770–79.

[3] National Research Council, N. Toxicity testing for assessment of environmental agents: Interim report. Washington, DC: The National Academies Press, 2006.

[4] Roggen EL. In vitro toxicity testing in the twenty-first century. Front Pharmacol. 2011;2:3.

[5] OECD, Draft guidance document on good in vitro method practices (GIVIMP) for the development and implementation of in vitro methods for regulatory use in human safety assessment, 2016.

[6] Eskes C, et al. Good cell culture practices & in vitro toxicology. Toxicol In Vitro. 2017;45 (Part 3):272–77.

[7] Coecke S, et al. Guidance on good cell culture practice. a report of the second ECVAM task force on good cell culture practice. Altern Lab Anim. 2005;33(3):261–87.

[8] Pamies D, et al. Good cell culture practice for stem cells and stem-cell-derived models. Altex. 2017;34(1):95–132.

[9] Moore GE, Hood DB. Modified RPMI 1640 culture medium. In Vitro Cell Dev Biol Anim. 1993;29A(4):268.

[10] Geraghty RJ, et al. Guidelines for the use of cell lines in biomedical research. Br J Cancer, 2014;111 (6):1021–46.

[11] Ryan JA. Evolution of cell cultures. BioFiles. 2008;3(8):21.

[12] Rubiolo JA, et al. Potassium currents inhibition by gambierol analogs prevents human T lymphocyte activation. Arch Toxicol. 2015;89(7):1119–34.

[13] Gordon J, Amini S, White MK. General overview of neuronal cell culture. Meth Mol Biol (Clifton, N J). 2013;1078:1–8.

[14] Martin V, et al. Differential effects of ciguatoxin and maitotoxin in primary cultures of cortical neurons. Chem Res Toxicol. 2014;27(8):1387–400.

[15] Martín V, et al. Chronic ciguatoxin treatment induces synaptic scaling through voltage gated sodium channels in cortical neurons. Chem Res Toxicol. 2015;28(6):1109–19.

[16] Mendez AG, et al. The marine guanidine alkaloid crambescidin 816 induces calcium influx and cytotoxicity in primary cultures of cortical neurons through glutamate receptors. ACS Chem Neurosci. 2017;8(7):1609–17.

[17] Anderson GW, et al. Characterisation of neurons derived from a cortical human neural stem cell line CTX0E16. Stem Cell Res Ther. 2015;6(149):015–0136.

[18] Eisenbrand G, et al. Methods of in vitro toxicology. Food Chem Toxicol. 2002;40(2):193–236.

[19] Elmore SA, et al. Recommendations from the INHAND Apoptosis/Necrosis Working Group. Toxicol Pathol. 2016;44(2):173–88.

[20] EPAA and D.s.o.A.M.D.-A. European Centre for the Validation of Alternative Methods (ECVAM). Revision and update of the sector Reproductive Toxicity testing of the publicly available "DataBase service on ALternative Methods to animal experimentation (DB-ALM), 2011.

[21] OECD (2018), Test No. 443: Extended One-Generation Reproductive Toxicity Study, OECD Guidelines for the Testing of Chemicals, Section 4, OECD Publishing, Paris, https://doi.org/10.1787/9789264185371-en.

[22] Vale C, et al. Cell volume decrease as a link between azaspiracid-induced cytotoxicity and c-Jun-N-terminal kinase activation in cultured neurons. Toxicol Sci. 2010;113(1):158–68.

[23] Watmuff B, et al. Human pluripotent stem cell derived midbrain PITX3(eGFP/w) neurons: A versatile tool for pharmacological screening and neurodegenerative modeling. Front Cell Neurosci. 2015;4:104.

[24]   Mahmood T, Yang P-C. Western Blot: Technique, theory, and trouble shooting. North Am J Med Sci. 2012;4(9):429–34.

[25]   Gorr TA, Vogel J. Western blotting revisited: Critical perusal of underappreciated technical issues. Proteomics Clin Appl. 2015;9(3–4):396–405.

[26]   Eaton SL, et al. Total protein analysis as a reliable loading control for quantitative fluorescent Western blotting. PLoS One. 2013;8(8):e72457.

[27]   Cagide E, et al. Production of functionally active palytoxin-like compounds by Mediterranean ostreopsis cf. siamensis. Cell Physiol Biochem. 2009;23(4–6):431–40.

[28]   Allen M, et al. Lactate dehydrogenase activity as a rapid and sensitive test for the quantification of cell numbers in vitro. Clin Mater. 1994;16(4):189–94.

[29]   Decker T, Lohmann-Matthes ML. A quick and simple method for the quantitation of lactate dehydrogenase release in measurements of cellular cytotoxicity and tumor necrosis factor (TNF) activity. J Immunol Methods. 1988;115(1):61–69.

[30]   Hiebl B, et al. Impact of serum in cell culture media on in vitro lactate dehydrogenase (LDH) release determination. J Cell Biotechnol. 2017;3(1):9–13.

[31]   Kaja S, et al. Quantification of lactate dehydrogenase for cell viability testing using cell lines and primary cultured astrocytes. Curr Protoc Toxicol. 2017;72:2 26 1–2 26 10.

[32]   Zuang V, Barroso J, Belz S, Berggren E, Bernasconi C, Bopp S, Bouhifd M, Bowe G, Campia I, Casati S, Coecke S, Corvi R, Dura A, Gribaldo L, Grignard E, Halder M, Holley T, Janusch Roi A, Kienzler A, Lostia A, Madia F, Milcamps A, Morath S, Munn S, Paini A, Pistollato F, Price A, Prieto-Peraita P, Richarz A, Triebe J, van der Linden S, Wittwehr C, Worth A, Whelan M. EURL ECVAM Status Report on the Development, Validation and Regulatory Acceptance of Alternative Methods and Approaches, 2017.

[33]   Sanchez JA, et al. Spongionella secondary metabolites regulate store operated calcium entry modulating mitochondrial functioning in SH-SY5Y neuroblastoma cells. Cell Physiol Biochem. 2015;37(2):779–92.

[34]   Sanchez JA, et al. Autumnalamide targeted proteins of the immunophilin family. Immunobiology. 2017;222(2):241–50.

[35]   Fang X, et al. Fluorescence detection and imaging of cytosolic calcium oscillations: A comparison of four equipment setups. Prog Nat Sci. 2009;19(4):479–87.

[36]   Tsien RY. Fluorescent probes of cell signaling. Annu Rev Neurosci. 1989;12:227–53.

[37]   Grienberger C, Konnerth A. Imaging calcium in neurons. Neuron. 2012;73(5):862–85.

[38]   Abdul Kadir L, Stacey M, Barrett-Jolley R. Emerging roles of the membrane potential: Action beyond the action potential. Front Physiol. 2018;9:1661.

[39]   Mustroph H. Oxonol dyes. Phys Sci Rev. 2021;7(1):37–43.

[40]   Adams DS, Levin M. Measuring resting membrane potential using the fluorescent voltage reporters DiBAC4(3) and CC2-DMPE. Cold Spring Harb Protoc. 2012;2012(4):459–64.

[41]   Maher MP, Wu NT, Ao H. pH-Insensitive FRET voltage dyes. J Biomol Screen. 2007;12(5):656–67.

Rebeca Alvariño

6 Marine toxins

6.1 Introduction

Out of the 5,000 known species of marine algae, marine phycotoxins ("phyco," from algae) are produced by approximately 75 microalgae species. Dinoflagellates are the main producers of these toxic compounds, but some diatoms and cyanobacteria are also involved in phycotoxin generation. These organisms are the first step of the food chain, so toxins accumulate in invertebrates, especially in filter-feeding mollusks, and some fish, finally reaching human consumers [1].

Dinoflagellates are unicellular protists, mainly planktonic, which can be found in fresh and salt water (Figure 6.1). Some species are autotrophic (photosynthetic) or heterotrophic, but most can combine both trophic modes (mixotrophic) [2]. They live as free organisms in the water column, but under environmental fluctuations, some dinoflagellate species can form cysts. These resting forms with thickened walls sink to the sea floor, where they can survive for decades until the restoration of optimal environmental conditions. Although a small proportion of dinoflagellates can form cysts, many toxic species are able to produce them during their life cycle, and these resting forms are considered the seeds of some algae blooms [3, 4]. Along with planktonic species, benthic dinoflagellates such as *Gambierdiscus* spp. or *Ostreopsis* spp. are also implicated in toxin production [5]. Moreover, some human poisoning has been related to macroalgae consumption. At least two edible red algae, *Acanthophora specifera* and *Gracilaria edulis*, were associated with human deaths in Asia. Then, the compound polycavernoside A was identified as the responsible for toxicity [6]. Under specific physical sea conditions (temperature, salinity, turbidity, light irradiation, and nutrients), explosive growths of dinoflagellates and diatoms may occur, causing staggering accumulations of toxins in seafood. This increase in unicellular algae concentration is known as "harmful algae blooms" (HABs), a phenomenon where microalgae accumulation draws big colored stains in the water. These blooms are commonly named "red tides," but depending on the organism type, they can also be green, brown, or orange (Figure 6.2).

Marine phycotoxins comprise a great diversity of molecules with complex chemical structures and different toxicities and mechanisms of action. The physiological or ecological function of these compounds remains unknown, and they seem to be harmless to those organisms that act as mere vectors. Phycotoxins enter the food chain through invertebrates (mollusks, gastropods, echinoderms, or crustaceans) and fishes, especially in tropical waters, but more and more frequently in temperate waters (Figure 6.3) [1]. Since these toxins are pharmacologically very potent, they pose an acute risk to consumers.

Rebeca Alvariño, Departamento de Fisiología, Facultad de Veterinaria, Lugo, Spain

https://doi.org/10.1515/9783111014449-006

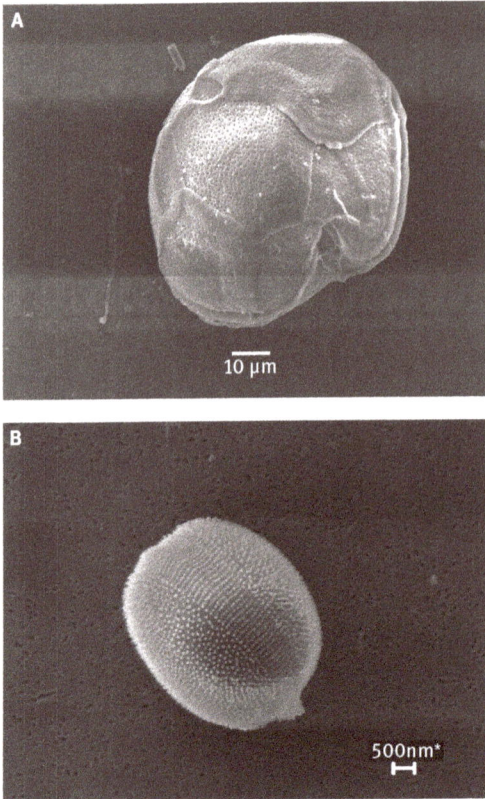

Figure 6.1: Dinoflagellates from *Gambierdiscus toxicus* (A) and *Prorocentrum minimum* (B).

Figure 6.2: Algae bloom: red tide.

Figure 6.3: Trophic chain.

Some of them, such as ciguatoxins (CTXs), cause food intoxications so frequently that they are considered as a potentially neglected third-world disease. Other phycotoxins, like palytoxin (PLTX) or maitotoxin (MTX), are among the deadliest compounds in nature [7]. Phycotoxin poisoning causes neuronal and gastrointestinal symptoms in humans. In fact, marine toxin classification used to be referred to human symptoms (neurotoxic, amnesic, and paralytic toxins) but nowadays, phycotoxins are classified by their chemical characteristics as lipophilic, hydrophilic, and amphiphilic toxins (Table 6.1). Interestingly, some toxins are produced by different species, but one species can produce several toxin groups (i.e., *Gambierdiscus* sp. may produce CTXs, gambierol, gambierona, and MTX [8, 9]).

Along with the unquestionable health risk, there is also a high economic cost for the fishery industry. HABs provoke the closure of fishing areas, increase the costs of toxin levels monitoring, and some of the toxins are harmful to fisheries and marine cultures (ichthyotoxins). Due to this, the control and legislation of marine toxins are of great importance. Countries have administrations that look after food safety and public health, and establish the legal levels of toxins in marine food. In Europe and the United States, the European Food Safety Authority (EFSA) and the Food and Drug Administration (FDA), respectively, issue the guidelines for marine toxin detection. For each legislated toxin, a maximum amount of the compound per kilogram of flesh and a detection method is settled. The international limits of marine biotoxins for international trade are set by the Codex Committee on Fish and Fishery Products. There is a "Standard for Live and Raw Bivalve Mollusks," where a maximum level in mollusks flesh for five groups of toxins is established (Table 6.2) [24].

Table 6.1: Classification of marine toxins.

Marine phycotoxins	Group	Reference compound [10]	Mechanism of action [10]	Analogues [10]	Producer organism (genus) [9]
Lipophilic toxins	Pectenotoxins (PTXs)	Pectenotoxin-2 (PTX2)	Actin inhibition [11]	PTX 1-14, PTX seco acid [11]	Dynophysis
	Yessotoxins (YTXs)	Yessotoxin (YTX)	Phosphodiesterase 4A activation [12]	More than 90 analogues, including hidroxiYTX, carboxyYTX, homoYTX, ketoYTX, and noroxoYTX [12]	Protoceratium Lyngulodinium Gonyaulux
	Azaspiracids (AZAs)	Azaspiracid-1 (AZA1)	Unknown	AZA 1-11	Azadinium Amphidoma
	Okadaic acid and dinophysistoxins (DTXs)	Okadaic acid	Protein phosphatases 1 and 2 inhibition	DTX 1-6	Dynophysis Prorocentrum
	Ciguatoxins (CTXs)	Pacific Ciguatoxin-1 (P-CTX1)	Voltage-dependent Na⁺ channel activation [13]	More than 25 analogues, classified into 4 groups: P-CTX I (or CTX4A), P-CTX II (or CTX3C), Caribbean-CTX (C-CTX), and Indian-CTX (I-CTX) [14]	Gambierdiscus
	Brevetoxins (BTXs)	Brevetoxin-1 (BTX1 or PbTx1) Brevetoxin- 2 (BTX2 or PbTx1)	Voltage-dependent Na⁺ channel activation	BTX 1-14 Brevenal	Karenia Chatonella
	Gambierol		Voltage-dependent K⁺ channel blockage [15]		Gambierdiscus
	Cyclic imines	Gymnodimine A (GYM A) Spirolide A (SPX A) Pinnatoxin A (PnTX A) Pteriatoxin A (PtTX A) Prorocentrolide A (PcTX A) [16]	Reversible blockage of cholinergic receptors [16]	GYM A-C, 12-methylgymnodimine, SPX A-I, desmethyl-SPX-C, desmethyl-SPX-G, pinnatoxins A–H, pteriatoxins A–C, pinnamine, prorocentrolides A–B, spiro-prorocentrimine, symbioimines, portimine [16, 17]	Gymnodinium Karenia Alexandrium Prorocentrum Vulcanodinium [16]

Hydrophilic toxins	Saxitoxins (STXs)	Saxitoxin (STX)	Site-1 voltage-dependent Na$^+$ channel inhibition	Carbamates (GTX 1-4, NeoSTX), N-sulfocarbamoil (C1-4, GTX 5-6), decarbamoil (dcGTX 1-4, dcSTX, dcNeo), benzoates (GC 1-6)	*Alexandrium Gymnodinium Pyrodinium Cyanobacteria* species
	Tetrodotoxins (TTXs)	Tetrodotoxin (TTX)	Site-1 voltage-dependent Na$^+$ channel inhibition	More than 30 analogues, divided into four groups: analogues chemically equivalent to TTX, deoxy compounds, one oxidized analogue, and C11-lacking analogues [18, 19]	Bacteria species of several genera (*Pseudomonas, Vibrio, Roseobacter,* etc.) [20]
	Domoic acid	Domoic acid	Kainate receptor activation		*Pseudo-nitzschia* and *Nitzschia*
Amphiphilic toxins	Maitotoxins (MTXs)	Maitotoxin (MTX or MTX1)	TRP channel activation [21] Ca^{2+} entry activation [15]	MTX 1-4, desulfo-MTX1, didehydro-demethyldesulfo-MTX1 [22]	*Gambierdiscus*
	Palytoxins (PLTXs)	Palytoxin (PLTX)	Na$^+$/K$^+$ ATPase blockage [23]	10^{21} possible isomers. Identified analogues are divided into four groups: PLTXs, ostreocins, ovatoxins, and mascarenotoxins [23]	*Ostreopsis*

Table 6.2: Toxin limits set by Codex Alimentarius.

Saxitoxin group	<0.8 mg saxitoxin equivalent/kg
Okadaic acid group	<0.16 mg okadaic acid equivalents/kg
Domoic acid group	<20 mg domoic acid/kg
Brevetoxin group	<200 mice units or equivalent/kg
Azaspiracid group	<0.16 mg/kg

Regarding tetrodotoxins (TTXs) and CTXs, they are not on the list of internationally regulated marine toxins, but the EU legislation states that poisonous fish derivatives and/or contaminated products should not be placed on the market. On the other hand, countries such as Japan or US have set limits for these toxins [19, 25]. Regardless of the lack of worldwide agreement about toxin limits and analytical methods for their detection, most of the countries tend to the consensus in the monitoring programs procedures. There are diverse methods for marine toxin detection and quantification that will be revised in this book (see related chapters). Mouse bioassay (MBA) was the official method for paralytic and lipophilic toxins for several years, but the guidelines for regulation and reduction in the use of experimental animals have tipped the balance in favor of analytical methods such as high-pressure liquid chromatography combined with mass spectrometry (MS/MS) or with ultraviolet detection. In MBA, seafood extracts are administered intraperitoneally to 20 g mice and the number of dead animals, death time, and symptoms allow a sample toxicity prediction. For analytical methods, a certified standard is needed to detect and quantify the toxins in samples. Moreover, advances in new technologies and understanding of the toxin's mechanism of action provide high-throughput functional detection methods based on the cellular target of these marine compounds [26]. Figure 6.4 shows the methods used for the monitoring of these toxins.

The following sections show a brief review about the origin, structure, and mechanism of action of the different marine phycotoxins.

6.2 Lipophilic toxins

The group of lipophilic toxins is divided into eight compound families: pectenotoxins (PTXs), yessotoxins (YTXs), azaspiracids (AZAs), okadaic acid and dinophysistoxins (DTXs), CTXs, brevetoxins (BTXs), gambierol, and cyclic imines. The representative toxins of each group are shown in Figure 6.5.

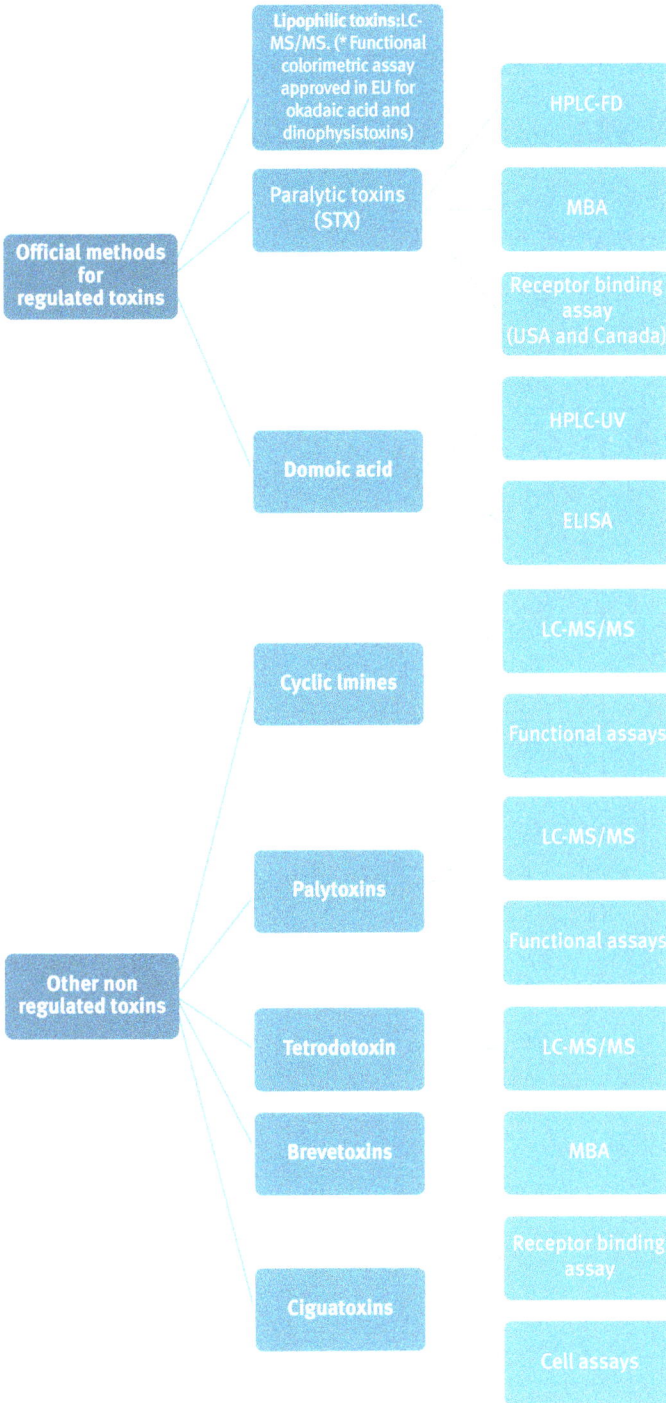

Figure 6.4: Methods used for the monitoring of regulated and non-regulated marine toxins.

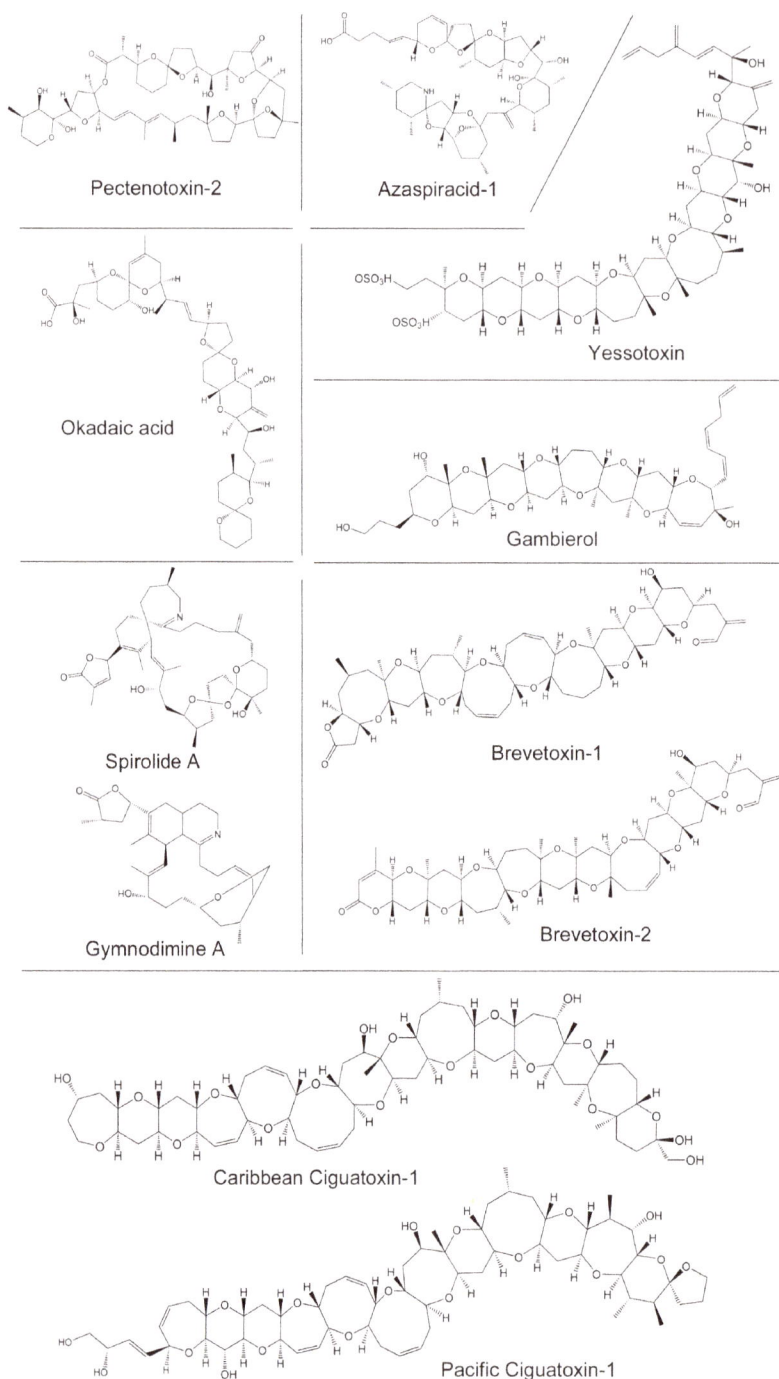

Pectenotoxin-2

Azaspiracid-1

Yessotoxin

Okadaic acid

Gambierol

Spirolide A

Brevetoxin-1

Gymnodimine A

Brevetoxin-2

Caribbean Ciguatoxin-1

Pacific Ciguatoxin-1

Figure 6.5: Chemical structures of the representative compounds from each group of lipophilic toxins.

6.2.1 Pectenotoxins

PTXs are polyether macrolides with a worldwide distribution produced by *Dinophysis* spp. Their name is due to *Patinopecten yessoensis*, the first organism from which these compounds were isolated. More than 20 analogues have been identified, but most of them are products of shellfish metabolism [27, 28]. PTXs -1, -2, -3, and -11 have similar toxicity in mice (LD_{50} 200–400 µg/kg), and the target organ seems to be the liver, where macroscopic damage can be observed after intraperitoneal injection. Their oral toxicity is lower due to poor intestinal absorption. No human poisonings have been associated with PTXs, some diarrheic episodes were related to PTX2 in the past, but the coexistence with okadaic acid was settled as the responsible for toxicity [11].

The mechanism of action of PTXs is the interaction with the actin skeleton of cells and the consequent destabilization of its structure. Particularly, in vitro studies have reported that PTX-2 inhibits actin polymerization. Actin is one of the most abundant cytoskeletal proteins and has a key role in important processes such as cell growth, cell signaling, and maintenance of cell shape [11, 29].

Until 2021, this group of toxins was regulated in Europe with a limit of 160 µg of okadaic acid equivalents/kg shellfish meat for the sum of okadaic acid, DTXs, and PTXs [30]. However, due to the lack of reports on adverse effects in humans associated with these toxins, PTXs have been removed from the list of biotoxins to be analyzed in bivalve mollusks in the EU [31].

6.2.2 Yessotoxins

YTX is a polyether compound that was described for the first time in Japanese waters after an outbreak of food intoxication with diarrheic symptoms. Due to the symptomatology and the fact that it was often isolated with okadaic acid, YTX was originally classified among diarrheic toxins. Lately, it was separated into its own group due to its biologic origin and effects [32]. This phycotoxin, with more than 90 analogues, is produced by different dinoflagellate species from the genera *Protoceratium, Lingulodinium,* and *Gonyaulax* [33–35]. Some analogues are directly generated by the dinoflagellate, while other ones are products of shellfish metabolism [36].

YTX is formed by 11 adjacent rings, an unsaturated lateral chain, and two sulfate groups. The carbonate skeleton is liposoluble, but the sulfate groups confer amphiphilic characteristics to the molecule, making it the most polar of the lipophilic toxins [37]. YTX is accumulated in the digestive gland of mussels and scallops but it can be also detected in the muscle [38]. MBA results in a high mortality when the toxin is administered intraperitoneally but no mortality is registered orally [32]. This behavior is due to the chemical characteristics of the compound, which modify intestinal absorption. Controversial results have been observed in vivo, with different LD_{50} and tissue alterations, being the most relevant effect in the heart [39–43]. The regulatory limit of YTX in Europe was originally 1 mg YTX equivalents/kg, but in 2013 it was set at 3.75 mg of YTX equivalents/kg shellfish meat due to the absence of human intoxications [44, 45].

Several intracellular pathways are modified by YTX, and different effects are observed depending on the cellular type. YTX affects, among other biological effects, calcium signaling, cAMP levels, phosphodiesterases (PDEs), A-kinase anchoring protein, and protein kinase C expression, and activates programmed cellular death [12]. For example, it is known that YTX activates PDE in human lymphocytes followed by a calcium-dependent cAMP decrease [46]. This binding of PDE and YTX has been also confirmed by biosensor techniques [47]. Interestingly, YTX potently induces death in several cell lines and primary cultures, and it has displayed promising antitumoral properties in vivo [12, 48, 49].

6.2.3 Azaspiracids

AZAs are polyether compounds synthesized by dinoflagellates from the genera *Azadinium* and *Amphidoma* that were first reported in mussels from Ireland [50]. From the approximately 50 analogues described, AZA1 and AZA2 are produced by the dinoflagellate itself, while other compounds such as AZAs 3-5 derive from shellfish biotransformation [51]. No human deaths have been reported and only diarrheic episodes (nausea, vomiting, diarrhea, and stomach cramps) are related to consumption of AZA-contaminated seafood.

A comparative acute oral toxicity assay in mice with AZAs 1-3 recently determined LD_{50}, resulting AZA1 as the most toxic compound, followed by AZA2 and AZA3 [52]. Toxicity equivalent factors (TEFs) derived from these data were 1.0, 0.7, and 0.5, respectively, which differ from the proposed by EFSA, pointing to the need for a re-evaluation [22]. Currently, the European limit is settled at 160 µg of AZA1 equivalents/kg shellfish meat [30, 44].

Organ damage is observed after AZA administration, with intestinal swelling, fatty acid deposition in the liver, lesions in the spleen and thymus, and even tumorigenic effects [53]. Moreover, neuro and cardiotoxic effects have been related to AZA exposure [54, 55]. The biological target of this group of compounds is still unknown, but they produce cytotoxicity in several cell lines, affecting the actin cytoskeleton and altering intracellular signaling molecules such as MAP kinases, Ca^{2+}, and cAMP [50]. Other works also link AZAs effect to the human *ether-á-go-go-related gene* potassium channel (hERG K^+), whose open state is inhibited by these toxins [56, 57]. Recently, the effects of AZAs on mitochondrial dehydrogenases and chloride channels have been also described, providing new insights into the pathways affected by this toxin family [58, 59].

6.2.4 Okadaic acid and dinophysistoxins

Okadaic acid and its analogues, DTXs, are isolated from several dinoflagellates of the genera *Prorocentrum* and *Dinophysis*. These polyether toxins have a wide distribution and can be found in the microorganism itself or in shellfish as acylated forms. Generally, okadaic acid, DTX1 and DTX2 are produced by the dinoflagellates, while DTX3 is found in seafood [22]. These compounds are responsible for the diarrhetic shellfish poisoning syndrome. Nausea, vomiting, diarrhea, and abdominal pain are observed after consumption of contaminated shellfish. The symptoms are usually minor, with a full recovery in 2–3 days [28].

These toxins are strong inhibitors of the serine/threonine phosphatases 1, 2A, and 3, with protein phosphatase 2A being the most affected. These enzymes are important signal-transducer proteins involved in the control of several cellular processes. The repetitive exposure to these phycotoxins has been also associated with a tumor-promoting effect in mice, and some studies have related human colorectal cancer risk to the consumption of shellfish contaminated with okadaic acid and analogues [60, 61]. Although the mode of action of this toxin group is well-defined, the diarrheic effect is not fully understood. It was originally thought to be caused by disruptions of the tight junctions on intestinal cells, but this was questioned by several works [62]. In addition, a neuroactive effect based on neuropeptide Y was proposed based on cellular assays [63, 64]. The in vivo administration of the peptide did not modify the toxicity of okadaic acid, however, an involvement of serotonin in the diarrhea induced by this toxin has been described, pointing to a role of the neurotransmitter in the intestinal toxicity produced by the compound [64, 65].

Okadaic acid and DTXs are regulated toxins, and the actual European limit is settled at 160 µg/kg shellfish meat. These compounds present different toxicities, and the TEFs proposed by EFSA are 1.0, 1.0, and 0.5 for okadaic acid, DTX1, and DTX2, respectively [66]. These values were calculated based on intraperitoneal administration, but a recent acute oral toxicity study reported TEF values of 1.0, 1.5, and 0.3, which agree with PP2A inhibitory potency [67, 68].

6.2.5 Ciguatoxins

Ciguatera is an endemic food intoxication observed after tropical and subtropical fish consumption in Indo-Pacific Oceans and Caribbean Sea areas. In recent years, the responsible dinoflagellates, *Gambierdiscus* spp., have been observed in Mediterranean and Atlantic areas in accordance with the expansion trend of these benthic organisms attributable to climate warming [69]. In Europe, 34 ciguatera outbreaks have been registered between 2012 and 2019, some of them due to the consumption of imported fish, but other ones related to autochthonous species [70]. Ciguatera symptoms include gastrointestinal and neurological manifestations such as diarrhea, nausea, vomiting, paresthesia, myalgia, headache, muscular weakness, and cold allodynia. Gastrointestinal manifestations usually appear before neurological symptoms but the latter are the most characteristic of this food intoxication. Specifically, cold allodynia is considered pathognomonic of ciguatera. Lethality is low, but in some cases, ciguatera symptoms persist or reappear several years after intoxication due to the accumulation of the toxin in adipose tissue and long-term effect in the nervous system [71–73].

Dinoflagellates from *Gambierdiscus* genus produce several polyether toxins. The most studied are CTXs, but these microalgae also generate gambieric acid, gambierol, or MTX. Different CTX structures are observed depending on the dinoflagellate location; therefore, the prefixes P-, C-, or I- are added to identify Pacific, Caribbean, or Indic molecules. These compounds are cyclic liposoluble heat-stable polyethers, highly oxygenated, which potently interact with sodium channels. They are strong activators of voltage-gated sodium channels through the competitive binding to the receptor site-5 at low nanomolar concentrations. The channel activation leads to cell depolarization, spontaneous nerve firing, elevation of intracellular free Ca^{2+} concentration, and even neurotransmitter release. Toxin binding to site-5 produces a shift in the voltage dependence activation of Na^+ channels, causing a Na^+ influx at membrane potentials at which these channels would be usually inactivated. However, other properties of these channels, such as the peak Na^+ current size, remain unaffected [13].

The differences in the chemical structure of CTXs result in toxicity variations. Pacific analogues are considered more potent than Caribbean CTXs, based on LD_{50} obtained in mice after intraperitoneal administration [74–76]. Taking into account these data, the FDA established guidance levels of 0.01 µg/kg P-CTX1B equivalents for Pacific CTXs and 0.1 µg/kg for Caribbean analogues [25]. In Europe, there is no regulated limit, but only CTX-free fish can reach the market [77]. However, the real potency of these toxins remains elusive due to the lack of certified reference standards and the few animal toxicity studies available [14]. In fact, a recent work reported a similar in vitro effect of Pacific and Caribbean analogues, which reinforces the need for further studies on CTXs toxicity [78].

6.2.6 Brevetoxins

Species from the genus *Karenia* are responsible for BTX production, mainly *Karenia brevis*. Although these dinoflagellates have a worldwide distribution, BTXs intoxications have been only reported in the United States, Mexico, and New Zealand. Their structure is similar to CTXs, with a ladder-like fused oxygen-containing rings skeleton. Based on their backbone structure, BTXs can be classified into A-type (BTXs -1, -7, and -10) and B-type (BTXs -2, -3, -5, -6, -8, and -9), being BTX2 the most common. This group of toxins binds to site-5 of voltage-gated sodium channels producing a Na^+ ion influx [79, 80].

BTXs are extensively studied for their ichthyotoxicity. They are absorbed through the gill membranes and produce important behavioral and developmental damage due to their interaction with nerve functioning and neuromuscular transmission [81]. BTXs intoxication is responsible for neurotoxic shellfish poisoning, leading to general gastrointestinal symptoms and neurological effects such as paresthesia, ataxia, disorientation, and reversal temperature sensation (similar to the observed with CTXs). Severe intoxication can end in limb paralysis and respiratory distress, however, no deaths have been reported and recovery is observed in a few days. Inhalation is the most common intoxication route; toxic effects in humans after ingestion are rare. Exposure to BTX-containing aerosols provokes respiratory tract and conjunctiva irritation [82–84].

Limited human data are available, but in some countries (the United States, New Zealand, and Australia), the regulatory limit is fixed at 200 mice units or 0.8 mg BTX-2 equivalents/kg shellfish meat. BTXs are not regulated in Europe, but in 2018, these toxins were detected for the first time in French mussels and, consequently, the French Agency for Food, Environmental, and Occupational Health and Safety proposed a guidance level of 180 µg BTX-3 equivalents/kg shellfish meat [80, 85].

6.2.7 Gambierol

Gambierol was isolated from the dinoflagellate *Gambierdiscus toxicus* as a part of CTX group. It is also a cyclic polyether, but with a different mechanism of action, symptoms, and potency. This toxin has a ladder-shaped octacyclic ether skeleton with a partially conjugated triene side chain and targets transmembrane proteins within the lipid bilayer [86, 87]. Gambierol elicits a high toxicity in mice, with a LD_{50} of 50 and 150 µg/kg after intraperitoneal and oral administration, respectively, and neurological symptoms [88].

Due to its structural similitude to CTXs, gambierol was believed to share their mechanism of action. However, CTXs are strong modulators of voltage-gated sodium channels with little effect in voltage-gated potassium channels [89–91], while gambierol does not block or affect sodium channels and it efficiently inhibits potassium channels at nanomolar concentrations in several cellular models [92–95]. Particularly, Kopljar et al. described that gambierol anchors the gating machinery of Kv in its resting state, requiring depolarization above the physiological range for the channel opening [94, 96].

6.2.8 Cyclic imines

This group of lipophilic toxins has a macrocyclic structure with an imino functional group. Among them, we can find spirolides (SPXs), gymnodimines (GYMs), pinnatoxins (PnTXs), pteriatoxins (PtTXs), prorocentrolides (PcTXs), spiro-prorocentrimines, symbioimines, and portimine. These compounds have similar intraperitoneal toxicity in mouse and characteristic neurological manifestations [97, 98]. SPXs are worldwide distributed, whereas PnTXs, PtTX, PcTX, and spiro-prorocentrimines have been reported only in Japan, China, and Taiwan areas [99].

SPXs have been detected in North American, Canadian, and European coasts, but no human intoxications have been reported [100, 101]. They were first isolated in New Scotland (Canada) in a routine monitoring and later, *Alexandrium ostenfeldii* was identified as the responsible dinoflagellate [102, 103]. Fourteen compounds have been identified until now and are classified into three different groups. The first one includes SPXs A–D, SPXs E and F are in the second group, and SPX-G forms the latest one [104]. These toxins act through the interaction with cholinergic receptors. Particularly, 13-desmethyl spirolide C, the most extended analogue, produces overexpression of muscarinic and nicotinic receptors in rats after high-dose administration and has a high affinity for muscular and neuronal nicotinic receptors at nanomolar concentrations [105, 106].

GYMs were detected for first time in oysters, in early 1990s. GYM A was the first compound isolated, produced by the dinoflagellate *Karenia selliformis* [107]. Similarly to SPXs, GYMs have not been linked to human intoxications, but their geographic distribution is more limited [108]. These toxins display high toxicity by the intraperitoneal route, which is followed by neurological symptoms and a fast animal death. The blockage of nicotinic receptors in the neuromuscular junction was settled as their mechanism of action and inhibition of acetylcholine currents in *Xenopus* oocytes has been also described [99, 106, 109].

Despite their frequent occurrence and toxicity in animals, no human intoxications with cyclic imines have been reported. In Europe, these toxins are not regulated, but EFSA concluded that the risk of exposure to toxins other than SPXs could not be assessed due to a lack of toxicological and epidemiological data [101]. In this sense, in 2019, due to the detection of high PnTX G levels in French shellfish, the French Agency for Food, Environmental, and Occupational Health and Safety established a limit of 23 µg PnTX G/kg [110].

6.3 Hydrophilic toxins

Hydrophilic toxins comprise three groups, saxitoxins (STXs), TTXs, and domoic acid, whose representative compounds are shown in Figure 6.6.

6.3.1 Saxitoxins

These non-peptide neurotoxins are produced by several members of the genus *Alexandrium*, as well as by the species *Gymnodinium catenatum* and *Pyrodinium bahamense*. Their poisoning is lethal and is characterized by neurological symptoms such as paralysis, dizziness, headache, and respiratory arrest. The mechanism of action of this group of toxins is the blockage of site-1 sodium channels in excitable cells, inter-

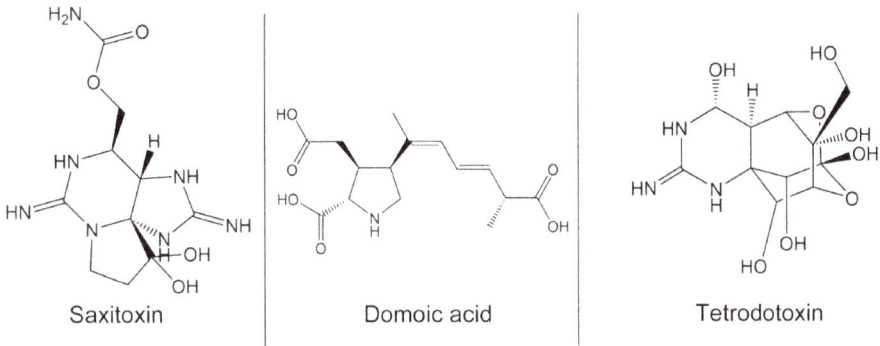

Saxitoxin   Domoic acid   Tetrodotoxin

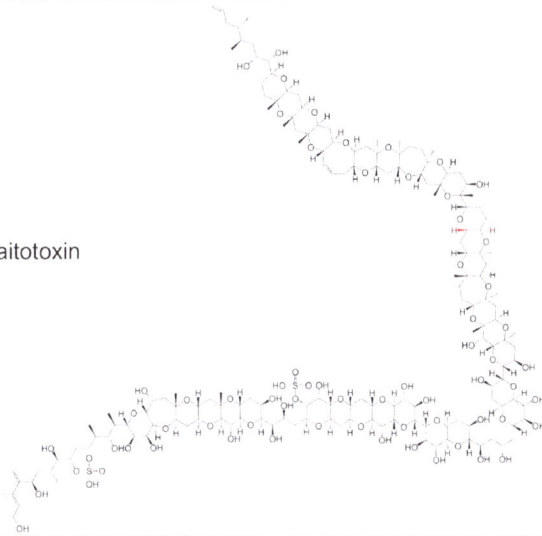

Maitotoxin

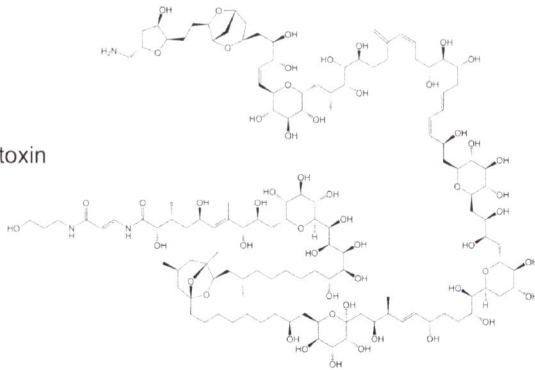

Palytoxin

Figure 6.6: Chemical structures of the representative compounds from each group of hydrophilic and amphiphilic toxins.

fering with normal nervous transmission. The intoxication symptoms appear rather fast, but the patients can recover in a few days with artificial ventilation, as the binding to the receptors is reversible [111].

They are included in the paralytic shellfish poisoning (PSP) toxins, a group formed by more than 57 analogues, closely related among them, which share a tetrahydropurine structure and are represented by STX [112]. STX basic structure is a 3,4-propinoperhydropurine that can be modified by the addition of hydroxyl, carbamyl, *N*-sulfocarbamoyl, or sulfate groups, producing several analogues with different chemical properties and potency, but with the same mechanism of action. The legal limit for this group of toxins is 800 µg of STX equivalents/kg or 4 mouse units/kg of meat, which corresponds to 18 µg of STX as the lethal amount needed for a 20 g mouse in 15 min [113].

Interestingly, these toxins also pose a hazard for marine animals, as STXs enter the food web through invertebrates like copepods or bivalves and larger organisms such as mammals or birds can be affected by their toxicity. Motor symptoms, respiratory paralysis, and death of these animals have been reported [22].

6.3.2 Tetrodotoxins

TTX is one of the most important neurotoxins. This heterocyclic compound consists of a guanidinium moiety connected to an oxygenated backbone with a 2,4-dioxaadamantane structure with six hydroxyl groups [114]. TTX has more than 30 analogues that are classified into four groups depending on their chemical properties: analogues chemically equivalent to TTX, deoxy compounds, one oxidized analogue, and C11 lacking analogues [18].

Pufferfish is the most commonly responsible for TTX poisoning, but the toxin has been also detected in arthropods, echinoderms, mollusks, and other fishes. The wide distribution of TTX among phyla is related to their origin, as bacteria of the genera *Pseudomonas* or *Vibrio* have been pointed out as the primary source of this compound. Different TTX-producing bacteria have been found in the mucus or gastrointestinal tract of marine organisms and are also associated with dinoflagellate blooms of *Alexandrium tamarense* or *Prorocentrum minimum* [19, 115, 116].

Intoxication symptoms appear between 10 and 45 min after TTX ingestion. They include perioral paresthesia, numbness and, in severe cases, respiratory distress, cyanosis, and paralysis that might lead to respiratory failure and death. TTX binds to receptor site-1 of sodium channels, inhibiting the propagation of action potentials in muscle and nerve cells. It has been used to identify and classify the voltage-dependent sodium channels subtypes with respect to their TTX sensitivity [117]. Consistently, sodium channels are classified into TTX-resistant subtypes, Nav 1.8 and Nav 1.9 (involved in neuropathic pain states) and TTX-sensitive subtypes, Nav 1.7, 1.3, 1.2, and 1.1, which are implicated in inflammation, epilepsy, or neuropathic pain [118]. Due to this, TTX is being used in several pharmacological researches for its therapeutic possibilities in migraines, addictions, or as an anesthetic agent for pain [119]. The median lethal dose for TTX in rodents is between 9–10 and 232 µg/kg by intraperitoneal and oral route [120, 121].

Although TTX presence was traditionally linked to Asian countries, in 2007, the first human intoxication was reported in Spain [122]. Since then, the neurotoxin has been detected in a great diversity of mollusks in European countries and TTX intoxications have been reported in Israel, Lebanon, Cyprus, or Malta [19, 123, 124]. Currently, Japan is the only country with a regulatory limit for TTX (2 mg/kg pufferfish tissue) [19]. In Europe, there is no established limit and the toxin is not regularly monitored, but products derived from pufferfish family cannot be marketed. However, due to the occurrence of the toxin in European coasts in recent years, EFSA proposed a safe concentration of 44 µg TTX equivalents/kg shellfish in 2017 [125]. In this report, the EFSA panel also pointed to a potential interaction between STX and TTX due to their similar mode of action. Recent toxicity studies in mice confirmed this point, since acute and chronical exposure to both toxins produced an additive effect, so their co-occurrence in food could pose a health risk [126, 127].

6.3.3 Domoic acid

Domoic acid is a worldwide-distributed toxin first isolated from the red macroalgae *Chondria armata* in 1958. Later, several diatom species, mainly from the genera *Pseudonitzschia* and *Nitzschia,* have been described as producers of this phycotoxin. Interestingly, bivalve species have diverse capacities to accumulate this toxin due to their different absorption and depuration rates. Most bivalves depurate domoic acid very fast, but scallops such as *Pecten maximus* or razor clams like *Siliqua patula* can accumulate great concentrations of this toxin [128].

This water-soluble compound is responsible for amnesic shellfish poisoning (ASP). The toxin is a glutamate analogue with a high affinity for neuronal glutamate ionotropic receptors. It binds a-amino-5-hydroxy-3-methyl-4-isoxazole propionic acid (AMPA) and kainate receptors and maintains them at an open state, inducing neuron excitability. Hippocampus has a high concentration of these receptors, so pyramidal hippocampal neurons are the main target of this toxin, specifically the CA3 area. The activation of these receptors leads to a decrease in ATP and an increase in intracellular calcium levels. These effects are followed by the generation of reactive oxygen and nitrogen species [129, 130].

In 1987, a serious food poisoning happened in Canada after consumption of mussels contaminated with domoic acid. Gastrointestinal, as well as serious and uncommon neurological symptoms were reported. Some patients showed confusion, disorientation, seizures, and even epileptic processes and coma. Three people died some days after the poisoning, whereas other deaths related to the intoxication were reported 2 months and 2 years later [131]. Due to these serious effects, the toxicity of domoic acid in animals has been extensively studied. After intraperitoneal injection, a characteristic behavior is observed in rodents. Lethargy is followed by forelimb paralysis, vigorous scratching, and circular movements. Progressively, animals lose postural control and in the final stages, seizures, and tremors are observed prior to death. The complexity of neurological damage also includes hypophysis injury and endocrine alterations [132]. Currently, the regulatory limit for this toxin is set at 20 mg/kg of shellfish meat [24, 30].

6.4 Amphiphilic toxins

The group of amphiphilic toxins includes two families of compounds: MTXs and PLTXs. The chemical structures of MTX or MTX1 and PLTX, the reference compounds, are shown in Figure 6.6.

6.4.1 Maitotoxins

MTX is a polyketide-derived polycyclic ether of 3422 Da, the largest and most complex natural non-biopolymer known, and is produced by species of the genus *Gambierdiscus*. Six analogues have been described, including MTXs -1,-2,-3, and-4 [22, 133, 134]. MTX1, the first identified, was discovered as one of the compounds involved in ciguatera poisoning [135, 136]. However, it is believed that CTXs are the main responsible for neurological symptoms and MTXs are less implicated due to their low oral absorption. MTXs have greater water solubility than CTXs, so their accumulation in fish and invertebrates is more difficult [136]. In fact, MTXs accumulate in carnivorous fishes, but mainly in their liver and viscera instead of in their flesh as CTXs [137].

MTX-1 is the most toxic compound isolated to date, with an intraperitoneal LD$_{50}$ of 50 ng/kg, but its oral toxicity is far lower [138]. However, recent studies have reported a great range of MTX in vivo toxicity [139, 140]. This compound family affects calcium homeostasis by inducing the ion entry, which leads to several cellular responses like depolarization, muscle contraction, neurotransmitter release, or sperm acrosome changes, among others [15, 133]. However, their exact mechanism of action remains unknown.

6.4.2 Palytoxins

PLTX was first isolated in 1971 from a specimen of the soft coral *Palythoa* sp. collected in Hawaii. After that, PLTX has been also obtained from corals of the genera *Zoanthus* and *Parazoanthus*, and from the red alga *Chondria armata*. Moreover, PLTX derivatives have been isolated in dinoflagellates from the genus *Ostreopsis*, a microalgae with a wide distribution in temperate and tropical waters that is propagating to new areas due to global warming [141].

PLTX is a large molecule with a polyketide structure of polyethers with both lipophilic and hydrophilic regions. It has 64 chiral centers, leading to a great number of possible stereoisomers. Currently, around 20 analogues have been identified and are classified into four groups: PLTXs, ostreocins, ovatoxins, and mascarenotoxins [23].

The main mechanism of action of PLTX is the disruption of the mammalian cell Na^+–K^+-adenosine triphosphatase (ATPase) pump. The toxin binds to the ATPase pump and turns it into a nonspecific permanently open ion channel. This pump is one of the most important transporters of the cellular membrane and its dysfunction leads to membrane potential alterations. In humans, PLTX poisoning is characterized by myalgia, myoglobinuria, respiratory distress, cyanosis, and alterations in creatine phosphokinase, AST, ALT, and lactate dehydrogenase levels. If death occurs, it may be due to respiratory failure. Furthermore, in coincidence with *Ostreopsis* blooms, several episodes of respiratory, ocular, and skin damage have been reported. In these cases, the exposure to the toxin in aerosols was by inhalation or skin contact [23]. Poisoning has been also reported due to the manipulation of aquarium corals [142, 143]. In rodents, ataxia and paralysis are the first symptoms observed after intraperitoneal administration. Mice death occurs by respiratory arrest and increased levels of creatine phosphokinase, AST, ALT, and lactate dehydrogenase, as in humans [144, 145]. In addition, after oral chronic treatment, mice present alterations in the digestive tract and in biochemical blood and urine parameters [146, 147].

Despite their toxicity, PLTXs are not currently regulated. The EFSA recommended a maximum of 30 µg/kg in 2009, but due to its chemical properties, the intestinal absorption rate of PLTX is poor [148]. Given a large number of potential analogues, their presence will be a matter of concern in the future. Since their occurrence is associated with warm waters and they seem to be affected by global warming, this family of compounds will need a follow-up in the future [149].

Keywords: Marine toxin, phycotoxin, dinoflagellate, microalgae, harmful algal bloom

References

[1]   Turner AD, et al. Marine invertebrate interactions with Harmful Algal Blooms – Implications for one health. J Invertebr Pathol. 2021;186:107555.
[2]   Stoecker DK, et al. Mixotrophy in the marine plankton. Ann Rev Mar Sci. 2017;9:311–35.
[3]   Wang ZH, et al. Distribution of dinoflagellate cysts in surface sediments from the Qingdao Coast, the Yellow Sea, China: The potential risk of harmful algal blooms. Front Mar Sci. 2022;9:910327.
[4]   Bravo I, Figueroa RI. Towards an ecological understanding of dinoflagellate cyst functions. Microorganisms. 2014;2(1):11–32.

[5]   Holmes MJ, Brust A, Lewis RJ. Dinoflagellate toxins: An overview. In: Botana LM, ed. Seafood and freshwater toxins: Pharmacology, physiology and detection. Taylor and Francis Group, Boca Ratón, 2014;3–38.

[6]   Louzao MC, Vilariño N, Yotsu-Yamashita M. Polycavernosides and other scarce new toxins. In: Botana LM, ed. Seafood and freshwater toxins: Pharmacology, physiology and detection. Taylor and Francis Group, Boca Ratón, 2014;857–71.

[7]   Katikou P, Vlamis A. Palytoxin and analogs: Ecobiology and origin, chemistry, and chemical analysis. In: Botana LM, ed. Seafood and freshwater toxins: Pharmacology, physiology and detection. Taylor and Francis Group, Boca Ratón, 2014;696–740.

[8]   Skinner MP, Lewis RJ, Morton S. Ecology of the ciguatera causing dinoflagellates from the Northern Great Barrier Reef: Changes in community distribution and coastal eutrophication. Mar Pollut Bull. 2013;77(1–2):210–19.

[9]   Vlamis A, Katikou P. Ecobiology and geographical distribution of potentially toxic marine dinoflagellates. In: Botana LM, ed. Seafood and freshwater toxins: Pharmacology, physiology and detection. Taylor and Francis Group, Boca Ratón, 2014;569–625.

[10]   Botana LM, et al. Derivation of toxicity equivalence factors for marine biotoxins associated with bivalve molluscs. Trends Food Sci Technol. 2017;59:15–24.

[11]   Espiña B, Rubiolo JA. Marine toxins and the cytoskeleton: Pectenotoxins, unusual macrolides that disrupt actin. FEBS J. 2008;275(24):6082–88.

[12]   Alfonso A, Vieytes MR, Botana LM. Yessotoxin, a promising therapeutic tool. Mar Drugs. 2016;14(2):30.

[13]   Vetter I, Zimmermann K, Lewis RJ. Ciguatera toxins: Pharmacology, toxicology and detection. In: Botana LM, ed. Seafood and freshwater toxins. Pharmacology, physiology and detection. Taylor and Francis Group, Boca Ratón, 2014;3–38.

[14]   FAO. Report of the expert meeting on ciguatera poisoning: Rome, 19–23 November 2018. 2020;9.

[15]   Shmukler YB, Nikishin DA. Ladder-shaped ion channel ligands: Current state of knowledge. Mar Drugs. 2017;15(7):232.

[16]   Molgo J, et al. Cyclic imine toxins from dinoflagellates: A growing family of potent antagonists of the nicotinic acetylcholine receptors. J Neurochem. 2017;142(Suppl 2):41–51.

[17]   Otero A, et al. Cyclic imines: Chemistry and mechanism of action: A review. Chem Res Toxicol. 2011;24(11):1817–29.

[18]   Yotsu-Yamashita M, et al. First identification of 5,11-dideoxytetrodotoxin in marine animals, and characterization of major fragment ions of tetrodotoxin and its analogs by high resolution ESI-MS/MS. Mar Drugs. 2013;11(8):2799–813.

[19]   Katikou P, et al. An updated review of tetrodotoxin and its peculiarities. Mar Drugs. 2022;20(1):37.

[20]   Magarlamov TY, Melnikova DI, Chernyshev AV. Tetrodotoxin-producing bacteria: Detection, distribution and migration of the toxin in aquatic systems. Toxins. 2017;9(5):166.

[21]   Flores PL, et al. Maitotoxin is a potential selective activator of the endogenous transient receptor potential canonical Type 1 channel in Xenopus laevis Oocytes. Mar Drugs. 2017;15(7):198.

[22]   Louzao MC, et al. Current trends and new challenges in marine phycotoxins. Mar Drugs. 2022;20(3):198.

[23]   Patocka J, et al. Palytoxin congeners. Arch Toxicol. 2018;92(1):143–56.

[24]   FAO and WHO. Codex Alimentarius International Food Standards. Standard for Live And raw Bivalve Molluscs Codex Standard 292–2008. Adopted in 2008. Amendment: 2013. Revision: 2014 and 2015.

[25]   FDA. Guidance for the industry: Fish and fishery products hazards and controls guidance. 2011.

[26]   Rodríguez I, Vieytes MR, Alfonso A. Analytical challenges for regulated marine toxins. Detection methods. Curr Opin Food. 2017;18(Supplement C):29–36.

[27]   Yasumoto T, et al. eds. Polyether toxins produced by dinoflagellates. Mycotoxins and phycotoxins ´88. Elsevier: Amsterdam, 1989;375–82.

[28] Reguera B, et al. Dinophysis toxins: Causative organisms, distribution and fate in shellfish. Mar Drugs. 2014;12(1):394–461.

[29] Ares IR, et al. Lactone ring of pectenotoxins: A key factor for their activity on cytoskeletal dynamics. Cell Physiol Biochem. 2007;19(5–6):283–92.

[30] EU. Regulation (EC) No 853/2004 of 29 April 2004 laying down specific hygiene rules for food of animal origin. Off J Eur Union. 2004;139:55–205.

[31] EU. Commission Delegated Regulation (EU) 2021/1374 of 12 April 2021 Amending Annex III to Regulation (EC) No 853/2004 of the European Parliament and of the Council on Specific Hygiene Requirements for Food of Animal Origin. Off J Eur Union, 2021;1–15.

[32] Ogino H, Kumagai M, Yasumoto T. Toxicologic evaluation of yessotoxin. Nat Toxins. 1997;5 (6):255–59.

[33] Miles CO, et al. Polyhydroxylated amide analogs of yessotoxin from Protoceratium reticulatum. Toxicon. 2005;45(1):61–71.

[34] Paz B, et al. Production and release of yessotoxins by the dinoflagellates Protoceratium reticulatum and Lingulodinium polyedrum in culture. Toxicon. 2004;44(3):251–58.

[35] Satake M, MacKenzie L, Yasumoto T. Identification of Protoceratium reticulatum as the biogenetic origin of yessotoxin. Nat Toxins. 1997;5(4):164–67.

[36] Draisci R, Lucentini L, Masciori A. Pectenotoxins and yessotoxins: Chemistry, toxicology, pharmacology, and analysis. In: Botana LM, ed. Seafood and freshwater toxins: Pharmacology, physiology, and detection. New York: Marcel Dekker, 2000;289–324.

[37] Yasumoto T, Takizawa A. Fluorometric measurement of yessotoxins in shellfish by high-pressure liquid chromatography. Biosci Biotechnol Biochem. 1997;61(10):1775–77.

[38] Murata M, et al. Isolation and structure of yessotoxin, a novel polyether compound implicated in diarrheic shellfish poisoning. Tetrahedron. 1987;28:5869–72.

[39] Ferreiro SF, et al. Acute cardiotoxicity evaluation of the marine biotoxins OA, DTX-1 and YTX. Toxins. 2015;7(4):1030–47.

[40] Tubaro A, et al. Ultrastructural damage to heart tissue from repeated oral exposure to yessotoxin resolves in 3 months. Toxicon. 2008;51(7):1225–35.

[41] Tubaro A, et al. Yessotoxins: A toxicological overview. Toxicon. 2010;56(2):163–72.

[42] Tubaro A, et al. Oral and intraperitoneal acute toxicity studies of yessotoxin and homoyessotoxins in mice. Toxicon. 2003;41(7):783–92.

[43] Ferreiro SF, et al. Subacute cardiotoxicity of yessotoxin: In vitro and in vivo studies. Chem Res Toxicol. 2016;29(6):981–90.

[44] EU. Regulation (EU) No 786/2013 of 16 August 2013 amending Annex III to Regulation (EC) No 853/2004 of the European Parliament and of the Council as regards the permitted limits of yessotoxins in live bivalve molluscs. Off J Eur Union. 2013;L220.

[45] EFSA. Yessotoxin group. Scientific opinion of the panel on contaminants in the food chain. EFSA J. 2008;907:1–62.

[46] Alfonso A, et al. Yessotoxin, a novel phycotoxin, activates phosphodiesterase activity. Effect of yessotoxin on cAMP levels in human lymphocytes. Biochem Pharmacol. 2003;65(2):193–208.

[47] Pazos MJ, et al. Study of the interaction between different phosphodiesterases and yessotoxin using a resonant mirror biosensor. Chem Res Toxicol. 2006;19(6):794–800.

[48] Korsnes MS, Espenes A. Yessotoxin as an apoptotic inducer. Toxicon. 2011;57(7–8):947–58.

[49] Tobio A, et al. Yessotoxin, a marine toxin, exhibits anti-allergic and anti-tumoural activities inhibiting melanoma tumour growth in a preclinical model. PLoS One. 2016;11(12):e0167572.

[50] Twiner MJ, Hess P, Doucette GJ. Azaspiracids: Toxicology, pharmacology and risk assessment. In: Botana LM, ed. Seafood and freshwater toxins: Pharmacology, physiology and detection. Taylor and Francis Group, Boca Ratón, 2014;823–55.

[51] Rossi R, et al. Mediterranean Azadinium dexteroporum (Dinophyceae) produces six novel azaspiracids and azaspiracid-35: A structural study by a multi-platform mass spectrometry approach. Anal Bioanal Chem. 2017;409(4):1121–34.

[52] Pelin M, et al. Toxic equivalency factors (TEFs) after acute oral exposure of azaspiracid 1, −2 and −3 in mice. Toxicol Lett. 2018;282:136–46.

[53] Ito E, et al. Chronic effects in mice caused by oral administration of sublethal doses of azaspiracid, a new marine toxin isolated from mussels. Toxicon. 2002;40(2):193–203.

[54] Ferreiro SF, et al. In vivo cardiomyocyte response to YTX- and AZA-1-induced damage: Autophagy versus apoptosis. Arch Toxicol. 2017;91(4):1859–70.

[55] Ferreiro SF, et al. In vivo arrhythmogenicity of the marine biotoxin azaspiracid-2 in rats. Arch Toxicol. 2014;88(2):425–34.

[56] Twiner MJ, et al. Marine algal toxin azaspiracid is an open-state blocker of hERG potassium channels. Chem Res Toxicol. 2012;25(9):1975–84.

[57] Ferreiro SF, et al. In vitro chronic effects on hERG channel caused by the marine biotoxin azaspiracid-2. Toxicon. 2014;91:69–75.

[58] Boente-Juncal A, et al. Targeting chloride ion channels: New insights into the mechanism of action of the marine toxin azaspiracid. Chem Res Toxicol. 2021;34(3):865–79.

[59] Pelin M, et al. Azaspiracids increase mitochondrial dehydrogenases activity in hepatocytes: Involvement of potassium and chloride ions. Mar Drugs. 2019;17(5):276.

[60] Jimenez-Carcamo D, Garcia C, Contreras HR. Toxins of okadaic acid-group increase malignant properties in cells of colon cancer. Toxins. 2020;12(3):179.

[61] Manerio E, et al. Shellfish consumption: A major risk factor for colorectal cancer. Med Hypotheses. 2008;70(2):409–12.

[62] Munday R. Is protein phosphatase inhibition responsible for the toxic effects of okadaic Acid in animals? Toxins. 2013;5(2):267–85.

[63] Louzao MC, et al. Diarrhetic effect of okadaic acid could be related with its neuronal action: Changes in neuropeptide Y. Toxicol Lett. 2015;237(2):151–60.

[64] Louzao MC, et al. Serotonin involvement in okadaic acid-induced diarrhoea in vivo. Arch Toxicol. 2021;95(8):2797–813.

[65] Costas C, et al. Intestinal secretory mechanisms in Okadaic acid induced diarrhoea. Food Chem Toxicol. 2022;169:113449.

[66] EFSA. Marine biotoxins in shellfish- okadaic acid and analogues, Scientific opinion of the Panel on Contaminants in the Food chain. EFSA J. 2008;589:1–62.

[67] Abal P, et al. Toxic action reevaluation of okadaic acid, dinophysistoxin-1 and dinophysistoxin-2: toxicity equivalency factors based on the oral toxicity study. Cell Physiol Biochem. 2018;49(2):743–57.

[68] Abal P, et al. Characterization of the dinophysistoxin-2 acute oral toxicity in mice to define the Toxicity Equivalency Factor. Food Chem Toxicol. 2017;102:166–75.

[69] Kohli GS, Farrell H, Murray SA. Gambierdiscus, the cause of ciguatera fish poisoning: An increased human health threat influenced by climate change. In: Botana LM, Louzao MC, Vilariño N, eds. Climate change and marine and freshwater toxins. De Gruyter: Berlin, 2015;273–312.

[70] Varela Martínez CLG, Martínez Sánchez I, Carmona Alférez EV, Nuñez Gallo R, Friedemann D, Oleastro M, Boziaris IM. Incidence and epidemiological characteristics of ciguatera cases in Europe. EFSA Supporting Publications. 2021;18:6650E.

[71] Bagnis R, Kuberski T, Laugier S. Clinical observations on 3,009 cases of ciguatera (fish poisoning) in the South Pacific. Am J Trop Med Hyg. 1979;28(6):1067–73.

[72] Dickey RW, Plakas SM. Ciguatera: A public health perspective. Toxicon. 2010;56(2):123–36.

[73] Swift AE, Swift TR. Ciguatera. J Toxicol Clin Toxicol. 1993;31(1):1–29.

[74]   Lewis RJ, Jones A, Vernoux JP. HPLC/tandem electrospray mass spectrometry for the determination of Sub-ppb levels of Pacific and Caribbean ciguatoxins in crude extracts of fish. Anal Chem. 1999;71 (1):247–50.

[75]   Vernoux JP, Lewis RJ. Isolation and characterisation of Caribbean ciguatoxins from the horse-eye jack (Caranx latus). Toxicon. 1997;35(6):889–900.

[76]   Lewis RJ, et al. Purification and characterization of ciguatoxins from moray eel (Lycodontis javanicus, Muraenidae). Toxicon. 1991;29(9):1115–27.

[77]   EU. Regulation (EU) No 627/2019 of 15 March 2019 laying down uniform practical arrangements for the performance of official controls on products of animal origin intended for human consumption in accordance with Regulation (EU) 2017/625 of the European Parliament and of the Council and amending Commission Regulation (EC) No 2074/2005 as regards official controls. Off J Eur Union. 2019;L 131/51.

[78]   Raposo-Garcia S, et al. In silico simulations and functional cell studies evidence similar potency and distinct binding of pacific and caribbean ciguatoxins. Expo Health, 2022.

[79]   Konoki K, et al. Molecular determinants of brevetoxin binding to voltage-gated sodium channels. Toxins. 2019;11(9):513.

[80]   EFSA. Scientific opinion on marine biotoxins in shellfish – Emerging toxins: Brevetoxin group. EFSA J. 2010;8:1677.

[81]   La Claire JW, Manning SR. Ichtyotoxins. In: Botana LM, Alfonso A, eds. Phycotoxins chemistry and biochemistry. Wiley Blackwell, Oxford, 2015;407–61.

[82]   Fleming LE, et al. Exposure and effect assessment of aerosolized red tide toxins (brevetoxins) and asthma. Environ Health Perspect. 2009;117(7):1095–100.

[83]   Fleming LE, et al. Review of Florida red tide and human health effects. Harmful Algae. 2011;10 (2):224–33.

[84]   Watkins SM, et al. Neurotoxic shellfish poisoning. Mar Drugs. 2008;6(3):431–55.

[85]   Amzil Z, et al. Monitoring the emergence of algal toxins in shellfish: First report on detection of brevetoxins in French mediterranean mussels. Mar Drugs. 2021;19(7):393.

[86]   Satake M, Murata M, Yasumoto T. Gambierol: A new toxic polyether compound isolated from the marine dinoflagellate Gambierdiscus toxicus. J Am Chem Soc. 1993;115(1):361–62.

[87]   Ujihara S, et al. Interaction of ladder-shaped polyethers with transmembrane alpha-helix of glycophorin A as evidenced by saturation transfer difference NMR and surface plasmon resonance. Bioorg Med Chem Lett. 2008;18(23):6115–18.

[88]   Ito E, et al. Pathological effects on mice by gambierol, possibly one of the ciguatera toxins. Toxicon. 2003;42(7):733–40.

[89]   Birinyi-Strachan LC, et al. Block of voltage-gated potassium channels by Pacific ciguatoxin-1 contributes to increased neuronal excitability in rat sensory neurons. Toxicol Appl Pharmacol. 2005;204(2):175–86.

[90]   Hidalgo J, et al. Pacific ciguatoxin-1b effect over Na+ and K+ currents, inositol 1,4,5-triphosphate content and intracellular Ca2+ signals in cultured rat myotubes. Br J Pharmacol. 2002;137 (7):1055–62.

[91]   Schlumberger S, et al. Dual action of a dinoflagellate-derived precursor of Pacific ciguatoxins (P-CTX -4B) on voltage-dependent K(+) and Na(+) channels of single myelinated axons. Toxicon. 2010;56 (5):768–75.

[92]   Cuypers E, et al. Gambierol, a toxin produced by the dinoflagellate Gambierdiscus toxicus, is a potent blocker of voltage-gated potassium channels. Toxicon. 2008;51(6):974–83.

[93]   Kopljar I, et al. A polyether biotoxin binding site on the lipid-exposed face of the pore domain of Kv channels revealed by the marine toxin gambierol. PNAS. 2009;106(24):9896–901.

[94]   Kopljar I, et al. The ladder-shaped polyether toxin gambierol anchors the gating machinery of Kv3.1 channels in the resting state. J Gen Physiol. 2013;141(3):359–69.

[95] Ghiaroni V, et al. Inhibition of voltage-gated potassium currents by gambierol in mouse taste cells. Toxicol Sci. 2005;85(1):657–65.

[96] Kopljar I, et al. Voltage-sensor conformation shapes the intra-membrane drug binding site that determines gambierol affinity in Kv channels. Neuropharmacology. 2016;107:160–67.

[97] Krock B, et al. LC-MS-MS aboard ship: Tandem mass spectrometry in the search for phycotoxins and novel toxigenic plankton from the North Sea. Anal Bioanal Chem. 2008;392(5):797–803.

[98] EFSA. EFSA report on data collections: Future directions. EFSA J. 2010;1533:1–53.

[99] Munday R. Toxicology of cyclic imines: Gymnodimine, spirolides, pinnatoxins, pteriatoxins, prorocentrolide, spiro-prorocentrimine, and symbioimines. In: Botana LM, ed. Seafood and freshwater toxins, pharmacology, physiology, and detection. Taylor and Francis group, Boca Ratón, 2008;581–94.

[100] Villar Gonzalez A, et al. First evidence of spirolides in Spanish shellfish. Toxicon. 2006;48(8):1068–74.

[101] EFSA. Scientific opinion on marine biotoxins in shellfish – Cyclic imines (spirolides, gymnodimines, pinnatoxins and pteriatoxins). EFSA J. 2010;8:1628.

[102] Cembella AD, Lewis NI, Quilliam MA. Spirolide composition of micro-extracted pooled cells isolated from natural plankton assemblages and from cultures of the dinoflagellate Alexandrium ostenfeldii. Nat Toxins. 1999;7(5):197–206.

[103] Hu T, et al. Spirolides B and D, two novel macrocycles isolated from the digestive glands of shellfish. J Chem Soc Chem Commun. 1995;20:2159–61.

[104] Cembella AK, Krock B. Cyclic imine toxins: Chemistry, biogeography, biosynthesis and pharmacology. In: Botana LM, ed. Seafood and freshwater toxins. CRC Press, Boca Ratón, 2008;561–78.

[105] Gill S, et al. Neural injury biomarkers of novel shellfish toxins, spirolides: A pilot study using immunochemical and transcriptional analysis. Neurotoxicology. 2003;24(4–5):593–604.

[106] Bourne Y, et al. Structural determinants in phycotoxins and AChBP conferring high affinity binding and nicotinic AChR antagonism. PNAS. 2010;107(13):6076–81.

[107] Miles CO, et al. Gymnodimine C, an isomer of gymnodimine B, from Karenia selliformis. J Agric Food Chem. 2003;51(16):4838–40.

[108] Meilert K, Brimble MA. Synthesis of the bis-spiroacetal moiety of the shellfish toxins spirolides B and D using an iterative oxidative radical cyclization strategy. Org Biomol Chem. 2006;4 (11):2184–92.

[109] Munday R, et al. Acute toxicity of gymnodimine to mice. Toxicon. 2004;44(2):173–78.

[110] Arnich N, et al. Health risk assessment related to pinnatoxins in French shellfish. Toxicon. 2020;180:1–10.

[111] Lehane L. Paralytic shellfish poisoning: A potential public health problem. Med J Aust. 2001;175 (1):29–31.

[112] Wiese M, et al. Neurotoxic alkaloids: Saxitoxin and its analogs. Mar Drugs. 2010;8(7):2185–211.

[113] Vale P. Saxitoxin and analogs: Ecobiology, origin, chemistry and detection. In: Botana LM, ed. Seafood and freshwater toxins: Pharmacology, physiology and detection. Taylor and Francis Group, Boca Ratón, 2014;991.

[114] Chau R, Kalaitzis JA, Neilan BA. On the origins and biosynthesis of tetrodotoxin. Aquat Toxicol. 2011;104(1–2):61–72.

[115] Pratheepa V, Vasconcelos V. Microbial diversity associated with tetrodotoxin production in marine organisms. Environ Toxicol Pharmacol. 2013;36(3):1046–54.

[116] Rodriguez I, et al. The association of bacterial C9-based TTX-like compounds with Prorocentrum minimum opens new uncertainties about shellfish seafood safety. Sci Rep. 2017;7:40880.

[117] Catterall WA, Goldin AL, Waxman SG. International Union of Pharmacology. XLVII. Nomenclature and structure-function relationships of voltage-gated sodium channels. Pharmacol Rev. 2005;57 (4):397–409.

[118] Wood JN, et al. Voltage-gated sodium channels and pain pathways. J Neurobiol. 2004;61(1):55–71.

[119] Jal S, Khora SS. An overview on the origin and production of tetrodotoxin, a potent neurotoxin. J Appl Microbiol. 2015;119(4):907–16.

[120] EFSA. Risks for public health related to the presence of tetrodotoxin (TTX) and TTX analogues in marine bivalves and gastropods. EFSA J. 2017;15(4):65.

[121] Abal P, et al. Acute oral toxicity of tetrodotoxin in mice: Determination of lethal dose 50 (LD50) and no observed adverse effect level (NOAEL). Toxins. 2017;9(3):75.

[122] Rodriguez P, et al. First toxicity report of tetrodotoxin and 5,6,11-trideoxyTTX in the trumpet shell Charonia lampas lampas in Europe. Anal Chem. 2008;80(14):5622–29.

[123] Silva M, et al. New gastropod vectors and tetrodotoxin potential expansion in temperate waters of the Atlantic Ocean. Mar Drugs. 2012;10(4):712–26.

[124] Turner AD, et al. Detection of the pufferfish toxin tetrodotoxin in European bivalves, England, 2013 to 2014. Euro Surveill. 2015;20(2):21009.

[125] EFSA. Scientific opinion. Risks for public health related to the presence of tetrodotoxin (TTX) and TTX analogues in marine bivalves and gastropods. EFSA J. 2017;15(4):4752–817.

[126] Finch SC, Boundy MJ, Harwood DT. The acute toxicity of tetrodotoxin and tetrodotoxin(-)saxitoxin mixtures to mice by various routes of administration. Toxins. 2018;10(11):423.

[127] Boente-Juncal A, et al. Oral chronic toxicity of the safe tetrodotoxin dose proposed by the European Food Safety Authority and its additive effect with saxitoxin. Toxins. 2020;12(5):312.

[128] Blanco J, et al. Twenty-five years of domoic acid monitoring in Galicia (NW Spain): Spatial, temporal and interspecific variations. Toxins. 2021;13(11):756.

[129] Petroff R, et al. Public health risks associated with chronic, low-level domoic acid exposure: A review of the evidence. Pharmacol Ther. 2021;227:107865.

[130] Pulido OM. Domoic acid toxicologic pathology: A review. Mar Drugs. 2008;6(2):180–219.

[131] Perl TM, et al. An outbreak of toxic encephalopathy caused by eating mussels contaminated with domoic acid. N Engl J Med. 1990;322(25):1775–80.

[132] Crespo A, et al. Dose-response and histopathological study, with special attention to the hypophysis, of the differential effects of domoic acid on rats and mice. Microsc Res Tech. 2015;78 (5):396–403.

[133] Pisapia F, et al. Maitotoxin-4, a Novel MTX Analog produced by Gambierdiscus excentricus. Mar Drugs. 2017;15(7):220.

[134] Estevez P, et al. Toxicity screening of a Gambierdiscus australes strain from the Western Mediterranean Sea and identification of a Novel Maitotoxin Analogue. Mar Drugs. 2021;19(8):460.

[135] Murata M, Yasumoto T. The structure elucidation and biological activities of high molecular weight algal toxins: Maitotoxin, prymnesins and zooxanthellatoxins. Nat Prod Rep. 2000;17(3):293–314.

[136] Wang DZ, Xin YH, Wang MH. Gambierdiscus and its associated toxins: A minireview. Toxins. 2022;14(7):485.

[137] Reyes JG, et al. Maitotoxin: An enigmatic toxic molecule with useful applications in the biomedical sciences. In: Botana LM, ed. Seafood and freshwater toxins: Pharmacology, physiology and detection. Taylor and Francis Group, Boca Ratón, 2014;677–94.

[138] Murata M, et al. Structure and partial stereochemical assignments for maitotoxin, the most toxic and largest natural non-biopolymer. J Am Chem Soc. 1994;116(16):7098–107.

[139] Raposo-Garcia S, et al. In vivo subchronic effects of ciguatoxin-related compounds, reevaluation of their toxicity. Arch Toxicol. 2022;96(9):2621–38.

[140] Munday R, et al. Ciguatoxins and Maitotoxins in Extracts of Sixteen Gambierdiscus isolates and one Fukuyoa isolate from the South Pacific and their toxicity to mice by intraperitoneal and oral administration. Mar Drugs. 2017;15(7):208.

[141] Munday R. Toxicology of seafood toxins: A critical review. In: Botana LM, ed. Seafood and freshwater toxins: Pharmacology, physiology and detection. Taylor and Francis Group, Boca Ratón, 2014;197–290.

[142] Schulz M, et al. Inhalation poisoning with palytoxin from aquarium coral: Case description and safety advice. Arh Hig Rada Toksikol. 2019;70(1):14–17.

[143] Pelin M, et al. Palytoxin-containing aquarium soft corals as an emerging sanitary problem. Mar Drugs. 2016;14(2):33.

[144] Sosa S, et al. Palytoxin toxicity after acute oral administration in mice. Toxicol Lett. 2009; 191(2–3):253–59.

[145] Taniyama S, et al. Ostreopsis sp., a possible origin of palytoxin (PTX) in parrotfish Scarus ovifrons. Toxicon. 2003;42(1):29–33.

[146] Boente-Juncal A, et al. In vivo evaluation of the chronic oral toxicity of the marine toxin palytoxin. Toxins. 2020;12(8):489.

[147] Boente-Juncal A, et al. Reevaluation of the acute toxicity of palytoxin in mice: Determination of lethal dose 50 (LD(50)) and No-observed-adverse-effect level (NOAEL). Toxicon. 2020;177:16–24.

[148] EFSA. Scientific opinion on marine biotoxins in shellfish – Palytoxin group. EFSA J. 2009;7:1–38.

[149] Fernández-Araujo A, et al. Warm seawater microalgae: Growth and toxic profile of Ostreopsis spp. from European coasts. Oceanography. 2013;1(1):104.

Alejandro Cao, Eva Cagide and Natalia Vilariño

7 Cyanobacterial toxins

7.1 Cyanobacteria and algal blooms

Cyanobacteria are an ancient bacterial phylum with unique characteristics and a wide environmental tolerance range. These microorganisms have been traditionally called "blue-green algae," due to their color, even though they are not related to algae. They belong to the *Bacteria* domain (Figure 7.1) [1–5], therefore they are not eukaryotes as algae are, but share some characteristic features resembling them, such as oxygenic photosynthesis and pigments; in fact, a common pigment is the blue-colored phycocyanin, which together with the green pigment chlorophyll a, is the reason for the former term.

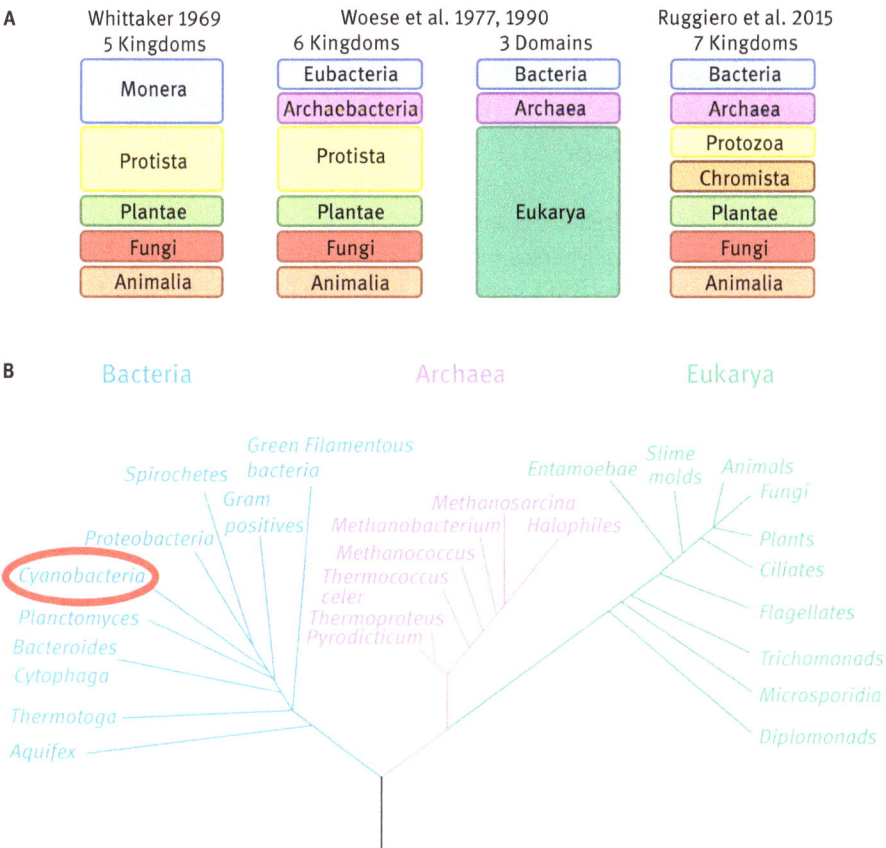

Figure 7.1: (A) Cyanobacteria classification according to Whittaker [3], Woese and Fox [4], Woese et al. [5], and Ruggiero et al. [2]. (B) Phylogenetic tree according to Woese's three domains.

https://doi.org/10.1515/9783111014449-007

Cyanobacteria are the only prokaryotes able to produce oxygen by photosynthesis, and are considered responsible for the increased production of oxygen in the Earth's atmosphere when the Great Oxidation Event occurred during the early Proterozoic, 2.5–2.3 billion years ago (2.5–2.3 Ga), contributing to the creation of the current atmosphere and allowing the great biodiversity of aerobic organisms [6].

In fact, the widely accepted endosymbiosis theory of Lynn Margulis explains the origin of eukaryotic cell organelles through the symbiosis of prokaryotic progenitors [7]. Specifically, the origins of plastids have been related to *Cyanobacteria*, probably nitrogen-fixing filamentous cyanobacteria [8]. This allowed the emergence of different lineages, making it one of the most significant evolutionary events with extraordinary importance for life on Earth.

The great morphological diversity of cyanobacteria is the basis for their classification in *Bergey's Manual of Systematic Bacteriology* [9–11], which has divided cyanobacteria into five subsections:

I (= Order Chroococcales)
II (= Order Pleurocapsales)
III (= Order Oscillatoriales)
IV (= Order Nostocales)
V (= Order Stigonematales)

Subsections I and II are unicellular cyanobacteria, usually aggregated in colonies by secreting mucopolysaccharides, whereas subsections III–V include filamentous cyanobacteria. Like most bacteria, subsection I reproduces by binary fission or budding [12], while subsection II reproduces by multiple fission and endospore release (baeocytes) [13].

Subsections III–V include filamentous cyanobacteria where reproduction takes place by fragmentation and formation of short motile filaments called hormogonia. Subsection III is formed by simple unidirectional filamentous cyanobacteria, and subsections IV and V have the ability to form differentiated cells: the metabolically specialized heterocysts (for nitrogen fixation, in anaerobic conditions, Figure 7.2) and the resting cells (akinetes) [14]. In addition, subsection V consists of the only cyanobacteria that produce "true branches," growing multiseriate filaments in several planes, and are the most evolved prokaryotes with regard to thallus organization and cell differentiation [15].

Heterocyst:
Nitrogen fixation
$$N_2 + 12ATP \xrightarrow{Nitrogenase} 2NH_3 + 12ADP + 12P_i$$

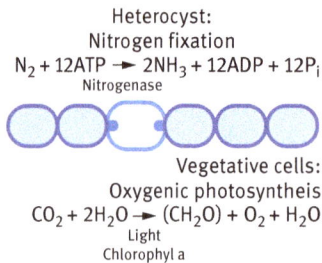

Vegetative cells:
Oxygenic photosyntheis
$$CO_2 + 2H_2O \xrightarrow[Chlorophyl\ a]{Light} (CH_2O) + O_2 + H_2O$$

Figure 7.2: Processes of nitrogen fixation and oxygen production in cyanobacteria.

Furthermore, cyanobacteria are named under Bacteriological and Botanical Codes, causing confusion due to the different rules that apply in both Codes. Advances in molecular techniques have allowed the evaluation of phylogenetic relationships, and efforts are being made to establish a consensus combining phenotype, ultrastructural, ecological, biochemical, and molecular approaches [16].

The cyanobacteria inhabit aquatic and terrestrial environments [17], surviving under a wide range of physicochemical conditions of pH, salinity, temperature, etc., and are naturally found in almost all environments on the Earth, from deserts to hot springs and ice-cold water [18]. Under specific conditions (mainly of temperature and nutrients), these organisms can massively proliferate and accumulate in the water (Figure 7.3), causing a negative impact on natural resources or humans [19]. These episodes are known as harmful algal blooms (HABs or CyanoHABs), which in some cases produce toxic compounds named cyanotoxins due to the producing organisms [20]. The main parameters favoring the occurrence of cyanobacterial blooms include [21, 22]:

- increased nutrients (nitrogen and phosphorus),
- warm water temperature,
- high intensity and duration of sunlight,
- quiescent surface water,
- water column stratification,
- changes in water pH, and
- occurrence of trace metals.

Figure 7.3: Cyanobacteria in culture, showing a high cellular proliferation.

Also, human activities leading to eutrophication and global climate change have been related to an increase in geographical distribution and frequency of HABs [23].

These HABs have a considerable impact on recreational water quality and have become a public health concern because the toxins that some species of cyanobacteria produce may provoke harmful effects on human consumers and wild or domestic animals, through drinking water, or recreational/direct exposure [24, 25].

7.2 Cyanotoxins

Cyanobacteria produce a wide range of compounds, over a hundred have been described, that vary in origin, chemical structure, and bioactivity, most of them considered secondary metabolites rather than primary products of cellular metabolism. Although some of these compounds may be beneficial, with diverse properties, including antiviral, anti-inflammatory, antitumor, anti-allergenic, and antioxidant properties [26], other molecules are toxic to humans and animals after oral, breathing, or skin contact exposure. These compounds are known as cyanotoxins (see Table 7.1) and they are predominantly found in the intracellular space of bacteria [27].

Despite extensive research, the biological function of cyanotoxins remains largely unknown. One possible role attributed to cyanotoxins is their potential as a defense mechanism against competition or predators. Certain types of cyanotoxins possess allelopathic properties, enabling them to inhibit the growth of other organisms, such as competing algae. Another hypothesis suggests that cyanotoxins may help to improve cellular physiology, promoting processes such as homeostasis, photosynthesis, or growth rates [28]. However, some researchers argue that cyanotoxins may not serve an ecological function or have not received sufficient scientific attention to determine their purpose [29].

These toxins are produced by certain species of cyanobacteria and within a species there are toxic and non-toxic strains [30], which cannot be identified separately by microscopic identification. Therefore, the presence of a toxin-producing species in a cyanobacteria bloom does not imply necessarily the presence of cyanotoxins, which must be determined by other means (see Section 7.3).

The adverse effects of cyanotoxins in vertebrates include the potential for mortality even at naturally occurring concentrations. The earliest recorded case of fatal poisoning in livestock following ingestion of water contaminated with cyanobacteria dates back to the nineteenth century. Subsequently, numerous cases of toxicity in humans have been reported through various routes of exposure, including drinking water, inhalation, contact with or ingestion of contaminated recreational water, and even the use of water in hemodialysis procedures. In some of these episodes, human casualties were reported. However, many cyanobacteria poisonings are likely to go unreported due to a lack of awareness of the relationship between symptoms and the toxicity of cyanobacteria [31].

Human exposure to cyanotoxins usually occurs through different routes due to the presence of cyanotoxins in freshwater, but their presence in vegetables watered with contaminated water and other food sources has also been occasionally reported.

The higher risk of poisoning through ingestion comes from drinking water and swallowing bathing water during recreational activities. In addition, caution should be exercised when considering the use of cyanobacteria-derived products. Dematotoxicity has also been reported due to bathing in freshwater reservoirs during cyanobacteria blooms. Respiratory problems have also been described due to the breathing of aerosols in the proximities of these blooms [32].

Table 7.1: General features of cyanotoxins.

Structure	Cyanotoxin	Primary target organ Mode of action	Cyanobacterial genera
Hepatotoxic cyclic peptides	Microcystins	Liver Protein phosphatase inhibition	*Microcystis, Anabaena, Planktothrix (Oscillatoria), Nostoc, Hapalosiphon, Anabaenopsis, Phormidium, Phanocapsa*
	Nodularins	Liver Protein phosphatase inhibition	*Nodularia*
Neurotoxic alkaloids	Anatoxin-a and homoanatoxin-a	Nerve synapse Agonist of nicotinic acetylcholine receptor	*Anabaena, Planktothrix (Oscillatoria), Aphanizomenon, Cylindrospermum*
	Guanitoxin	Nerve synapse Acetylcholinesterase inhibition	*Anabaena*
	Saxitoxins	Nerve axons Blockage of voltage-gated sodium channels	*Anabaena, Aphanizomenon, Lyngbya, Cylindrospermopsis, Oscillatoria, Phormidium, Planktothrix*
Cytotoxic alkaloids	Cylindrospermopsins	Liver, kidneys, lungs, spleen, intestine Protein synthesis inhibition	*Cylindrospermopsis, Anabaena, Aphanizomenon, Umezakia, Lyngbya, Oscillatoria, Raphidiopsis*
Dermatoxic alkaloids	Lyngbyatoxin-a	Skin, gastrointestinal tract Potentiation of protein kinase C	*Lyngbya*
	Aplysiatoxins	Skin Potentiation of protein kinase C	*Lyngbya, Schizothrix, Planktothrix (Oscillatoria)*

Table 7.1 (continued)

Structure	Cyanotoxin	Primary target organ Mode of action	Cyanobacterial genera
Lipopolysaccharides		Potential irritant; affects any exposed tissue	*All*
Amino acids	BMAA (β-*N*-methylamino-L-alanine)	Potentially related to neurodegenerative diseases	*Anabaena, Cylindrospermopsis, Microcystis, Nostoc, Planktothrix* (*Oscillatoria*)

Cyanotoxins are a diverse group of natural toxins that have been classified attending to different criteria. From a toxicological point of view, cyanotoxins can be classified into five major classes [33]:

- *Neurotoxins*: Anatoxins, saxitoxins (STXs), and β-*N*-methylamino-L-alanine (BMAA).
- *Hepatotoxins*: Microcystins (MCs) and nodularins (NODs).
- *Cytotoxins*: Cylindrospermopsins (CYNs).
- *Dermatoxins*: Aplysiatoxins and lyngbyatoxin.
- *Irritant and gastrointestinal toxins*: Lipopolysaccharide (LPS) endotoxins.

Structurally, the cyanotoxins are quite diverse [17, 34–36], and they can be categorized into four large groups according to their chemical structures:

- *Cyclic peptides*: MCs and NODs.
- *Alkaloids*: Anatoxins, STXs, CYNs, lyngbyatoxins, and aplysiatoxins.
- *LPSs*.
- *Other bioactive compounds*: Neurotoxic amino acids, different from the 20 standard amino acids that comprise proteins (BMAA).

7.2.1 Microcystins

7.2.1.1 Origin and chemical structure

Microcystins (MCs) have received considerable attention in the last decades as the most extensively studied class of freshwater cyanotoxins, largely due to their widespread distribution and potent toxicity. These toxins have been identified on all continents and account for 69% of global cyanotoxin detections [37, 38]. The name "microcystins" derives from the first cyanobacterial genus associated with their biosynthesis, *Microcystis*. However, it is now known that these toxins are also produced by other genera of cyanobacteria, including *Chrysosporum, Dolichospermum, Limnothrix, Nostoc, Phormidium,* and *Planktothrix* [39].

The first chemical structures of cyclic peptide cyanotoxins were identified in the early 1980s and the number of fully characterized toxin variants has greatly increased during the 1990s. The first of such compounds found in freshwater cyanobacteria were microcystins (see Table 7.2) [40].

Table 7.2: Some structural variations of microcystins reported in the scientific literature.

Microcystin	MW (g/mol)	$LD_{50}{}^{a}$	1	2	$3/R^2$	4	$5/R^3$	6	$7/R^1$
MC-LR	994	50	Ala	Leu	Me	Arg	Me	Glu	Mdha
[D-Asp³]MC-LR	980	160–300	Ala	Leu	H	Arg	Me	Glu	Mdha
[Dha⁷]MC-LR	980	250	Ala	Leu	Me	Arg	Me	Glu	Dha
[D-Asp³, Dha⁷]MC-LR	966	+	Ala	Leu	H	Arg	Me	Glu	Dha
[DMAdda⁵]MC-LR	980	90–100	Ala	Leu	Me	Arg	H	Glu	Mdha
[(6Z)-Adda⁵]MC-LR	994	>1,200	Ala	Leu	Me	Arg	Me	Glu	Mdha
[L-Ser⁷]MC-LR	998	+	Ala	Leu	Me	Arg	Me	Glu	Ser
[D-Glu(OCH₃)⁶]MC-LR	1,008	>1,000	Ala	Leu	Me	Arg	Me	Glu-OMe	Mdha
[DAsp³, D-Glu(OCH₃)⁶]MC-LR	994	NR	Ala	Leu	H	Arg	Me	Glu-OMe	Mdha
[ADMAdda⁵]MC-LR	1,022	60	Ala	Leu	Me	Arg	COCH₃	Glu	Mdha
[D-Asp³, ADMAdda⁵, Dhb⁷]MC-LR	1,009	+	Ala	Leu	H	Arg	COCH₃	Glu	Dhb
[L-MeSer⁷]MC-LR	1,012	150	Ala	Leu	Me	Arg	Me	Glu	Mser
[D-Asp³, ADMAdda⁵]MC-LR	1,008	160	Ala	Leu	H	Arg	COCH₃	Glu	Mdha
[D-Ser¹, ADMAdda⁵]MC-LR	1,038	+	Ser	Leu	Me	Arg	COCH₃	Glu	Mdha
[ADMAdda⁵, MeSer⁷]MC-LR	1,040	+	Ala	Leu	Me	Arg	COCH₃	Glu	Mser
[L-MeLan⁷]MC-LR	1,115	1,000	Ala	Leu	M	Arg	Me	Glu	MeLan
[D-Leu¹]MC-LR	1,036		Leu	Leu	Me	Arg	Me	Glu	Mdha
[D-Asp³]MC-RR	1,023	250	Ala	Arg	H	Arg	Me	Glu	Mdha
[Dha⁷]MC-RR	1,023	180	Ala	Arg	Me	Arg	Me	Glu	Dha

Table 7.2 (continued)

Microcystin	MW (g/mol)	LD$_{50}$[a]	1	2	3/R^2	4	5/R^3	6	7/R^1
[D-Asp^3, Dha^7]MC-RR	1,009	+	Ala	Arg	H	Arg	Me	Glu	Dha
[(6Z)-Adda^5]MC-RR	1,037	>1,200	Ala	Arg	Me	Arg	Me	Glu	Mdha
[L-Ser^7]MC-RR	1,041	+	Ala	Arg	Me	Arg	Me	Glu	Ser
[D-Asp^3, MeSer^7]MC-RR	1,041	+	Ala	Arg	H	Arg	Me	Glu	Mser
[D-Asp^3, ADMAdda^5, Dhb^7]MC-RR	1,052	+	Ala	Arg	H	Arg	COCH$_3$	Glu	Dhb
MC-YR	1,044	70	Ala	Tyr	Me	Arg	Me	Glu	Mdha
[Dha^7]MC-YR	1,030	+	Ala	Tyr	Me	Arg	Me	Glu	Dha
[D-Asp^3]MC-YR	1,030	+	Ala	Tyr	H	Arg	Me	Glu	Mdha
MC-LA	909	50	Ala	Leu	Me	Ala	Me	Glu	Mdha
MC-LF	985	+	Ala	Leu	Me	Phe	Me	Glu	Mdha
MC-LY	1,001	90	Ala	Leu	Me	Tyr	Me	Glu	Mdha
MC-LW	1,024	NR	Ala	Leu	Me	Trp	Me	Glu	Mdha
MC-WR	1,067	150–200	Ala	Trp	Me	Arg	Me	Glu	Mdha
MC-HilR	1,008	100	Ala	Hil	Me	Arg	Me	Glu	Mdha
MC-HtyR	1,058	80–100	Ala	Homo-tyr	Me	Arg	Me	Glu	Mdha
MC-LAba	923	NR	Ala	Leu	Me	Aba	Me	Glu	Mdha
MC-LL	951	+	Ala	Leu	Me	Leu	Me	Glu	Mdha
MC-AR	953	250	Ala	Ala	Me	Arg	Me	Glu	Mdha
MC-YA	959	NR	Ala	Tyr	Me	Ala	Me	Glu	Mdha
MC-VF	971	NR	Ala	–	Me	Phe	Me	Glu	Mdha
MC-FR	1,028	250	Ala	Phe	Me	Arg	Me	Glu	Mdha
[Dha^7]MC-FR	1,014	NR	Ala	Phe	Me	Arg	Me	Glu	Dha
[D-Asp^3, ADMAdda^5]MC-LHar	1,022	+	Ala	Leu	H	Homo-Arg	COCH$_3$	Glu	Mdha
[D-Asp^3, Dha^7]MC-E(OMe)E(OMe)	983	+	Ala	Glu(OMe)	H	Glu(OMe)	Me	Glu	Dha
[D-Asp^3, Dha^7]MC-EE(OMe)	969	+	Ala	Glu	H	Glu(OMe)	Me	Glu	Dha

Table 7.2 (continued)

Microcystin	MW (g/mol)	LD$_{50}$[a]	1	2	3/R²	4	5/R³	6	7/R¹
[Dha⁷]MC-HphR	1,028	+	Ala	Homo-Phe	Me	Arg	Me	Glu	Dha
MC-M(O)R	1,028	700–800	Ala	Met(O)	Me	Arg	Me	Glu	Mdha
MC-XR	–	NR	Ala	X	Me	Arg	Me	Glu	Mdha
MC-LZ	–	NR	Ala	Leu	Me	Z	Me	Glu	Mdha

[a]Toxicity determined intraperitoneal mouse (µg/kg); the LD$_{50}$ value is the dose of toxin that kills 50% of exposed animals; a " + " denotes a toxic result in a non-quantitative mouse bioassay or inhibition of protein phosphatase and "NR" denotes "Not reported."

Aba	Aminoisobutyric acid	E(OMe)	Glutamic acid methyl ester Δ
ADMAdda	O-Acetyl-O-demethyl-Adda	(H₄)Y	1,2,3,4,-Tetrahydrotyrosine
Dha	Dehydroalanine	Har	Homoarginine
Dhb	Dehydrobutyrine	Hil	Homoisoleucine
DMAdda	O-Demethyl-Adda	Hph	Homophenylalanine
Hty	Homotyrosine	MeSer	N-Methylserine
MeLan	N-Methyllanthionine	(6Z)-Adda	Stereoisomer of Adda at the Δ⁶ double bond
M(O)	Methionine-S-oxide	X, Z	Unknown amino acids

This table does not intend to be a full description of all microcystins described in the literature. Revision of more than 200 analogs of microcystins can be checked in Appendix 3, Tables of Microcystins and Nodularins of [41].

Amino acid variations in the structures were present at sites 1-2-3-4-5-6-7.

MCs are cyclic heptapeptides composed of protein and nonprotein amino acids with molecular masses around 1,000 Da (Figure 7.4). The seven amino acids that form the core structure of a MC include a unique β-amino acid (Adda, (2S,3S,8S,9S)-3-amino-9-methoxy-2,6,8-trimethyl-10-phenyldeca-4,6-dienoic acid)), as well as alanine (D-Ala), D-erythro-β-methylaspartic acid (D-MeAsp), glutamic acid (D-Glu), and N-methyl dehydroalanine (Mdha).

Structural variations have been reported in all seven amino acids, but the most frequently found are substitution of L-amino acids in the two variable positions 2 and 4, demethylation of amino acids at positions 3 (Masp) and/or 7 (Mdha), and methylesterification of D-Glu at position 6 (Figure 7.4). To date, more than 300 different microcystin variants have been discovered [42], see Table 7.2. The general structure consists of:

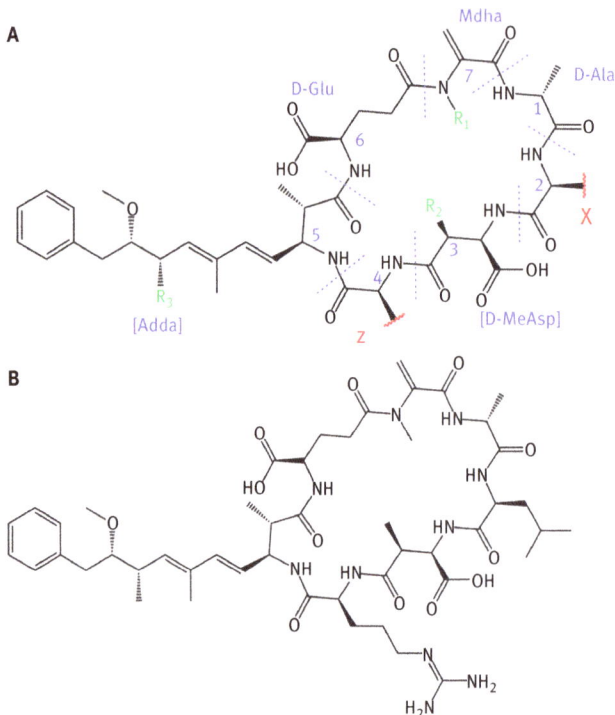

Figure 7.4: (A) General structure of microcystin. General numbering of residues is indicated. (B) 2D structure of microcystin-LR where **X** is L-Leu, **Z** is L-Arg, and R_1, R_2, R_3 are, respectively, methyl groups.

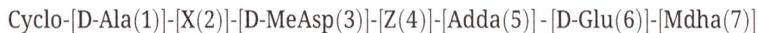

Cyclo-[D-Ala(1)]-[X(2)]-[D-MeAsp(3)]-[Z(4)]-[Adda(5)]-[D-Glu(6)]-[Mdha(7)]

where X and Z are the variable residues. The most common isoform is microcystin-LR, where X and Z are, respectively, leucine and arginine:

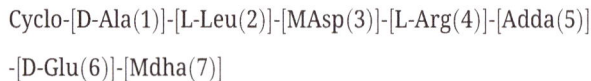

Cyclo-[D-Ala(1)]-[L-Leu(2)]-[MAsp(3)]-[L-Arg(4)]-[Adda(5)]

-[D-Glu(6)]-[Mdha(7)]

MCs are relatively polar molecules due to the presence of carboxylic acids at 3 and 6 positions and the frequent presence of arginine at positions 2 and 4 (see Figure 7.5 for surface polar regions of MC-RR). However, some variants contain hydrophobic nonpolar amino acids (Ala, Ile, and Val) as amino acid residues in the highly variable parts of the molecule. In MC-LF and MC-LW, the more hydrophobic phenylalanine (F) and tryptophan (W), respectively, have replaced arginine (R) in MC-LR (see Figure 7.5 to compare polarity). Depending on the structure, MCs are expected to have different in vivo toxicity and bioavailability, but only few studies have considered the toxic properties of the more hydrophobic variants.

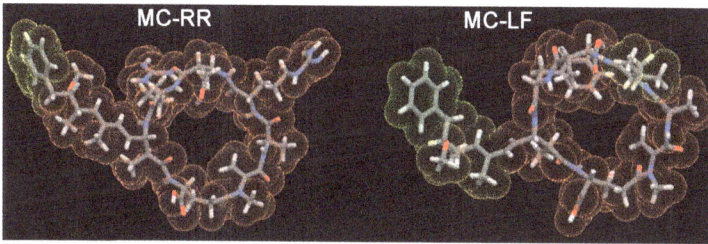

Figure 7.5: Molecular lipophilicity potential (MLP) surface of polar microcystin-RR and more hydrophobic microcystin-LF obtained by the online tool Molinspiration Cheminformatics software. Orange and red color represents molar polar region, while the more lipophilic regions are encoded by violet and blue colors. Although all MCs are soluble in water, the hydrophobicity varies considerably, which is primarily important for the uptake of the toxins by organisms and cells.

MCs are very stable compounds: spontaneous hydrolysis occurs at negligible rates, boiling at neutral pH does not lead to considerable decay for weeks, and their half-life time at pH 1 and 40 °C is around 3 weeks. Moreover, MCs were found to be resistant to enzymatic cleavage by common proteases like trypsins. Because MCs do not absorb UV light in the sunlight spectrum, no photolysis happens upon exposure of pure compounds. However, indirect photodegradation occurs in the presence of photosynthetic pigments or humic substances under natural conditions [22].

7.2.1.2 Mechanism of action and toxicity

Exposure to microcystin toxicity occurs usually through ingestion. After entering the bloodstream through the small intestine, MC-LR is transported to the liver via the portal blood system [43]. In the liver, it is carried across the cell membranes of hepatocytes using the multispecific bile acid transport system facilitated by organic anion-transporting polypeptides (OATPs). In mice, the most commonly involved OATP is OATP1B2, whereas in humans the orthologues, OATP1B1 and OATP1B3, play a major role in this process [44]. These transporting proteins are more abundant in the liver because they are responsible for the transmembrane transport of bile acids and hormones in hepatocytes, but they are also expressed in the gastrointestinal tract, kidney, and brain; therefore, these are also vulnerable organs as well, although the main organ target is the liver [17, 45, 46].

In the cell, MC-LR initiates cellular toxicity by inhibiting protein phosphatases 1 (PP1) and 2A (PP2A), leading to excessive phosphorylation [44]. This abnormal phosphorylation mediated by MCs disrupts the cytoskeleton, leading to cellular disintegration, deformation of hepatocytes, and ultimately cell death and organ failure [47]. In addition, the cellular alterations that MCs cause may provoke genomic instability. It has been demonstrated that the inhibitory action of microcystin-LR on ser/thr protein phosphatases leads to the production of reactive oxygen substances inducing DNA damage with genotoxic effects. Moreover, this toxin also interferes with DNA damage repair pathways, contributing to their carcinogenicity. In fact, the International Agency for Research on Cancer (IARC, a branch of the World Health Organization (WHO)) has included Microcystin-LR in Group 2B, as a possible carcinogenic substance to humans [48].

Cases of acute poisoning due to MC exposure have been reported in both humans and animals [49–51]. In humans, symptoms of acute poisoning range from liver damage to gastrointestinal distress and fever. The most important and serious case of MC intoxication occurred at a hemodialysis clinic in Caruaru, Pernambuco State (Brazil), in February 1996, where 116 patients were intoxicated and experienced visual disturbances, nausea, vomiting, and muscle weakness after routine dialysis, and 52 died of a syndrome now called Caruaru syndrome [52–54]. Since then, several human and animal poisonings due to MCs have been described, with gastroenteritis being the major clinical manifestation.

To conclude, as shown in Table 7.2, MC-LR and MC-LA are the most toxic congeners of the group, with a LD_{50} of 50 µg/kg, determined by intraperitoneal injection in mice.

7.2.2 Nodularins

7.2.2.1 Origin and chemical structure

Nodularins (NODs), which are structurally similar to microcystins, are also hepatotoxins. They are mainly synthesized by *Nodularia spumigena* [55]. However, the production of certain variants has also been attributed to another cyanobacterial species, *Nodularia sphaerocarpa*, and the sponge *Theonella swinhoei* [56]. Nevertheless, the origin of its synthesis in this sponge remains controversial, with suggestions that it may be produced by symbiotic cyanobacteria [57].

This cyanotoxin class has a global distribution, although cases of its occurrence have predominantly been documented in specific regions such as New Zealand, the Baltic Sea, and tropical areas including Australia and South Africa. Notably, its prevalence is relatively low compared to other cyanotoxins [38, 58].

Structurally, NODs are cyclic pentapeptides composed of protein and nonprotein amino acids with molecular masses around 800 Da (Figure 7.6). The five amino acids that are involved in the NOD structure include: Adda, D-glutamic acid (D-Glu), *N*-methyldehydrobutyrine (MeDhb), D-erythro-β-methylaspartic acid (D-MeAsp), and L-arginine (L-Arg).

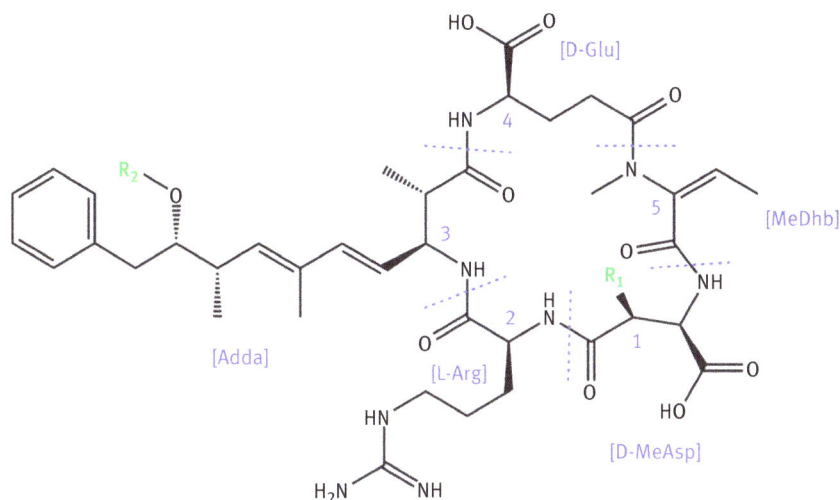

Figure 7.6: Structure of nodularin (NOD), $R_1 = R_2 = CH_3$. General numbering of residues is indicated. The L-Arg residue of NOD may be replaced with a homoarginine (nodularin-Har) or valine residue (motuporin). The molecular weight of NOD variants ranges between 760 and 840 Da.

The structure of NOD is

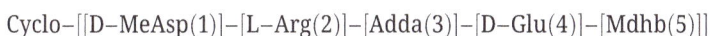

$$\text{Cyclo}-[[\text{D}-\text{MeAsp}(1)]-[\text{L}-\text{Arg}(2)]-[\text{Adda}(3)]-[\text{D}-\text{Glu}(4)]-[\text{Mdhb}(5)]]$$

Structural variations have been reported in all five amino acids, two of these variants have alterations within the Adda residue, which reduces or abolishes the toxicity of the compound. The D-Glu residue is essential for the toxicity of NOD, so the esterification of the free carboxyl abolishes toxicity. On the contrary, the substitution at position 1 has little effect on toxicity. The other two isoforms, nodularin-Har and motuporin, vary at position 2 (Figure 7.6). Over 10 naturally occurring analogs of NOD have been reported to date versus more than 200 structural variants of MCs (see Table 7.3). NODs, similarly to MCs, are chemically very stable.

Table 7.3: Structural variants of NOD reported in the scientific literature. Amino acid variations in the structures were present at sites 1-2-3-4-5.

Nodularin	MW/g / mol	LD$_{50}$[a]	1/R^1	2	3/R^2	4	5
Nodularin (NOD-R)	825	50	Me	Arg	Me	Glu	Mdhb
Motuporin (NOD-V) ([l-Val^2]NOD)	768	NR	Ala	Val	Me	Glu	Mdhb
[d-Asp^1]NOD	811	NR	H	Arg	Me	Glu	Mdhb
[DMAdda^3]NOD	811	NR	Me	Arg	H	Glu	Mdhb
[dhb^5]NOD	811	NR	Me	Arg	Me	Glu	Dhb
[6(Z)-Adda^3]NOD	825	nontoxic	Me	Arg	Me	Glu	Mdhb

Table 7.3 (continued)

Nodularin	MW/g / mol	LD$_{50}$[a]	1/R^1	2	3/R^2	4	5
[D-Glu(OCH$_3$)^4]NOD	839		NR Me	Arg	Me	Glu-OMe	Mdhb
[l-Har^2]NOD	839		NR Me	Har	Me	Glu	Mdhb
[MeAdda^3]NOD	839		NR Me	Arg	Me + extraMe	Glu	Mdhb
Linear nodularin	843	Adda-d-Glu(γ)-Mdhb-d-MeAsp(β)-l-Arg-OH					

[a]The mean dose that kills 50% of the animals.

7.2.2.2 Mechanism of action and toxicity

After ingestion, nodularins, like microcystins, are absorbed from the ileum and enter the bloodstream. Liver uptake of the toxin is facilitated by a non-specific organic anion transporter, such as bile acid transporters [59], as described for MCs. The amino acid Adda has been identified as the component responsible for the toxic activity, and their mechanism of action involves inhibition of ser/thr protein phosphatases PP1, PP2A, and PP3.

The damage caused to the liver includes disruption of cytoskeletal structures, oxidation of lipids, impaired integrity of cell membranes, fragmentation of DNA strands, formation of cellular blebs, induction of programmed cell death (apoptosis), impairment of cellular functions, tissue necrosis, and internal hemorrhage within the liver [60]. This cascade of adverse effects can potentially culminate in death from hemorrhagic shock [60]. Since nodularin mechanism of action is similar to MCs, the LD$_{50}$ measured in mice is similar as well (see Table 7.3).

In addition to their hepatotoxic properties, nodularins also are considered capable of inducing and promoting tumors. This makes them a potential source of carcinogenic risk [61]. However, the precise mechanisms underlying their hepatotoxic and carcinogenic properties are not fully understood. Several theories suggest that they are involved in inhibiting protein phosphatases or triggering the formation of reactive oxygen species, ultimately leading to DNA damage [55, 61].

Nodularins may have been involved in the earliest documented case of cyanotoxin poisoning, reported by George Francis in 1878 at Lake Alexandrina in South Australia. This incident involved a bloom of cyanobacteria with the characteristic features of *Nodularia spumigena*, which may have led to the deaths of sheep, horses, dogs, and pigs [62]. Over time, various cases of acute intoxication in animals have been recorded, with symptoms varying depending on the species [63, 64]. However, there are no confirmed cases of human poisoning directly attributed to nodularin.

7.2.3 Anatoxin-a and homoanatoxin-a

7.2.3.1 Origin and chemical structure

Anatoxins were discovered as a result of their involvement, during the 1960s decade, in the poisoning of cattle during a massive bloom of *Anabaena flos-aquae* in Saskatchewan Lake, Canada [65]. Later, the production of anatoxins was linked to other genera of cyanobacteria, namely *Aphanizomenon, Planktothrix, Oscillatoria,* or *Phormidium* [66, 67]. Anatoxin-a (ATXa) is present globally, accounting for 9% of cyanotoxin detections [38].

ATXa is a low-molecular weight alkaloid ($C_{10}H_{15}NO$, MW = 165 Da), a secondary amine, 2-acetyl-9-azabicyclo(4-2-1)non-2-ene (Figure 7.7). Anatoxin-a is found in nature as the (+)-anatoxin-a enantiomer, whereas much research has been done to investigate the synthesis of both (+)- and (−)-anatoxin-a enantiomers [68]. Commercially available products are usually a racemic mixture.

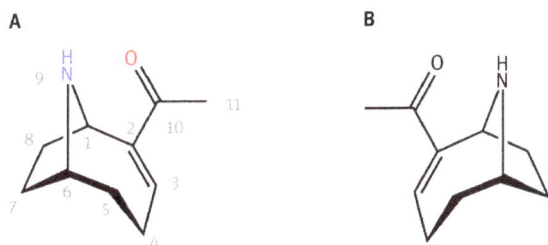

Figure 7.7: Structure of (A) natural (+)-anatoxin-a and structure of (B) synthetic (−)-anatoxin-a.

Homoanatoxin-a ($C_{11}H_{17}NO$, MW = 179) is an anatoxin-a homolog with a propionyl group at C-2 instead of the acetyl group in anatoxin-a. Both are highly soluble in water. Anatoxin and homoanatoxin are unstable and are converted into epoxy and dihydro degradation products, processes that are dependent on several environmental parameters, especially pH and light (see Figure 7.8) [69].

Although these reaction products have reduced toxicity or are nontoxic, the detection of dihydroanatoxin-a, epoxyanatoxin-a, dihydrohomoanatoxin-a, and/or epoxyhomoanatoxin-a during routine screening would be useful in alerting the presence of lower, but still toxic, levels of parent toxins, or in forensic investigations to determine the previous presence of parent toxins.

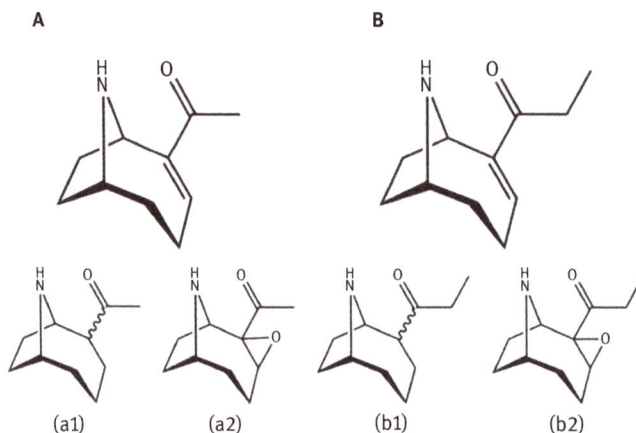

Figure 7.8: Structure of (A) (+)-anatoxin-a, (B) homoanatoxin-a, and degradation products: a1) dihydroanatoxin-a, a2) epoxyanatoxin-a, b1) dihydrohomoanatoxin-a, and b2) epoxyhomoanatoxin-a.

7.2.3.2 Mechanism of action and toxicity

Anatoxins are primarily neurotoxic substances, exerting their toxicity at the cellular level by binding irreversibly to nicotinic acetylcholine receptors, which causes a continuous nerve and muscle depolarization, resulting in the blockage of further impulse transmission. The evidenced clinical signs of neurotoxicity include loss of coordination, cyanosis, convulsions, and death from respiratory paralysis [70]. Ataxia, dizziness and blurred vision were described in humans [133]. Many case reports of wildlife, dogs, and livestock poisoning have been related to anatoxins since its discovery.

Regarding their toxicological potency, the LD_{50} in mice of anatoxin-a administrated intraperitoneally is 200 µg/kg, whereas homoanatoxin-a possesses one-tenth the toxicity of anatoxin-a [71].

7.2.4 Guanitoxin

7.2.4.1 Origin and chemical structure

Guanitoxin is a unique phosphate ester of a cyclic *N*-hydroxyguanine (MW = 252) (Figure 7.9). The production of this toxic compound has been attributed to *Anabaena flos-aquae* [72], although little information is available on this toxin. This scarcity may be due to the similarity of poisoning symptoms caused by guanitoxin and anatoxin-a, which can lead to miss guanitoxin implications in toxic blooms.

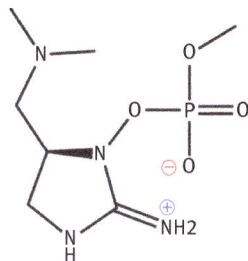

Figure 7.9: Structure of guanitoxin.

7.2.4.2 Mechanism of action and toxicity

Guanitoxin causes neurological effects such as paralysis, tremors, and convulsions [73], similar to ATX-a. However, guanitoxin induces intense salivation and lacrimation in mice, which distinguishes it from anatoxin-a [74]. Owed to this excessive salivation, guanitoxin was previously known by the name of anatoxin-a(s).

Its mechanism of action is similar to organophosphate and carbamate insecticides, acting through irreversible inhibition of acetylcholinesterase [75]. The toxicity of guanitoxin remains relatively unclear and understudied. One study suggests that the LD_{50} of this toxin administered intraperitoneally to mice in a single dose is 20 µg/kg, more potent than anatoxin-a [76].

7.2.5 Saxitoxins

7.2.5.1 Origin and chemical structure

Saxitoxin (STX) receives its name from the mollusk, *Saxidomus giganteus*, from which it was first isolated and identified. Although the production of these toxins by marine dinoflagellate species belonging to the genus *Alexandrium, Gymnodinium*, and *Pyrodinium* was initially published, synthesis by cyanobacteria also occurs. The main cyanobacterial organisms related to their production are *Anabaena, Aphanizomenon, Lyngbya, Planktothrix*, or *Cylindrospermopsis* [70].

STX is the parent molecule in a class of compounds, collectively termed as paralytic shellfish poisons (PSPs). Structurally, PSPs are a group of carbamate alkaloids sharing a common trialkyl tetrahydropurine tricyclic skeleton with two guanidinium moieties, which are responsible for their high polarity (Figure 7.10) [77].

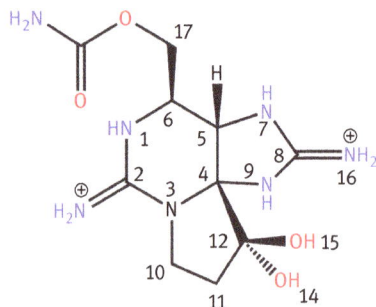

Figure 7.10: Chemical structure of saxitoxin. It possesses two pK_as of 8.22 and 11.28, which belong to the 7,8,9- and 1,2,3-guanidinium groups, respectively [56].

Variations in functional moieties at five defined positions around the ring categorize these molecules into subgroups according to the functional group at the R4 position (Table 7.4):

- Carbamate toxins have a carbamoyl moiety.
- Decarbamoyl toxins only have a hydroxyl group.
- N-Sulfocarbamate toxins possess an N-sulfocarbamoyl.
- Deoxydecarbamoyl toxins with no functional group.
- Hydroxybenzoate toxins are 4-hydroxybenzoate ester derivatives.
- Acetate toxins are acetate ester derivatives.

Table 7.4: Saxitoxins and analogues reported from cyanobacterial strains and bloom samples.

Toxin	Variable chemical groups in toxins					Cyanobacteria				TEF[e]
	R_1	R_2	R_3	R_4	R_5	Aph[a]	Ana[b]	Lyn[c]	Cyl[d]	
STX	H	H	H	$OCONH_2$	OH	+	+		+	1.0
GTX2	H	H	OSO_3^-	$OCONH_2$	OH		+			0.4
GTX3	H	OSO_3^-	H	$OCONH_2$	OH		+			0.6
GTX5	H	H	H	$OCONHSO_3^-$	OH		+			0.1
C1	H	H	OSO_3^-	$OCONHSO_3^-$	OH		+			–
C2	H	OSO_3^-	H	$OCONHSO_3^-$	OH		+			0.1
NEO	OH	H	H	$OCONH_2$	OH	+			+	1.0
GTX1	OH	H	OSO_3^-	$OCONH_2$	OH		*			1.0
GTX4	OH	OSO_3^-	H	$OCONH_2$	OH		*			0.7
GTX6	OH	H	H	$OCONHSO_3^-$	OH		*			0.1
dcSTX	H	H	H	H	OH		+	+		1.0
dcGTX2	H	H	OSO_3^-	H	OH		+	+		0.2
dcGTX3	H	OSO_3^-	H	H	OH		+	+		0.4
LWTX1	H	OSO_3^-	H	$OCOCH_3$	H			+		–
LWTX2	H	OSO_3^-	H	$OCOCH_3$	OH			+		–
LWTX3	H	H	OSO_3^-	$OCOCH_3$	OH			+		–
LWTX4	H	H	H	H	H			+		–

Table 7.4 (continued)

Toxin	Variable chemical groups in toxins					Cyanobacteria				TEF[e]
	R_1	R_2	R_3	R_4	R_5	Aph[a]	Ana[b]	Lyn[c]	Cyl[d]	
LWTX5	H	H	H	$OCOCH_3$	OH			+		–
LWTX6	H	H	H	$OCOCH_3$	H			+		–

STX, saxitoxin; GTX, gonyautoxins; C, C-toxins; dcSTX, decarbamoylsaxitoxin; LWTX, Lyngbya-wollei-toxins.
[a]Toxins found in *Aphanizomenon flos-aquae*, New Hampshire, USA [60, 61].
[b]Toxins reported in an *Anabaena circinalis* strain and bloom samples, Australia [62–64]. dcGTX2 and dcGTX3 are probably break down products of C1 and C2 in this species [65]. An asterisk in this column denotes toxins reported by Humpage et al. [62] for *Anabaena circinalis* based on retention time data, but not confirmed by mass spectrometry, and not found in subsequent studies.
[c]Toxins detected in *Lyngbya wollei*, USA [66].
[d]Toxins thus far found in *Cylindrospermopsis raciborskii*, Brazil [67].
[e]Toxicity equivalency factor expresses the toxicity of PSPs in terms of the saxitoxin dihydrochloride form recommended by the European Food Safety Authority (EFSA).

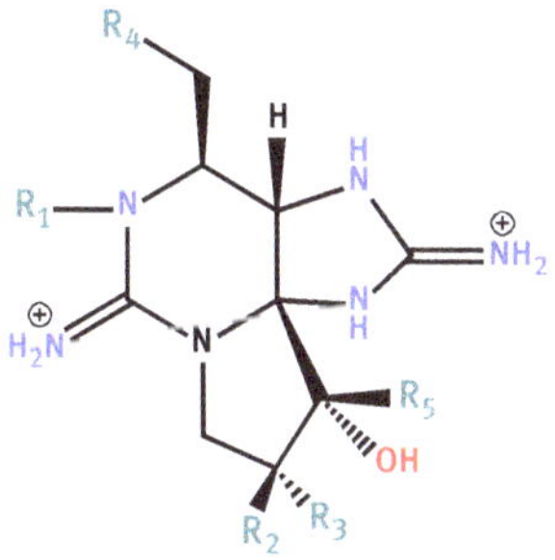

Other variations are N-oxidation at N-1 and sulfate substitution at C-11. The latter results in α/β epimeric pairs, that is, C1/C2, GTX1/GTX4, GTX2/GTX3, and dcGTX2/dcGTX3, which are in equilibrium through keto-enol taumerism. Both variations result in a decrease in toxicity relative to STX except GTX1, which exhibits toxicity comparable to that of STX. The toxicity of the derivatives varies by approximately two orders of magnitude, with STX being the most toxic, followed by neosaxitoxin and gonyautoxins 1 and 3. Different values have been reported in the literature; as these values are dependent in part upon the purity of the compounds, it is likely these differences are simply a result of differences in the purities of the toxin preparations.

Currently, PSPs include more than 50 components [78–87]. Due to structural differences, each analogue has a slightly different affinity to the binding to target sites, and thus may justify variations in individual toxic potency.

Lyngbya-wollei-toxins (LWTXs) do not appear to be as toxic as other STXs, the presence of acetate in the side chain resulted in a 7-fold to 17-fold decrease in mouse toxicity compared to their carbamoyl counterparts, while the reduction at C-12 resulted in a complete loss of mouse toxicity [66]. However, given their structural similarity to STX, the abiotic or biotic transformation of LWTXs to other more toxic analogs cannot be ruled out without additional bioassays. This type of transformation has been observed for N-sulfocarbamoyl STXs (C1 and C2) that can be converted to the more toxic decarbamoylgonyautoxins [80].

7.2.5.2 Mechanism of action and toxicity

This group of toxins binds selectively and reversibly to the extracellular side of the voltage-dependent Na^+ channels present in excitable cells, more specifically, to site 1. Although STX could also act on other targets, such as Ca^{2+} and K^+ channels. Binding to voltage-dependent Na^+ channels prevents the entry of Na^+ ions into the cell triggered by depolarization, which inhibits the transmission of nervous impulses in the peripheral nerves and skeletal muscle, causing the interruption of a wide variety of cellular functions, including muscle functions and leading to death by respiratory arrest [88, 89].

The main poisoning produced by STXs is known as paralytic shellfish poisoning (PSP), caused by these neurotoxins affecting the peripheral nervous system. Symptoms start with tingling, burning, and numbness of the lips and fingertips, which spread across the face and neck to the extremities, and continue with loss of muscle coordination, respiratory distress, and, in the most severe cases, total paralysis, which may lead to death due to respiratory muscle paralysis [17]. Each analogue has a different toxicological potency, expressed in the toxicity equivalency factor (TEF) in Table 7.4.

7.2.6 Cylindrospermopsin

7.2.6.1 Origin and chemical structure

Cylindrospermopsin (CYN) was identified for the first time in 1979 after the hospitalization of 148 people with symptoms of hepato-enteritis in Isla Palmera, Australia, after a HAB in the local drinking water supply where the predominant species was *Cylindrospermopsis raciborskii* [35, 90]. Currently, other genera of cyanobacteria are known to produce CYN: *Anabaena, Aphanizomenon, Lyngbya, Raphidiopsis, Umezakia*, and *Oscillatoria* [17]. Toxic blooms of CYN-producing genera are most commonly found in tropical, subtropical, and arid zone water bodies, and have recently been found in Australia, Europe, Israel, Japan, and the USA [38]. It is considered the second most commonly identified group of cyanotoxins after microcystins, accounting for 10% of identified cases [38].

CYN is a polyketide-derived alkaloid with a central functional guanidino moiety and a hydroxymethyluracil attached to the tricyclic carbon skeleton with a molecular weight of 415 Da (Figure 7.11). The compound is zwitterionic and highly water soluble. CYN is stable to extreme temperatures (no degradation at 100 °C for 15 min) and pH [91].

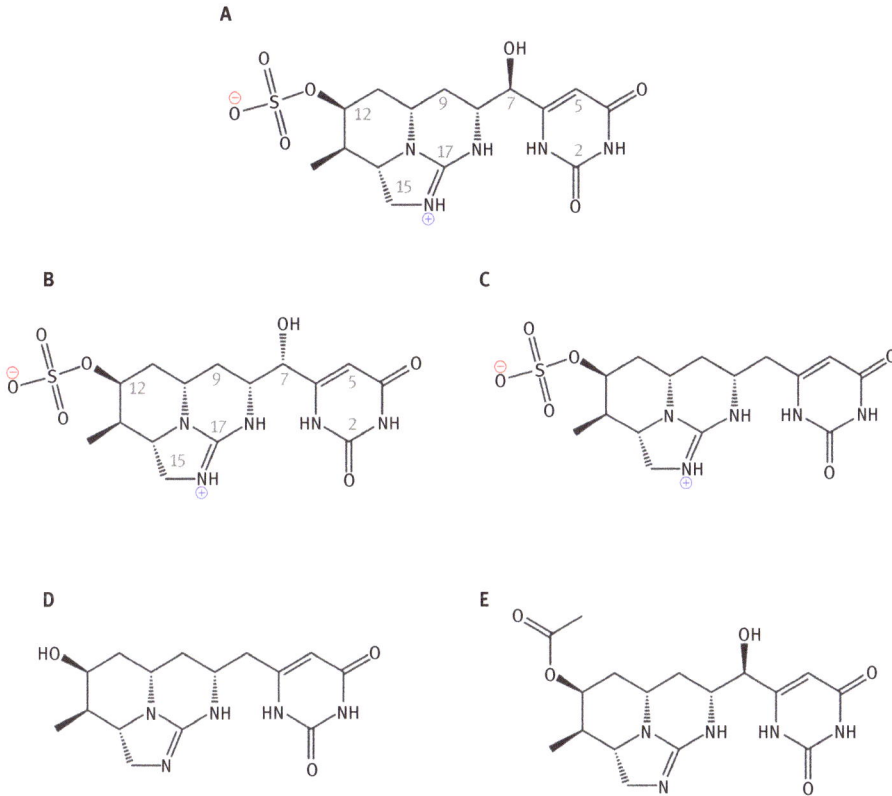

Figure 7.11: Structure of (A) cylindrospermopsin, (B) 7-epicylindrospermopsin, (C) deoxy-cylindrospermopsin, (D) 7-deoxy-desulfo-cylindrospermopsin, and (E) 7-deoxy-desulfo-12-acetylcylindrospermopsin.

Four naturally occurring analogs of CYN are 7-epicylindrospermopsin (7-Epi-CYN), deoxy-cylindrospermopsin (7-deoxy-CYN), 7-deoxy-desulfo-cylindrospermopsin, and 7-deoxy-desulfo-12-acetylcylindrospermopsin [92, 93]. Based on their structural features, it is likely that these new analogs also possess the harmful biological activities displayed by the representative molecule of the CYN family.

The uracil moiety as well as the hydroxyl at C7 are crucial for toxicity, making CYNs very important because of their potent toxicity as well as potential chronic effects at exposure levels below the acute toxicity threshold.

7.2.6.2 Mechanism of action and toxicity

CYN has important toxic effects on the liver, but it also affects the kidneys, adrenals, lungs, heart, spleen, and thymus. CYN mechanism of action has not been fully elucidated, with inhibition of protein synthesis, inhibition of glutathione synthesis leading to oxidative stress, and activation of cytochrome P450 being proposed as relevant for toxicity of this compound [94, 95].

The first clinical signs of poisoning include liver and kidney failure. In fact, it can cause general cytotoxicity and, depending on the injured cells, provoke hepatitis, renal malfunction, gastroenteritis, and hemorrhage from blood vessels. Finally, CYN is also genotoxic and can cause chromosome loss and DNA strand breakage, so it has been suggested to be a potential carcinogen [70].

About its toxic potency, when CYN is administered to mice by intraperitoneal injection, an estimated 24-hour LD_{50} of 2,100 µg/kg was obtained [96]. However, with prolonged exposure, the LD_{50} for mice decreases to 200 µg/kg body weight over 5–6 days [96]. This suggests that the toxicity of CYN escalates with prolonged exposure.

7.2.7 Aplysiatoxins and lyngbyatoxins

7.2.7.1 Origin and chemical structure

Aplysiatoxins and lyngbyatoxins were originally isolated from the *Aplysiidae* sea hare *Stylocheilus longicauda* [97], although the cyanobacteria *Lyngbya majuscule* was later reported to be the producing organism [98–100]. Since then, they have been isolated from other cyanobacteria, such as *Schizothrix* and *Oscillatoria* [101].

The chemical structures of these molecules are shown in Figure 7.12.

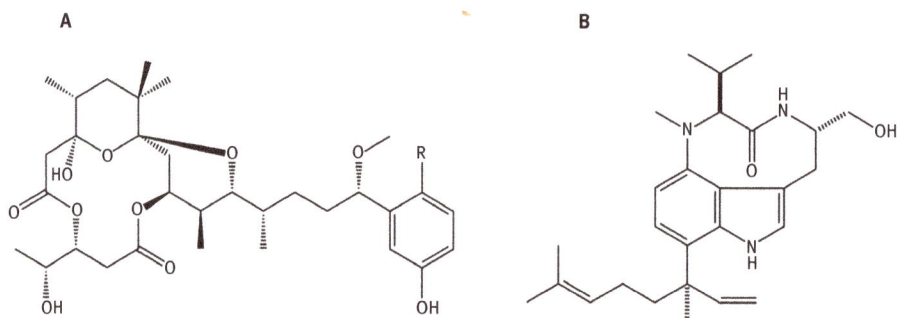

Figure 7.12: (A) Structure of aplysiatoxin (R = Br) and debromoaplysiatoxin (R = H). (B) Structure of lyngbyatoxin-a.

7.2.7.2 Mechanism of action and toxicity

Aplysiatoxins and lyngbyatoxins are responsible for severe cases of contact dermatitis, causing the "swimmers itch" syndrome due to their inflammatory activity [98]. Moreover, it has been proved that these toxins have tumor-promoting activity through the activation of protein kinase C, in a similar way as 12-O-tetradecanoylphorbol-13-acetate (TPA), binding to the phorbol receptor on the cell membrane [102].

7.2.8 Lipopolysaccharides

7.2.8.1 Origin and chemical structure

LPSs are common irritants that usually form part of the outer membrane of the cell wall in Gram-negative bacteria. Therefore, they are also found in cyanobacteria, although cyanobacterial LPSs usually exert less toxic potency [30]. LPSs, as the name implies, are condensed products of a sugar, usually a hexose, and a lipid, normally a hydroxy C14-C18 fatty acid.

7.2.8.2 Mechanism of action and toxicity

These substances contribute to the membrane stability and integrity of bacteria, but they also act as a protection system. They have been described to activate the innate immune response, binding and activating the Toll-like receptor 4 (TLR-4), therefore triggering the proinflammatory cytokine cascade [103].

It is generally the fatty acid component of the LPS molecule that elicits an irritant or allergic response in humans and mammals. Though not as potent as other cyanotoxins, some researchers have claimed that all LPSs in cyanobacteria can irritate the skin, while other researchers doubt the toxic effects are common.

The effects provoked by LPSs include gastrointestinal symptoms, fever, headache, cutaneous signs, allergy, or respiratory disease, contributing to skin lesions and gastrointestinal incidents in humans exposed to cyanobacteria-contaminated freshwaters [104].

7.2.9 BMAA

7.2.9.1 Origin and chemical structure

Most groups of cyanobacteria, including cyanobacterial symbionts of the genus *Nostoc*, and free-living cyanobacteria from the five cyanobacterial subsections may produce the neurotoxic amino acid BMAA [105]. Though there are other organisms as diatoms that can produce this neurotoxin as well [106].

BMAA is a nonproteinogenic amino acid ubiquitously produced by cyanobacteria in marine, freshwater, brackish, and terrestrial environments. The chemical structure is shown in Figure 7.13.

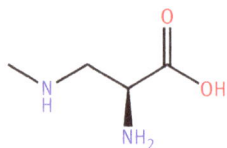

Figure 7.13: Structure of BMAA.

7.2.9.2 Mechanism of action and toxicity

This cyanotoxin acts as a glutamate agonist at AMPA, kainate, and NMDA receptors, increasing the intracellular concentration of Ca^{2+} and inducing neuronal activity by hyperexcitation. In addition, it has been pointed out that BMAA could mimic the amino acid L-serine, and cause misfolded proteins in the brain after chronic exposure, causing neurological illness [107]. The neurotoxicity of BMAA was first suggested during the 1960s [108], when this compound was proposed to lead to amyotrophic lateral sclerosis or Parkinsonism dementia in Guam, due to the consumption of cycad flour (*Cycas micronesica*) [109, 110].

This compound is being investigated for its potential as an environmental risk factor for neurodegenerative diseases, including amyotrophic lateral sclerosis, Parkinson's disease, and Alzheimer's disease [111]. The relationship between BMAA exposure and the onset of neurodegenerative disease is difficult to demonstrate, and despite increasing evidence, more studies are needed to establish the implication of this toxin in these pathologies.

7.3 Detection methods

Several methodologies have been described for the analysis of cyanotoxins in water that are based on different principles. The use of one or another methodology is linked to the use to be made of the analytical results [112, 113]. Table 7.5 summarizes the main characteristics of the methodologies used in the cyanotoxins analysis. Table 7.6 shows different approaches detecting the most relevant cyanotoxins in freshwater.

Developing analytical methods to detect cyanotoxins reliably and routinely in natural matrices must be based on three practical criteria: (a) reliability (accurate and precise), (b) the speed and consequently also to a large degree the cost of the analysis, and (c) the possibility to conduct an "on-site" analysis. Based on the above-mentioned criteria, techniques for analysis can be divided into *screening methods* (mouse bioassay, ELISA, receptor assays, and qPCR) and *confirmatory methods* (allow an unambiguous identification of the analyte): instrumental methods such as chromatography (GC/MS, HPLC-UV/PDA, HPLC-MS/MS).

Table 7.5: ☑ Advantages and ☒ disadvantages of principal analytical methods to detect cyanotoxins.

Analytical method	Performance
Mouse bioassay (MBA)	☑ Measurement of real toxicity. Biological test system used to directly demonstrate the presence of unknown toxins. ☒ Not specific. ☒ Conflict with Animal Welfare Directive. Death of animals is method endpoint. Therefore, it is not respectful of animal welfare. ☒ Test cannot be validated. Large variation in results between laboratories. Sex, weight, and strain of mice influence the test results.
Receptor-based assay (RBA)	☑ The fastest detection method possible with high sample throughput. ☑ Specific, sensitive, and reproducible. ☑ Measurement of real toxicity. ☑ Good comparability with chromatographic methods and mouse bioassay. ☒ May overestimate the toxin concentration. Not specific to each analog, and may indicate the presence of other substances interfering with receptor. ☒ Does not show the same sensitivity for all cyanotoxin variants.
LC-UV, FLD LC-MS/MS	☑ Rapid and accurate identification and quantitative determination of toxins: suitable for detection of all cyanotoxins. ☑ More sensitive than mouse bioassay. ☑ High selectivity, suitable for monitoring as a preventive measure. ☑ Specific. ☒ Reference materials not available for all toxins. ☒ Indirect measurement of toxin analogues: concentrations calculated by reference to a standard corresponding to the parental compound. In UV, absorption reflects overall concentration as toxin equivalents. ☒ NO measurement of real toxicity. In general, structurally unrelated, unknown toxins cannot be detected. ☒ Expensive and requires highly skilled analysts. ☒ No standard methods, expensive, requires complex data interpretation, time-consuming.
ELISA	☑ Relatively inexpensive. ☑ Kits available, relatively easy to use. ☑ Fast and sensitive. ☒ Indirect measurement of the toxin. ☒ Irregular cross-reactivity between congeners. ELISA assay might overestimate or underestimate the amount of cyanotoxins present in the sample resulting in both false positives and false negatives. ☒ NO measurement of real toxicity. Total results reported as "Toxin equivalents" irrespective of the congeners present.
Molecular methods	☑ Time efficiency. ☑ Comparability of results for toxic versus toxin-producing cells. ☑ Specific – no gene = no toxin. ☒ Irregular false-positive and false-negative results.

Table 7.6: Typical detection methods for cyanotoxins.

	ATXs	CYNs	MCs	NODs	STXs
Biological assays					
MBA	Y	Y	Y	Y	Y
RBA	N	N	Y	Y	N
Neurochemical	Y	N	N	N	Y
ELISA	Y	Y	Y	Y	Y
Molecular methods (nontoxin methods)					
qPCR	Y	Y	Y	Y	Y
DNA microarray chips	N	Y	Y	Y	?
Chromatographic methods					
GC/MS	Y	N	N	N	N
LC/UV	Y	Y	Y	Y	Y
LC/FLD	Y	N	N	N	Y
LC/MS	Y	Y	Y	Y	Y
LC-MS/MS	Y	Y	Y	Y	Y

Y, implemented method; N, no implemented method; ?, not confirmed.

The first group of methods mainly aimed at the qualitative detection of toxins as fast as possible, preferably already in the field, that is, have a "yes" or "no" answer.

The methods in the second group are predominantly used to confirm a positive result by means of a rapid method and/or to provide a more accurate quantitative result ("how much of the toxin is actually in there?").

As a premise, the screening methods must be sufficiently sensitive to avoid false negative results, in addition, they must be simple and fast, and allow the analysis of a large number of samples. While, the main disadvantage of instrumental methods is that they require the existence of certified reference material for each molecule to be unequivocally identified and this is not always possible, especially in the case of MCs (there are more than 100 identified variants). Often more than one toxin may be present in a sample. The consensus among those using analytical methods is that a single method will not suffice. The best approach for monitoring is to use a combination of screening and more sophisticated and cost-specific chemical methods.

In any case, it is clear that both strategies are perfectly complementary, allowing an initial screening of the samples by a rapid method based on the toxigenic potential, that is, in a detection proportional to the toxicity of the molecule and not to its structural properties. The samples whose results have been positive for the biological screening test can be later analyzed for the identification of the specific toxin responsible for water contamination by an instrumental method. In some cases, positive samples in the screening stage may not be confirmed with the instrumental method due to the aforementioned problem of the absence of reference patterns for many of the cyanotoxins discovered so far [114].

Both methods have in common that they require a prior sample preparation step. Cyanotoxins can be found intracellularly and must be released prior to their analysis by an appropriate procedure. Thus, sample preparation may need to include sonication (to break up cells) and a variety of extraction procedures in order to isolate the different (i.e., more lipophilic or polar) compounds. In the scientific literature, there are many protocols for lysis and extraction of toxins. So far, published studies on the levels of cyanotoxins in water supplies have generally not indicated whether total or free toxins were measured. Moreover, many of them have not been perfectly evaluated in terms of their capacity to recover and therefore their application can generate an underestimation of the toxin content that can put the health of the water at risk. In addition, the extraction method must be compatible with the analysis technique used.

On the other hand, the methods for the detection of cyanotoxins can also be classified as *structural methods, functional methods,* and *molecular methods.*

The first group is based on the physicochemical properties of the molecule and among them are HPLC [114] and immunochemical methods. Functional methods use the biological target on which the toxin acts, and therefore, they are an indication of the toxigenic potential of the molecule [35, 115]. The latter includes quantitative polymerase chain reaction (qPCR) which simultaneously quantifies total cyanobacteria along with genes responsible for toxin production. Positive detections indicate that the gene is available in the bloom material and may produce toxins (the test does not indicate that the cells are actively producing the toxin). Negative detections indicate that the gene for a particular toxin is not in the bloom material (captured by the sample) and likely not produced.

7.4 Health aspects: guidelines and legislation

Toxic cyanobacteria are encountered around the world, and problems related to safe drinking water production are common, presenting an expanding global threat to Public and Ecosystem Health [116]. The presence of cyanobacterial toxins in drinking and bathing waters has been recognized as a human health hazard by the WHO and a provisional guideline value (GV) for the common hepatotoxin, microcystin-LR in drinking water has been established [117]. National legislation has been recently introduced in some European countries and elsewhere to control MC levels [30, 118, 119].

MC-LR is the most toxic and most frequently found derivative of MCs in water resources. Health-related episodes in humans and animals caused by MC-LR contamination have been reported in several countries, including the United States, Australia, China, Great Britain, and Brazil.

Sufficient oral toxicity data on cyanotoxins in mammals to permit GVs [120, 121] to be derived with some confidence for human drinking water are only available for microcystin-LR [25] and, more recently, for CYN [122] (Table 7.7) but are insufficient for the derivation of concentration limits. However, it is recommended that the total MCs as gravimetric or molar equivalents are evaluated against the GVs [123], and that the CYN analogs 7-deoxy-CYN and 7-epi-CYN are included in calculations of total CYN [124]. Although GVs for ATX-a cannot be derived yet, a provisional health-based reference value has been suggested for short-term exposure [125].

After the first human fatal incident occurred in Brazil in 1996, the WHO set the provisional microcystin-LR GV in potable water to 1 µg/L [25, 126]. The WHO has also established the tolerable daily intake to 0.04 µg/kg day since cyanobacterial toxins bio-accumulate in aquatic microorganisms that humans consume and because of their use as dietary supplements. A similar GV for CYN in drinking water may be appropriate [122]. Since microcystin-LR appears to be one of the most toxic of known cyanotoxins to mammals, and adequate GVs for other cyanotoxins are currently lacking, it seems prudent to apply the microcystin-LR GV to other cyanotoxins until further data are available. Inevitably, the significance of cyanotoxin data is being interpreted in terms of the microcystin-LR GV. However, the interpretation of such data for risk management is a developing practice. Several countries throughout the world have already adopted or adapted the WHO GV for microcystin-LR into national water legislation. Others prefer to use the GV for guidance only, taking into account that safety factors are built into GV derivation [25, 33]. It is not appropriate here to favor the regulatory or guidance approach, but rather to emphasize that the purpose and scope of a GV for drinking water [126] needs to be recognized. Thus for a cyanotoxin in drinking water, the GV is:

- an estimate of the concentration of cyanotoxin which would not result in a significant risk to a consumer over a lifetime of drinking water consumption;
- advisory;
- derived to accommodate uncertainties and safety factors in its derivation;
- provisional and subject to revision in response to further advances in basic knowledge and practical experience;
- not intended as a recommended concentration to which cyanotoxin-containing water can be allowed to degrade; and
- a tool for use in the development and application of cyanotoxin risk management approaches, taking into account practicality, feasibility, and the protection of health and water resources.

Legislation governing the occurrence and concentrations of specific cyanotoxins for health protection is not only limited to MCs but also to STXs. Regulations governing STX concentrations were first introduced for health protection from eating shellfish potentially contaminated with STXs. The EU specifies that shellfish should not contain over 80 µg STX per 100 g of mussel meat [127]. However, these neurotoxins are known to be produced by several cyanobacteria genera, and the GV has been retrieved from poisoning events caused by mixtures of STXs, considering the STX concentration equivalents and therefore GV applies to the total STXs in the sample [128].

Table 7.7: Guideline values (GVs) and provisional health-based reference values (RFs) for cyanotoxins in drinking water according to WHO (2020) [123–125], recommending that the GVs are applied to total MCs, total CYNs, and total STXs as gravimetric or molar equivalents.

Compound	Value (µg/L)	Observations
MC-LR	1	GV
CYN	0.7	GV
ATX-a	30	RV
STX	3	GV

For STX toxicity equivalents, see WHO 2020 [128].

It is therefore required to regulate limits of cyanotoxins in the environment, and efforts are being made to develop sensitive methods. Furthermore, sensitive methods using liquid chromatography and tandem mass spectrometry (LC-MS/MS) have been standardized by the International Organization for Standardization (ISO) and the U.S. Environmental Protection Agency (EPA) [129–132].

Keywords: BMAA, anatoxin, aplysiatoxin, blue-green algae, cyanobacteria, cyanotoxin, cylindrospermopsin, freshwater toxins, guideline values, legislation, lipopolysaccharides, lyngbyatoxin, microcystin, nodularin, PSP, saxitoxin

Abbreviations: Adda, (2S,3S,8S,9S)-3-amino-9-methoxy-2,6,8-trimethyl-10-phenyldeca -4,6-dienoic acid; BMAA, β-N-methylamino-L-alanine; Ala, alanine; ATX-a, anatoxin-a; Arg, arginine; CYNs, cylindrospermopsins; GC, gas chromatography; Glu, glutamic acid; GV, guideline value; HABs, harmful algal blooms; HPLC, high-performance liquid chromatography; Har, homoarginin; IARC, International Agency for Research on Cancer; Leu, leucine; LPS, lipopolysaccharide; LWTXs, Lyngbya-wollei toxins; MS, mass spectrometry; Mdha, methyl dehydroalanine; MeAsp, methylaspartic acid; MeDhb, methyldehydrobutyrine; MCs, microcystins; MLP, molecular lipophilicity potential; MBA, mouse bioassay; NODs, nodularins; PSPs, paralytic shellfish poisons; PDA, photodiode array; PP, protein phosphatase; RBA, receptor-based assay; STXs, saxitoxins; Tyr, tyrosine; UV, ultraviolet; US EPA, US Environmental Protection Agency; WHO, World Health Organization

Acknowledgments: This research, leading to these results, has received funding from the following grants: Ministerio de Ciencia e Innovación PID 2020-11262RB-C21, IISCIII/PI19/001248, Grant CPP2021-008447 funded by MCIN/AEI/10.13039/501100011033 and by the European Union NextGenerationEU/PRT; Campus Terra (USC), BreveRiesgo (2022-PU011) CLIMIGAL (2022-PU016); Conselleria de Cultura, Educacion e Ordenación Universitaria, Xunta de Galicia, GRC (ED431C 2021/01); and European Union, Interreg EAPA-0032/2022 – BEAP-MAR, HORIZON-MSCA-2022-DN-01-MSCA Doctoral Networks 2022 101119901-BIOTOXDoc, and HORIZON-CL6-2023-CIRCBIO-01 COMBO-101135438.

References

[1] Dextro RB, Delbaje E, Cotta SR, Zehr JP, Fiore MF. Trends in free-access genomic data accelerate advances in cyanobacteria taxonomy. J Phycol. 2021;57(5):1392–402.

[2] Ruggiero MA, Gordon DP, Orrell TM, Bailly N, Bourgoin T, Brusca RC, et al. A higher level classification of all living organisms. PloS One. 2015;10(4):e0119248.

[3] Whittaker RH. New Concepts of Kingdoms of Organisms: Evolutionary relations are better represented by new classifications than by the traditional two kingdoms. Science. 1969;163(3863):150–60.

[4] Woese CR, Fox GE. Phylogenetic structure of the prokaryotic domain: The primary kingdoms. Proc Natl Acad Sci. 1977;74(11):5088–90.

[5] Woese CR, Kandler O, Wheelis ML. Towards a natural system of organisms: Proposal for the domains Archaea, Bacteria, and Eucarya. Proc Natl Acad Sci. 1990;87(12):4576–79.

[6] Schirrmeister BE, Gugger M, Donoghue PC. Cyanobacteria and the Great Oxidation Event: Evidence from genes and fossils. Palaeontology. 2015;58(5):769–85.

[7] Sagan L. On the origin of mitosing cells. J Theor Biol. 1967;14(3):225–IN6.

[8] Ochoa de Alda JA, Esteban R, Diago ML, Houmard J. The plastid ancestor originated among one of the major cyanobacterial lineages. Nat Commun. 2014;5(1):4937.

[9] Castenholz RW, Wilmotte A, Herdman M, Rippka R, Waterbury JB, Iteman I, et al. Phylum BX. cyanobacteria. Bergey's manual® of systematic bacteriology. Springer, New York, 2001;473–599.

[10] Garrity GM, Bell JA, Lilburn TG. Taxonomic outline of the prokaryotes. Bergey's manual of systematic bacteriology. New York, Berlin, Heidelberg: Springer, 2004.

[11] Rippka R, Deruelles J, Waterbury JB, Herdman M, Stanier RY. Generic assignments, strain histories and properties of pure cultures of cyanobacteria. Microbiology. 1979;111(1):1–61.

[12] Rippka R, Herdman M, eds. Division patterns and cellular differentiation in cyanobacteria. Annales de l'Institut Pasteur/Microbiologie. Elsevier Masson. 1985;136(1):33–39.

[13] Waterbury JB, Stanier RY. Patterns of growth and development in pleurocapsalean cyanobacteria. Microbiol Rev. 1978;42(1):2–44.

[14] Adams DG, Duggan PS. Tansley Review No. 107. Heterocyst and akinete differentiation in cyanobacteria. New Phytol. 1999;144(1):3–33.

[15] Takai Aah K. Freshwater hepatoxins: Ecobiology and classification. In: Botana LM, ed. Seafood and freshwater toxins: Pharmacology, physiology, and detection. New York: CRC Press (Marcel Dekker, Inc.), 2000;715–26.

[16] Palinska KA, Surosz W. Taxonomy of cyanobacteria: A contribution to consensus approach. Hydrobiologia. 2014;740:1–11.

[17] Pearson L, Mihali T, Moffitt M, Kellmann R, Neilan B. On the chemistry, toxicology and genetics of the cyanobacterial toxins, microcystin, nodularin, saxitoxin and cylindrospermopsin. Mar Drugs. 2010;8(5):1650–80.

[18] Rastogi RP, Madamwar D, Incharoensakdi A. Bloom dynamics of cyanobacteria and their toxins: Environmental health impacts and mitigation strategies. Front Microbiol. 2015;6:1254.

[19] Caruso G, Caruso G, Laganà PL, Santi Delia A, Parisi S, Barone C, et al. Biological toxins from marine and freshwater microalgae. In: Microbial toxins and related contamination in the food industry. Springer, Cham. 2015;13–55.

[20] Cheung MY, Liang S, Lee J. Toxin-producing cyanobacteria in freshwater: A review of the problems, impact on drinking water safety, and efforts for protecting public health. J Microbiol. 2013;51:1–10.

[21] Agency EUSEP. Algal toxin risk assessment and management strategic plan for drinking water. Strategy Submitted to Congress to Meet the Requirements of P.L., 2015:114–45.

[22] Botana LM. Seafood and freshwater toxins: Pharmacology, physiology, and detection. Crc Press, Boca Raton (USA). 2014.

[23] Paerl HW, Paul VJ. Climate change: Links to global expansion of harmful cyanobacteria. Water Res. 2012;46(5):1349–63.

[24] Mulvenna V, Dale K, Priestly B, Mueller U, Humpage A, Shaw G, et al. Health risk assessment for cyanobacterial toxins in seafood. Int J Environ Res Public Health. 2012;9(3):807–20.

[25] Zhang W, Liu J, Xiao Y, Zhang Y, Yu Y, Zheng Z, et al. The impact of cyanobacteria blooms on the aquatic environment and human health. Toxins. 2022;14(10):658.

[26] Ku CS, Yang Y, Park Y, Lee J. Health benefits of blue-green algae: Prevention of cardiovascular disease and nonalcoholic fatty liver disease. J Med Food. 2013;16(2):103–11.

[27] Moreira C, Vasconcelos V, Antunes A. Cyanobacterial blooms: Current knowledge and new perspectives. Earth. 2022;3(1):127–35.

[28] Holland A, Kinnear S. Interpreting the possible ecological role (s) of cyanotoxins: Compounds for competitive advantage and/or physiological aide? Mar Drugs. 2013;11(7):2239–58.

[29] Janssen EM-L. Cyanobacterial peptides beyond microcystins – A review on co-occurrence, toxicity, and challenges for risk assessment. Water Res. 2019;151:488–99.

[30] Chorus I, Welker M. Toxic cyanobacteria in water: A guide to their public health consequences, monitoring and management. Taylor & Francis, Boca Raton (USA). 2021.

[31] WHO. Guidelines for safe recreational water environments: Coastal and fresh waters. World Health Organization, 2003.

[32] Weralupitiya C, Wanigatunge RP, Gunawardana D, Vithanage M, Magana-Arachchi D. Cyanotoxins uptake and accumulation in crops: Phytotoxicity and implications on human health. Toxicon. 2022;211:21–35.

[33] Codd GA, Morrison LF, Metcalf JS. Cyanobacterial toxins: Risk management for health protection. Toxicol Appl Pharmacol. 2005;203(3):264–72.

[34] Miles CO, Sandvik M, Nonga HE, Rundberget T, Wilkins AL, Rise F, et al. Identification of microcystins in a Lake Victoria cyanobacterial bloom using LC–MS with thiol derivatization. Toxicon. 2013;70:21–31.

[35] Sivonen K, Jones G. Cyanobacterial toxins. In: Toxic cyanobacteria in water: A guide to their public health consequences, monitoring and management. Elsevier Inc., London. 1999;1:43–112.

[36] Westrick JA, Szlag DC, Southwell BJ, Sinclair J. A review of cyanobacteria and cyanotoxins removal/ inactivation in drinking water treatment. Anal Bioanal Chem. 2010;397:1705–14.

[37] Mowe MA, Mitrovic SM, Lim RP, Furey A, Yeo DC. Tropical cyanobacterial blooms: A review of prevalence, problem taxa, toxins and influencing environmental factors. J Limnol. 2015; 74/2):205–224.

[38] Svirčev Z, Lalić D, Bojadžija Savić G, Tokodi N, Drobac Backović D, Chen L, et al. Global geographical and historical overview of cyanotoxin distribution and cyanobacterial poisonings. Arch Toxicol. 2019;93:2429–81.

[39] Catherine A, Bernard C, Spoof L, Bruno M. Microcystins and nodularins. In: Handbook of cyanobacterial monitoring and cyanotoxin analysis (eds Meriluoto J, Spoof L and G.A. Codd). John Wiley & Sons, Ltd, Toronto. 2016;107–26.

[40] Miller TR, Beversdorf LJ, Weirich CA, Bartlett SL. Cyanobacterial toxins of the Laurentian Great Lakes, their toxicological effects, and numerical limits in drinking water. Mar Drugs. 2017;15(6):160.

[41] Meriluoto J, Spoof L, Codd GA. Handbook of cyanobacterial monitoring and cyanotoxin analysis. John Wiley & Sons, Toronto. 2017.

[42] Baliu-Rodriguez D, Peraino NJ, Premathilaka SH, Birbeck JA, Baliu-Rodriguez T, Westrick JA, et al. Identification of novel microcystins using high-resolution MS and MS n with Python Code. Environ Sci Technol. 2022;56(3):1652–63.

[43] Svirčev Z, Lujić J, Marinović Z, Drobac D, Tokodi N, Stojiljković B, et al. Toxicopathology induced by microcystins and nodularin: A histopathological review. J Environ Sci Health, Part C. 2015;33(2):125–67.

[44] Lu H, Choudhuri S, Ogura K, Csanaky IL, Lei X, Cheng X, et al. Characterization of organic anion transporting polypeptide 1b2-null mice: Essential role in hepatic uptake/toxicity of phalloidin and microcystin-LR. Toxicol Sci. 2008;103(1):35–45.

[45] MacKintosh C, Beattie KA, Klumpp S, Cohen P, Codd GA. Cyanobacterial microcystin-LR is a potent and specific inhibitor of protein phosphatases 1 and 2A from both mammals and higher plants. FEBS Lett. 1990;264(2):187–92.

[46] Runnegar M, Berndt N, Kong S-M, Lee EY, Zhang L. In vivo and in vitro binding of microcystin to protein phosphatase 1 and 2A. Biochem Biophys Res Commun. 1995;216(1):162–69.

[47] Zhou M, Tu W-w, Xu J. Mechanisms of microcystin-LR-induced cytoskeletal disruption in animal cells. Toxicon. 2015;101:92–100.

[48] IARC Working Group on the Evaluation of Carcinogenic Risks to Humans. Ingested nitrate and nitrite, and cyanobacterial peptide toxins. IARC monographs on the evaluation of carcinogenic risks to humans, Lyon (France). 2010;94:v–vii, 1–412

[49] Foss AJ, Aubel MT, Gallagher B, Mettee N, Miller A, Fogelson SB. Diagnosing microcystin intoxication of canines: Clinicopathological indications, pathological characteristics, and analytical detection in postmortem and antemortem samples. Toxins. 2019;11(8):456.

[50] Menezes C, Nova R, Vale M, Azevdo J, Vasconcelos V, Pinto C. First description of an outbreak of cattle intoxication by cyanobacteria (blue-green algae) in the South of Portugal. Bov Pract. 2019;53(1):66–70.

[51] Mez K, Beattie KA, Codd GA, Hanselmann K, Hauser B, Naegeli H, et al. Identification of a microcystin in benthic cyanobacteria linked to cattle deaths on alpine pastures in Switzerland. Euro J Phycol. 1997;32(2):111–17.

[52] Azevedo SM, Carmichael WW, Jochimsen EM, Rinehart KL, Lau S, Shaw GR, et al. Human intoxication by microcystins during renal dialysis treatment in Caruaru – Brazil. Toxicology. 2002;181:441–46.

[53] Pouria S, de Andrade A, Barbosa J, Cavalcanti R, Barreto V, Ward C, et al. Fatal microcystin intoxication in haemodialysis unit in Caruaru, Brazil. Lancet. 1998;352(9121):21–26.

[54] Yuan M, Carmichael WW, Hilborn ED. Microcystin analysis in human sera and liver from human fatalities in Caruaru, Brazil 1996. Toxicon. 2006;48(6):627–40.

[55] Vankova D, Pasheva M, Kiselova-Kaneva Y, Ivanov D, Ivanova D. Mechanisms of cyanotoxin toxicity – Carcinogenicity, anticancer potential, and clinical toxicology. In: Medical toxicology (eds Erkekoglu P and Ogawa T). IntechOpen, 2019;38.

[56] De Silva ED, Williams DE, Andersen RJ, Klix H, Holmes CF, Allen TM. Motuporin, a potent protein phosphatase inhibitor isolated from the Papua New Guinea sponge Theonella swinhoei Gray. Tetrahedron Lett. 1992;33(12):1561–64.

[57] Huang I-S, Zimba PV. Cyanobacterial bioactive metabolites – A review of their chemistry and biology. Harmful Algae. 2019;86:139–209.

[58] Mashile PP, Mashile GP, Dimpe KM, Nomngongo PN. Occurrence, quantification, and adsorptive removal of nodularin in seawater, wastewater and river water. Toxicon. 2020;180:18–27.

[59] Ufelmann H, Krüger T, Luckas B, Schrenk D. Human and rat hepatocyte toxicity and protein phosphatase 1 and 2A inhibitory activity of naturally occurring desmethyl-microcystins and nodularins. Toxicology. 2012;293(1–3):59–67.

[60] Gácsi M, Antal O, Vasas G, Máthé C, Borbély G, Saker ML, et al. Comparative study of cyanotoxins affecting cytoskeletal and chromatin structures in CHO-K1 cells. Toxicol In Vitro. 2009;23(4):710–18.

[61] Žegura B, Štraser A, Filipič M. Genotoxicity and potential carcinogenicity of cyanobacterial toxins – A review. Mutat Res/Rev Mutat Res. 2011;727(1–2):16–41.

[62] Francis G. Poisonous australian lake. Nature. 1878;18(444):11–12.

[63] Algermissen D, Mischke R, Seehusen F, Göbel J, Beineke A. Lymphoid depletion in two dogs with nodularin intoxication. Vet Record-English Edition. 2011;169(1):15.

[64] Simola O, Wiberg M, Jokela J, Wahlsten M, Sivonen K, Syrjä P. Pathologic findings and toxin identification in cyanobacterial (Nodularia spumigena) intoxication in a dog. Vet Pathol. 2012;49(5):755–59.

[65] Dillenberg H, Dehnel M. Toxic waterbloom in Saskatchewan, 1959. Can Med Assoc J. 1960;83 (22):1151.

[66] Devlin J, Edwards O, Gorham P, Hunter N, Pike R, Stavric B. Anatoxin-a, a toxic alkaloid from Anabaena flos-aquae NRC-44h. Can J Chem. 1977;55(8):1367–71.

[67] Méjean A, Paci G, Gautier V, Ploux O. Biosynthesis of anatoxin-a and analogues (anatoxins) in cyanobacteria. Toxicon. 2014;91:15–22.

[68] Petersen JS, Fels G, Rapoport H. Chirospecific syntheses of (+)-and (-)-anatoxin a. J Am Chem Soc. 1984;106(16):4539–47.

[69] James KJ, Furey A, Sherlock IR, Stack MA, Twohig M, Caudwell FB, et al. Sensitive determination of anatoxin-a, homoanatoxin-a and their degradation products by liquid chromatography with fluorimetric detection. J Chromatogr A. 1998;798(1–2):147–57.

[70] Van Apeldoorn ME, Van Egmond HP, Speijers GJ, Bakker GJ. Toxins of cyanobacteria. Mol Nutr Food Res. 2007;51(1):7–60.

[71] Fawell J, Mitchell R, Hill R, Everett D. The toxicity of cyanobacterial toxins in the mouse: II anatoxin-a. Hum Exp Toxicol. 1999;18(3):168–73.

[72] Carmichael WW, Gorham PR. Anatoxins from clones of Anabaena flos-aquae isolated from lakes of western Canada: With 3 figures and 2 tables in the text. Internationale Vereinigung Für Theoretische Und Angewandte Limnologie: Mitteilungen. 1978;21(1):285–95.

[73] Carmichael WW. Health effects of toxin-producing cyanobacteria:"The CyanoHABs". Human and ecological risk assessment. Int J. 2001;7(5):1393–407.

[74] Matsunaga S, Moore RE, Niemczura WP, Carmichael WW. Anatoxin-a (s), a potent anticholinesterase from Anabaena flos-aquae. J Am Chem Soc. 1989;111(20):8021–23.

[75] Metcalf JS, Bruno M. Anatoxin-a (S). Handbook of cyanobacterial monitoring and cyanotoxin analysis. John Wiley and Sons, New Jersey. 2016;155–59.

[76] Carmichael WW, Azevedo S, An JS, Molica R, Jochimsen EM, Lau S, et al. Human fatalities from cyanobacteria: Chemical and biological evidence for cyanotoxins. Environ Health Perspect. 2001;109(7):663–68.

[77] Rogers RS, Rapoport H. The pKa's of saxitoxin. J Am Chem Soc. 1980;102(24):7335–39.

[78] Humpage A, Rositano J, Bretag A, Brown R, Baker P, Nicholson B, et al. Paralytic shellfish poisons from Australian cyanobacterial blooms. Marine Freshwater Res. 1994;45(5):761–71.

[79] Ikawa M, Wegener K, Foxall TL, Sasner JJ Jr. Comparison of the toxins of the blue-green alga Aphanizomenon flos-aquae with the Gonyaulax toxins. Toxicon. 1982;20(4):747–52.

[80] Jones GJ, Negri AP. Persistence and degradation of cyanobacterial paralytic shellfish poisons (PSPs) in freshwaters. Water Res. 1997;31(3):525–33.

[81]   Lagos N, Onodera H, Zagatto PA, Andrinolo D, Azevedo SM, Oshima Y. The first evidence of paralytic shellfish toxins in the freshwater cyanobacterium Cylindrospermopsis raciborskii, isolated from Brazil. Toxicon. 1999;37(10):1359–73.

[82]   Mahmood NA, Carmichael WW. Paralytic shellfish poisons produced by the freshwater cyanobacterium Aphanizomenon flos-aquae NH-5. Toxicon. 1986;24(2):175–86.

[83]   Negri AP, Jones GJ. Bioaccumulation of paralytic shellfish poisoning (PSP) toxins from the cyanobacterium Anabaena circinalis by the freshwater mussel Alathyria condola. Toxicon. 1995;33(5):667–78.

[84]   Negri AP, Jones GJ, Blackburn SI, Oshima Y, Onodera H. Effect of culture and bloom development and of sample storage on paralytic shellfish poisons in the cyanobacterium Anabaena circinalis. J Phycol. 1997;33(1):26–35.

[85]   Onodera H, Satake M, Oshima Y, Yasumoto T, Carmichael WW. New saxitoxin analogues from the freshwater filamentous cyanobacterium Lyngbya wollei. Nat Toxins. 1997;5(4):146–51.

[86]   Rey V, Botana AM, Alvarez M, Antelo A, Botana LM. Liquid chromatography with a fluorimetric detection method for analysis of paralytic shellfish toxins and tetrodotoxin based on a porous graphitic carbon column. Toxins. 2016;8(7):196.

[87]   Wiese M, D'Agostino PM, Mihali TK, Moffitt MC, Neilan BA. Neurotoxic alkaloids: Saxitoxin and its analogs. Mar Drugs. 2010;8(7):2185–211.

[88]   Cusick KD, Sayler GS. An overview on the marine neurotoxin, saxitoxin: Genetics, molecular targets, methods of detection and ecological functions. Mar Drugs. 2013;11(4):991–1018.

[89]   Farabegoli F, Blanco L, Rodríguez LP, Vieites JM, Cabado AG. Phycotoxins in marine shellfish: Origin, occurrence and effects on humans. Mar Drugs. 2018;16(6):188.

[90]   Bourke A, Hawes R, Neilson A, Stallman N. An outbreak of hepato-enteritis (the Palm Island mystery disease) possibly caused by algal intoxication. Toxicon. 1983;21:45–48.

[91]   Chiswell RK, Shaw GR, Eaglesham G, Smith MJ, Norris RL, Seawright AA, et al. Stability of cylindrospermopsin, the toxin from the cyanobacterium, Cylindrospermopsis raciborskii: Effect of pH, temperature, and sunlight on decomposition. Environ Toxicol: Int J. 1999;14(1):155–61.

[92]   Kinnear S. Cylindrospermopsin: A decade of progress on bioaccumulation research. Mar Drugs. 2010;8(3):542–64.

[93]   Wimmer KM, Strangman WK, Wright JL. 7-Deoxy-desulfo-cylindrospermopsin and 7-deoxy-desulfo-12-acetylcylindrospermopsin: Two new cylindrospermopsin analogs isolated from a Thai strain of Cylindrospermopsis raciborskii. Harmful Algae. 2014;37:203–06.

[94]   Norris R, Seawright A, Shaw G, Senogles P, Eaglesham G, Smith M, et al. Hepatic xenobiotic metabolism of cylindrospermopsin in vivo in the mouse. Toxicon. 2002;40(4):471–76.

[95]   Runnegar MT, Kong S-M, Zhong Y-Z, Ge J-L, Lu SC. The role of glutathione in the toxicity of a novel cyanobacterial alkaloid cylindrospermopsin in cultured rat hepatocytes. Biochem Biophys Res Commun. 1994;201(1):235–41.

[96]   Ohtani I, Moore RE, Runnegar MT. Cylindrospermopsin: A potent hepatotoxin from the blue-green alga Cylindrospermopsis raciborskii. J Am Chem Soc. 1992;114(20):7941–42.

[97]   Kato Y, Scheuer PJ. Aplysiatoxin and debromoaplysiatoxin, constituents of the marine mollusk Stylocheilus longicauda. J Am Chem Soc. 1974;96(7):2245–46.

[98]   Cardellina JH, Marner F-J, Moore RE. Seaweed dermatitis: Structure of lyngbyatoxin A. Science. 1979;204(4389):193–95.

[99]   Chlipala GE, Tri PH, Hung NV, Krunic A, Shim SH, Soejarto DD, et al. Nhatrangins A and B, aplysiatoxin-related metabolites from the marine cyanobacterium Lyngbya majuscula from Vietnam. J Nat Prod. 2010;73(4):784–87.

[100] Mynderse JS, Moore RE, Kashiwagi M, Norton TR. Antileukemia activity in the Oscillatoriaceae: Isolation of debromoaplysiatoxin from Lyngbya. Science. 1977;196(4289):538–40.

[101] Moore RE, Blackman AJ, Cheuk CE, Mynderse JS, Matsumoto GK, Clardy J, et al. Absolute stereochemistries of the aplysiatoxins and oscillatoxin A. J Org Chem. 1984;49(13):2484–89.

[102] Fujiki H, Mori M, Nakayasu M, Terada M, Sugimura T, Moore RE. Indole alkaloids: Dihydroteleocidin B, teleocidin, and lyngbyatoxin A as members of a new class of tumor promoters. Proc Natl Acad Sci. 1981;78(6):3872–76.

[103] Gemma S, Molteni M, Rossetti C. Lipopolysaccharides in cyanobacteria: A brief overview. Adv Microbiol. 2016;6(5):391–97.

[104] Stewart I, Schluter PJ, Shaw GR. Cyanobacterial lipopolysaccharides and human health – A review. Environ Health. 2006;5(1):1–23.

[105] Cox PA, Banack SA, Murch SJ, Rasmussen U, Tien G, Bidigare RR, et al. Diverse taxa of cyanobacteria produce β-N-methylamino-L-alanine, a neurotoxic amino acid. Proc Natl Acad Sci. 2005;102 (14):5074–78.

[106] Violi JP, Facey JA, Mitrovic SM, Colville A, Rodgers KJ. Production of β-methylamino-L-alanine (BMAA) and its isomers by freshwater diatoms. Toxins. 2019;11(9):512.

[107] Dunlop RA, Cox PA, Banack SA, Rodgers KJ. The non-protein amino acid BMAA is misincorporated into human proteins in place of L-serine causing protein misfolding and aggregation. PloS One. 2013;8(9):e75376.

[108] Vega A, Bell E. α-Amino-β-methylaminopropionic acid, a new amino acid from seeds of Cycas circinalis. Phytochemistry. 1967;6(5):759–62.

[109] Duncan MW, Steele JC, Kopin IJ, Markey SP. 2-Amino-3-(methylamino)-propanoic acid (BMAA) in cycad flour: An unlikely cause of amyotrophic lateral sclerosis and parkinsonism-dementia of Guam. Neurology. 1990;40(5):767-.

[110] Glover WB, Liberto CM, McNeil WS, Banack SA, Shipley PR, Murch SJ. Reactivity of β-methylamino-L-alanine in complex sample matrixes complicating detection and quantification by mass spectrometry. Anal Chem. 2012;84(18):7946–53.

[111] Ra D, Sa B, Sl B, Js M, Sj M, DA D, et al. Is exposure to BMAA a risk factor for neurodegenerative diseases? A response to a critical review of the BMAA hypothesis. Neurotox Res. 2021;39:81–106.

[112] Kaushik R, Balasubramanian R. Methods and approaches used for detection of cyanotoxins in environmental samples: A review. Crit Rev Environ Sci Technol. 2013;43(13):1349–83.

[113] Zhang C. Current techniques for detecting and monitoring algal toxins and causative harmful algal blooms. Journal of Environmental Analytical Chemistry, 2015;2(1).

[114] Trifirò G, Barbaro E, Gambaro A, Vita V, Clausi MT, Franchino C, et al. Quantitative determination by screening ELISA and HPLC-MS/MS of microcystins LR, LY, LA, YR, RR, LF, LW, and nodularin in the water of Occhito lake and crops. Anal Bioanal Chem. 2016;408:7699–708.

[115] Harada K-I, Murata H, Qiang Z, Suzuki M, Kondo F. Mass spectrometric screening method for microcystins in cyanobacteria. Toxicon. 1996;34(6):701–10.

[116] Kaloudis T, Hiskia A, Triantis TM. Cyanotoxins in bloom: Ever-increasing occurrence and global distribution of freshwater cyanotoxins from Planktic and Benthic cyanobacteria. Toxins (MDPI). 2022;14:264.

[117] Gordon BC, Vickers P, C. Guidelines for drinking-water quality. WHO Chron. 2011;38(4):104–08.

[118] Antoniou MG, De La Cruz AA, Dionysiou DD. Cyanotoxins: New generation of water contaminants. Journal of Environmental Engineering, Reston (USA). 2005;131(9):1239–1243.

[119] Gupta S. Cyanobacterial toxins: microcystin-LR. Guidelines for Drinking-water Quality, Addendum to. 1998;2:95–110.

[120] Ibelings BW, Backer LC, Kardinaal WEA, Chorus I. Current approaches to cyanotoxin risk assessment and risk management around the globe. Harmful Algae. 2014;40:63–74.

[121] Metcalf J, Meriluoto J, Codd G. Legal and security requirements for the air transportation of cyanotoxins and toxigenic cyanobacterial cells for legitimate research and analytical purposes. Toxicol Lett. 2006;163(2):85–90.

[122] Humpage A, Falconer I. Oral toxicity of the cyanobacterial toxin cylindrospermopsin in male Swiss albino mice: Determination of no observed adverse effect level for deriving a drinking water guideline value. Environ Toxicol: Int J. 2003;18(2):94–103.

[123] WHO. Cyanobacterial toxins: Microcystins. World Health Organization, 2020.

[124] WHO. Cyanobacterial toxins: Cylindrospermopsins. World Health Organization, 2020.

[125] WHO. Cyanobacterial toxins: Anatoxin-a and analogues. World Health Organization, 2020.

[126] WHO. Guidelines for drinking-water quality. World Health Organization, 2004.

[127] Council Directive 91/492/EEC of 15 July 1991 laying down the health conditions for the production and the placing on the market of live bivalve molluscs, 1991.

[128] WHO. Cyanobacterial toxins: Saxitoxins. World Health Organization, 2020.

[129] EPA. Method 544 – Determination of microcystins and nodularin in drinking water by solid phase extraction and liquid chromatography/tandem mass spectrometry (LC/MS/MS). Cincinnati, OH, USA: National Exposure Research Laboratory, Office of Research and Development, 2015.

[130] EPA. Method 545 – Determination of Cylindrospermopsin and Anatoxin-a in Drinking Water by Liquid Chromatography Electrospray Ionization Tandem Mass Spectrometry (LC/ESI-MS/MS). Cincinnati, OH, USA: National Exposure Research Laboratory, Office of Research and Development, 2015.

[131] EPA. Method 546 – Determination of total microcystins and nodularins in drinking water and ambient water by adda enzyme-linked immunosorbent assay. Standards and Risk Management Division. Cincinnati, OH, USA: Technical Support Center, Office of Ground Water and Drinking Water, 2016.

[132] International Organization for Standardization. Water Quality – Determination of Microcystins – Method using liquid chromatography and tandem mass spectrometry (LC-MS/MS). ISO, 2021.

[133] Biré R, Bertin T, Dom I, Hort V, Schmitt C, Diogène J, et al. First evidence of the presence of anatoxin-A in sea figs associated with human food poisonings in France. Marine Drugs. 2020;18(6):285.

María J. Sainz, Jesús M. González-Jartín, Olga Aguín, Vanesa Ferreiroa,
Nadia Pérez-Fuentes, J. Pedro Mansilla, and Luis M. Botana

8 Isolation, characterization, and identification of mycotoxin-producing fungi

8.1 Introduction

Filamentous fungi produce a wide diversity of compounds with biological activity. Some of these compounds may be toxic to humans and other vertebrates (other mammals, poultry, and fish) and are called mycotoxins [1, 2]. The term mycotoxin is restricted to toxic low-molecular-weight metabolites produced by species of microscopic filamentous fungi (commonly known as molds) belonging to several genera in the phylum Ascomycota, primarily *Fusarium, Aspergillus*, and *Penicillium*, which can colonize plants cultivated for human and animal consumption [3]. Some species of Basidiomycota, such as *Amanita phalloides, A. muscaria, Clytocybe fragrans, Cortinarius orellanus*, and *Paxillus involutus*, and Ascomycota, as *Gyromitra esculenta*, that form mushrooms, also produce toxic metabolites in their fruiting bodies, but they are not considered mycotoxins. The distinction between a mycotoxin and a mushroom toxin is mainly based on human intention: mushrooms poisons are usually deliberately ingested by humans, frequently as a result of mistaken identity [4], and mycotoxin exposure is almost always accidental [2].

Mycotoxins are produced by the fungal mycelium and secreted into the substrate, which is usually an agricultural commodity susceptible to mold colonization, but can also be formed in the reproductive structures and then be present in asexual and sexual spores.

Exposure to mycotoxins is mostly by ingestion of contaminated food and feed [5]. The main source of mycotoxin contamination in the human food chain are cereals and their by-products (Table 8.1) either directly through the consumption of contaminated cereal-based food or indirectly through the intake of residues and metabolites of mycotoxins present in milk and other animal products obtained from livestock given contaminated feeds [1, 6]. It should be noted that cereals are the most important energy source in animal feed. Among cereals, maize is widely considered to be one of the most susceptible crops to mycotoxins and rice among the least ones.

María J. Sainz, Vanesa Ferreiroa, Departamento de Producción Vegetal y Proyectos de Ingeniería, Facultad de Veterinaria, Universidade de Santiago de Compostela, Campus s/n, 27002 Lugo, Spain
Jesús M. González-Jartín, Nadia Pérez-Fuentes, Luis M. Botana, Departamento de Farmacología, Facultad de Veterinaria, Universidade de Santiago de Compostela, Campus s/n, 27002 Lugo, Spain
Olga Aguín, J. Pedro Mansilla, Estación Fitopatolóxica Areeiro, Deputación de Pontevedra, Subida á Carballeira 26, 36153 Pontevedra, Spain

https://doi.org/10.1515/9783111014449-008

Mycotoxins can be also found in grapes, coffee, cocoa, groundnuts, tree nuts, some fruits, and other food commodities, and in animal feeds, as spoiled stored fodder (like silage), cereal by-products used in feed processing, etc. [7].

Table 8.1: Major mycotoxins in foodstuffs, important mycotoxigenic fungi that produce them [8, 9], and EU limits and EU indicative limits for legislated and nonlegislated mycotoxins, respectively.

Mycotoxin	Food/feed commodity	Main producing fungi	EU limits* (µg/kg)
Aflatoxins B_1, B_2, G_1, G_2	Cereals, dried fruits, dried figs, peanuts, tree nuts, almonds, pistachios, apricot kernels, hazelnuts, Brazil nuts, dried spices (chilies, chili powder, cayenne, paprika, pepper, nutmeg, turmeric), ginger (dried), baby food, processed cereal-based food, and food for special medical purposes for infants and young children	*Aspergillus flavus* *Aspergillus parasiticus*	0.1–12 for B_1 4–15 for sum of B_1, B_2, G_1, and G_2
Aflatoxin M_1	Milk, milk-based products, infant and young-child formulae, food for special medical purposes for infants and young children		0.05 in milk and milk products 0.025 in infant and young-child formulae and food for special medical purposes
Ochratoxin A	Cereals, cereal-based products, soybeans, coffee, cocoa powder, sunflower seeds, pumpkin seeds, (water) melon seeds, hempseeds, dried fruits, date syrup, pistachios, dried herbs, dried ginger roots, dried marshmallow and dandelion roots, dried orange blossoms, nonalcoholic beverages, wheat gluten, dried spices, liquorice, products containing liquorice, wine, wine-based products, grape juice, grape nectar, grape must, baby food, processed cereal-based food, and food for special medical purposes for infants and young children	*Aspergillus ochraceus* *Aspergillus carbonarius* *Penicillium verrucosum*	0.5–80
Deoxynivalenol	Maize, other cereals and cereal-based food (except rice and rice products), baby food, processed cereal-based food for infants, and young children	*Fusarium graminearum* *Fusarium culmorum*	200–1,750

Mycotoxin	Food/feed commodity	Main producing fungi	EU limits[*] (µg/kg)
Zearalenone	Maize, other cereals, and cereal-based food (except rice and rice products), refine maize oil, baby food, and processed cereal-based food for infants and young children	*Fusarium graminearum Fusarium culmorum*	20–400
Fumonisins B$_1$ + B$_2$	Maize, maize products, baby food containing maize, and processed maize-based food for infants and young children	*Fusarium verticillioides Fusarium proliferatum*	200–4,000
Patulin	Apple juice, apple products, fruit juices, fruit nectars, drinks derived from apples, and baby food	*Penicillium expansum*	10–50
Citrinin	Food supplements based on rice fermented with red yeast *Monascus purpureus*	*Monascus purpureus*	100
Ergot alkaloids	Grains and milling products of rye and other cereals (barley, wheat, spelt, oats), wheat gluten, and processed cereal-based food for infants and young children	*Claviceps purpurea*	20–500 (20–400 as from 1 July 2024)
			EU indicative limits[**] (µg/kg)
T-2 + HT-2 toxins	Cereals (maize, barley, wheat, and oats)	*Fusarium sporotrichioides Fusarium poae*	15–1,000

[*]Commission Regulation (EC) No. 2023/915 of 25 April 2023 on maximum levels for certain contaminants in food and repealing Regulation (EC) No. 1881/2006; Commission Regulation (EC) No. 1126/2007 of 28 September 2007 amending Regulation (EC) No. 1881/2006 setting maximum levels for certain contaminants in foodstuffs as regards *Fusarium* toxins in maize and maize products.
[**]Commission Recommendation of 27 March 2013 on the presence of T-2 and HT-2 toxins in cereals and cereal products.
For all mycotoxins, the lowest limits are for baby food and food for infants and young children.

Contamination of food and feed with mycotoxins is a threat to human and animal health [10]. This process can occur either preharvest, when grain and forage crops are growing in the field, or postharvest during handling, transportation, storage, and processing of raw materials. Mycotoxins are thermostable, have great chemical stability, and withstand industrial processing so that all products made from contaminated raw materials are likely to contain these compounds.

Fusarium species are field fungi, as they colonize crops and produce mycotoxins before harvest, some of them living endophytically and other causing serious plant diseases. However, *Aspergillus* and *Penicillium* species commonly invade cereals and other food and feed commodities and produce mycotoxins, after harvest under inadequate handling, transportation, and storage conditions [11]. Apart from *Fusarium, Aspergillus,* and *Penicillium,* some other microscopic ascomycetes (e.g., species of *Alternaria* and *Claviceps*) also produce mycotoxins in the field and/or during storage and are receiving increasing attention as food safety hazards, even though they generally do not accumulate to toxic levels in agricultural products.

Most mycotoxins are produced in the field [12]. Key factors that influence the natural infection of plants by mycotoxigenic fungi are environmental conditions during plant growth: temperature and/or water stress, high moisture content (water activity), and insect attacks [13]. These environmental conditions may vary from year to year, resulting in differences in the fungal species that colonize the plants and thus in the type and amount of mycotoxins which contaminate the whole food chain [14].

Cereals, and other crops, may be invaded either preharvest or postharvest by several mycotoxigenic fungi (Figure 8.1), each frequently able to produce several mycotoxins, a fact that increases significantly the potential for multitoxin contamination [7, 15].

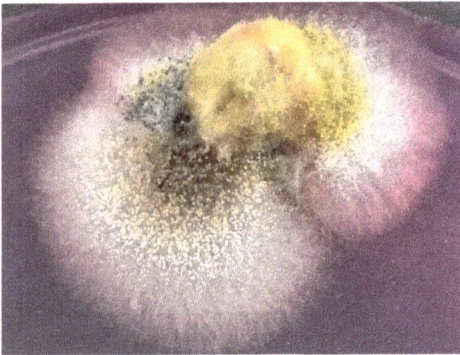

Figure 8.1: Different *Aspergillus* species, as noted by different colors of their asexual reproductive structures, growing from a kernel of maize on dichloran glycerol agar (DG18) medium.

Mycotoxins can also be produced by indoor molds growing on building materials. The most toxic are those from species of *Stachybotrys.* Adverse health effects may appear when mycotoxins are released from fungal spores and colony fragments after inhalation [16].

8.2 Major mycotoxins and their toxicity

More than 400 mycotoxins and derivatives have been described, but around 20 can be present in foods and feeds at significant levels and often enough to constitute a food safety problem and be considered important in human and animal health [17]. In most countries, there are specific regulations, those of European Union being presented in Table 8.1, that set limits to the presence of the mycotoxins most frequently detected in products intended for human consumption: aflatoxins (B_1, B_2, G_1, G_2, M_1), ochratoxin A (OTA), the trichothecenes deoxynivalenol (DON) and T2 and HT-2 toxins, zearalenone (ZEN), fumonisins (FB_1, FB_2), patulin (PAT), citrinin, and ergot alkaloids (EAs). Aflatoxins, OTA, and PAT are produced by *Aspergillus* and *Penicillium* species, trichothecenes, ZEN and fumonisins by *Fusarium*, citrinin by *Aspergillus*, *Penicillium*, and *Monascus*, and EAs by *Claviceps*. The trichothecenes T2 and HT-2 toxins are not regulated; however, indicative limits for the sum of both toxins in cereals and cereal products, either in food or feed, are set out in the Commission Recommendation 2013/165/EU.

The only major regulated mycotoxin not produced by fungi is aflatoxin M_1, which is the principal hydroxylated metabolite of aflatoxin B_1 produced in the liver of dairy cows fed with aflatoxin B_1-contaminated feedstuffs. Aflatoxin M_1 is secreted in the milk and can be found in milk and other dairy products [18].

Animal feed legislation usually only regulates the maximum permitted levels of aflatoxin B_1 and rye ergot (as an indirect level of EAs), while makes recommendations for other mycotoxins [3] (Table 8.2).

Table 8.2: EU limits for major mycotoxins in feeds.

Mycotoxin	Animal feeds	EU limit (mg/kg)
Aflatoxin B_1[*]	All feed materials	0.02
	Complementary and complete feed	0.01
	With the exception of:	
	– Compound feed for dairy cattle and calves, dairy sheep and lambs, dairy goats and kids, piglets, and young poultry animals	0.005
	– Compound feed for other cattle, sheep, goats, pigs, and poultry	0.02
Rye ergot[*] (***Claviceps purpurea***)	Feed materials and compound feed containing unground cereals	1,000
Deoxynivalenol[**]	Cereals and cereal products	8
	Maize by-products	12
	Complementary and complete feeding stuffs with the exception of:	5
	– Complementary and complete feeding stuffs for pigs	0.9
	– Complementary and complete feeding stuffs for calves, lambs, and kids	2

Table 8.2 (continued)

Mycotoxin	Animal feeds	EU limit (mg/kg)
Zearalenone**	Cereals and cereal products	2
	Maize by-products	3
	Complementary and complete feeding stuffs for piglets and gilts	0.1
	Complementary and complete feeding stuffs for sows and fattening pigs	0.25
	Complementary and complete feeding stuffs for calves, dairy cattle, sheep, and goats	0.5
Ochratoxin A**	Cereals and cereal products	0.25
	Complementary and complete feeding stuffs for pigs	0.05
	Complementary and complete feeding stuffs for poultry	0.1
Fumonisins B_1 + B_2**	Maize and maize products	60
	Complementary and complete feeding stuffs for:	
	– Pigs, horses, rabbits, and pet animals	5
	– Fish	10
	– Poultry, calves, lambs, and kids	20
	– Adult ruminants and mink	50
T-2 + HT-2 toxins***	Oat milling products (husks)	2
	Other cereal products	0.5
	Compound feed, with the exception of feed for cats	0.25

*Commission Regulation (EU) No. 574/2011 of 16 June 2011 amending Annex I to Directive 2002/32/EC of 7 May 2002 on undesirable substances in animal feed.
**Commission Recommendation 2006/576/EC of 17 August 2006 on the presence of deoxynivalenol, zearalenone, ochratoxin A, T-2 and HT-2, and fumonisins in products intended for animal feeding.
***Commission Recommendation of 27 March 2013 in the presence of T-2 and HT-2 toxins in cereals and cereal products.
Legislated limits are only for aflatoxin B_1 and rye ergot (ergot alkaloids indirectly). For the remaining mycotoxins, limits are recommendations. All levels are relative to a feed with a moisture content of 12%.

In recent years, increasing attention is being paid to other mycotoxins as food safety hazards because, even though they do not accumulate to toxic levels in agricultural products, their incidence is rapidly increasing: the so-called emerging mycotoxins and the modified and matrix-associated forms of DON, ZEN, fumonisins, and T-2 and HT-2 toxins, initially called masked mycotoxins, most mainly produced on cereal crops by *Fusarium* species and not subjected to any regulation [19]. The term "masked mycotoxin" was coined for metabolites of a parent mycotoxin conjugated or attached to certain molecules and formed in the plant or fungus.

Emerging mycotoxins include enniatins (ENNs), beauvericin (BEA), fusaproliferin (FUS), and moniliformin (MON), produced by *Fusarium*; alternariol (AOH), alternariol monomethyl ether (AME), and tenuazonic acid (TeA), produced by *Alternaria*; and sterigmatocystin (STE), produced by *Aspergillus* [3, 20–23].

Many works have shown that some emerging mycotoxins are present in high concentrations in food and feed, which increases the interest in studying their toxicity, whereas the existence of modified and matrix-associated mycotoxins means that actual amounts of the parent mycotoxins in food and feedstuffs might be higher than currently determined by routine and conventional analytical methods [11, 24, 25]. New conjugated mycotoxins and analogs are continually being described [26], raising the need of studying whether they might contribute to toxicity either directly or indirectly through the release of the parent mycotoxins.

Carcinogenicity, nephrotoxicity, hepatotoxicity, reproductive problems, gastrointestinal effects, immunosuppression, dermal effects, and central nervous system disorders are toxic effects associated to these fungal metabolites (Table 8.3). The highest acute and chronic toxicity in humans and animals is produced by aflatoxins, especially the most predominant aflatoxin B_1, which are potent carcinogens [14]. Aflatoxins B_1, B_2, G_1, G_2, and M_1 have been classified as carcinogenic to humans by the International Agency for Research on Cancer [27]. OTA, FB_1, FB_2, and STE are possibly carcinogenic to humans [28]. Zearalenone has oestrogenic and anabolic activity, and DON, the most abundant trichothecene, has been related to immunosuppression, reproductive disorders, vomiting, and other symptoms in humans and animals [29].

Table 8.3: Major toxic effects of major mycotoxins found in food and feed [8, 9, 27, 28].

Mycotoxin	Toxic effects
Aflatoxins B_1, B_2, G_1, G_2	Liver cirrhosis, immunosuppression, and cancer
Aflatoxin M_1	
Ochratoxin A	Liver and kidney damage
Deoxynivalenol	Emesis, feed refusal, and immunosuppression
Zearalenone	Estrogenic activity
T-2 and HT-2 toxins	Immunosuppression, causative agent of alimentary toxic aleukia (ATA)
Fumonisins B_1, B_2, B_3	Neural tube defects, causative agent of leukoencephalomalacia in horses
Patulin	Inflammatory alterations of the gastrointestinal tract
Citrinin	Kidney damage
Ergot alkaloids	Vasoconstriction and neural disorders

Diseases caused by exposure to mycotoxins are collectively called mycotoxicoses. They are mostly the result of eating contaminated food, although skin contact with mold-infected substrates and inhalation of spore-borne mycotoxins are also sources of exposure. Mycotoxin exposure is more likely in parts of the world where poor methods of food handling and storage are common, where problems of malnutrition are important, in buildings harboring high levels of molds and in countries with few mycotoxin regulations [2].

Toxic effects associated to mycotoxin exposure may be acute or chronic. These effects depend on the mycotoxin type, the level, and duration of exposure, and in production animals, the animal species that is exposed and the age of the animal [30]. Although acute mycotoxicoses produced by high doses of some mycotoxins in humans and animals (e.g., turkey X disease and human ergotism) are the best known, they are rare. Ingestion of low to moderate amounts of mycotoxins, particularly those produced by *Fusarium* in cereals in the field, is common and generally does not result in obvious intoxication. However, the greatest risk to human and animal health is related to chronic exposure (e.g., cancer induction and immune suppression) [2].

8.2.1 Aflatoxins

Aflatoxins are difuro-coumarin derivatives produced by *Aspergillus* species, predominantly *A. flavus*, and *A. parasiticus* [31]. The aflatoxins group was identified as causative agent of "turkey X" disease in the 1960s after the death of thousands of turkey poults, ducklings, and chicks fed with contaminated peanut meal in England [3].

Aflatoxins are readily absorbed from the gastrointestinal tract and reach the systemic circulation [32]. In this way, aflatoxins are distributed to various body organs, especially the liver in which they are biotransformed through hydroxylation, hydration, demethylation, and epoxidation. The epoxidation of aflatoxin B_1 is more efficient than that of the others naturally occurring aflatoxins and, therefore, is the most toxic compound [33]. The resultant metabolites are excreted in bile and urine after conjugation [34]. Acute effects of aflatoxins are mediated by one of these metabolites, namely aflatoxin B_1 exo-8,9-epoxide, which causes impairment of protein synthesis and binds to proteins in the blood serum. Consequently, a reduction in protein content in body tissues is observed which has been associated with liver and kidney necrosis [32]. Carcinogenic activity of aflatoxins is also mediated by aflatoxin B_1 exo-8,9-epoxide, which reacts with the guanine bases of liver cells forming a DNA adduct (aflatoxin-N7-guanine). If the DNA adduct is not repaired, it causes mutations in p53 tumor suppressor gene which results in carcinogenesis [35].

The mycotoxicosis that results from the ingestion of aflatoxins is called aflatoxicosis. Acute aflatoxicosis is caused by the intake of large doses of aflatoxins, which result in direct damage to the liver, usually through liver cirrhosis [36]. In some cases, the illness leads to death. In this sense, 125 people died during one of the latest aflatoxicosis outbreaks happened in Kenya in 2004 [37]. Chronic exposure to sublethal doses of aflatoxins has been related with immunosuppression and nutritional alterations [36]. Long exposure to aflatoxins in the diet causes cancer in many animal species.

The carcinogenic activity of aflatoxin B_1 is well established, the liver being the primary target; however, tumors have been found in other sites such as kidney or colon.

8.2.2 Ochratoxins

Ochratoxins are produced by some *Aspergillus* and *Penicillium* species, including *A. ochraceus*, *A. carbonarius*, and *P. verrucosum* [38]. Members of the ochratoxin group are dihydrocoumarines linked to a molecule of L-β-phenylalanine via an amide bond. Ochratoxin A is the major produced and most toxic analog of the ochratoxin family. It has been implicated in a diverse range of toxicological effects in both animals and humans [39].

The main target organ of OTA is the kidney. In mammalians, the ingestion of OTA causes nephropathy in a dose- and time-dependent way [40]. Kidney alterations can be consequence of both acute and chronic exposure to this environmental-toxicant. In animals, the administration of acute lethal doses of OTA produces hemorrhages, intravascular coagulation, and necrosis of liver, kidney, and lymphoid organs [41]. In humans, the inhalation of OTA produced by *Aspergillus ochraceus* was related with the development of acute renal failure [42]. Toxicity resulting from chronic exposure seems to be the most significant since acute exposure is rare. In this regard, exposure to low doses of this toxin over long periods of time leads to characteristic toxic effects, which are manifested through kidney damage. Ocharoxin A is also an immunotoxic and hepatotoxic agent. Immunosuppressant activity of OTA is characterized by a significant reduction of immune organs and changes in the number and functions of immune cells [43]. Hepatotoxicity is mediated by the production of reactive oxygen species (ROS), which cause DNA damage leading to apoptosis in hepatic cells [44].

In addition to the described effects, OTA can cross the placenta from mother to fetus, resulting in embryo-toxic and teratogenic effects in animals. The most common symptoms are reduced birth weight and craniofacial abnormalities [45].

8.2.3 Trichothecenes

Trichothecenes are a group of sesquiterpenes which include more than 200 analogs, all of them containing an epoxide group. According to their structure, these compounds are classified into four groups, namely types A, B, C, and D [46]. Trichothecenes are produced by *Fusarium* species yet only types A and B; both the types are commonly found in cereals and cereal-based products.

Type-A trichothecenes include T-2 toxin, its deacetylated form HT-2 toxin, monoacetoxyscirpenol, diacetoxyscirpenol, and neosolaniol, T-2 toxin being considered as the most toxic trichothecene. Strains of *F. sporotrichioides* and *F. poae* are the main producers of these toxins [47].

Type-B trichothecenes include deoxynivalenol and its derivative forms, that is, 3-acetyl DON (3-ADON) and 15-acetyl DON (15-ADON), nivalenol (NIV), and fusarenone X (FUS-X). Strains of *F. culmorum* and *F. graminearum* are the major producers [48].

Within the group of type-A trichothecenes, most data on toxicokinetics and mechanism of action deal with T-2 toxin. Absorption, distribution, and excretion of T-2 toxin are rapid. These mycotoxins are metabolized in the liver and other tissues through hydroxylation, de-epoxidation, acetylation, and conjugation [49]. This mycotoxin is metabolized in the liver and other organs. Although many products have been reported from T-2 metabolism, HT-2 toxin is the major one. T-2 toxin and its metabolites are excreted in urine and feces [50]. The toxicity of trichothecenes primarily arises from their binding to the 60S subunit of ribosomes which lead to an inhibition of protein synthesis and subsequently leads to secondary DNA disruption and RNA synthesis. In this way, type-A trichothecenes inhibit hematopoiesis, disrupting the differentiation of monocytes into macrophages and dendritic cells. The acute intoxication caused by type-A trichothecenes, particularly T-2 toxin, results in the development of a condition known as alimentary toxic aleukia (ATA). Notably, during World War II, an outbreak of ATA caused the death of thousands of individuals in Russia. On the other hand, chronic exposure to type A trichothecenes is linked to symptoms such as anorexia, reduced body weight gain, and the occurrence of lesions in the upper digestive tract [51].

Available data on type-B trichothecenes are focused on DON in production animals. The oral absorption is low for ruminants and poultry, while swine and rodents readily absorb this mycotoxin. In pigs, up to 82% of the orally administered DON reaches the systemic circulation [52].

Gastrointestinal tract toxicity is the most characteristic one of type-B trichothecenes. Acute intoxication with DON (also known as vomitoxin) leads to anorexia and emesis. Other symptoms include abdominal distress, increased salivation, and diarrhea. Only exposure to extremely high concentrations produces mortality or marked tissue injury [53, 54]. Deoxynivalenol hinders the function of sodium-glucose transport protein 1, located in the brush border membrane of the small intestine, leading to a decrease in glucose uptake. Moreover, when present in high concentrations, this toxin can diminish the digestibility of crucial amino acids in pigs. All together negatively impacts in pig growth. In addition, the chronic exposure to deoxynivalenol provokes immunotoxicity and increases the expression of cytokines involved in the inflammatory response in the intestine [55].

NIV presents a chemical structure similar to DON, and it is synthesized by several *Fusarium* species, including *F. cortaderiae, F. crookwellense, F. culmorum, F. graminearum,* and *F. poae.* NIV is a naturally occurring contaminant that can be found in a wide range of foodstuffs worldwide, including wheat, barley, and maize [56], but NIV contamination is less frequent, and when it does occur, it occurs at lower levels than DON [57].

NIV has been observed to inhibit cell proliferation in various mammalian cells, particularly in tissues with high rates of cell turnover, such as intestinal epithelial cells. It is known to induce gastrointestinal toxicity and genotoxicity. However, its primary mechanism of action appears to be immunotoxicity. In this sense, NIV influences the maturation process of murine bone marrow-derived dendritic cells in response to lipopolysaccharide, and it also leads to an increase in expression of inflammation-associated genes leading to a stronger proinflammatory response [58].

8.2.4 Zearalenone

Zearalenone is a phenolic resorcylic acid lactone produced by *Fusarium* species, among which are *F. culmorum* and *F. graminearum* [59]. Zearalenone and its analogs alpha-zearalenol (α-ZOL) and beta-zearalenol (β-ZOL) are found in the field, usually as cereals contaminants. These mycotoxins are, in many cases, responsible for estrogen-related diseases observed in farm animals [60].

Zearalenone is readily absorbed after oral ingestion. However, this toxin is subject to an extensive presystemic metabolism and just low amounts reach the systemic circulation [61]. Zearalenone is biotransformed in the liver and intestines, where it is converted to α-ZOL and β-ZOL by an enzymatic reaction catalyzed by hydroxysteroid dehydrogenases. Thereafter, α-ZOL and β-ZOL are transformed in α-zearalanol and β-zearalanol (α-ZAL and β-ZAL) [62]. These metabolites are subsequently conjugated by UDP-glucuronyl transferases and afterward excreted in urine and bile [63].

The mechanism of action of ZEN and its metabolites is mediated by their binding affinity to estrogen receptors which yields estrogenic effects by the activation of gene transcription [64]. Not all compounds have the same affinity for estrogen receptors, being as follows: α-ZOL > ZEN > β-ZOL [65]. Swine is the most sensitive species to ZEN exposure since it shows a preferential conversion of ZEN to α-ZOL, which is the metabolite with the highest affinity for estrogen receptors [66].

Zearalenone exhibits low acute toxicity after oral ingestion. Chronic exposure to ZEN has well-established effects on the endocrine and reproductive systems. This mycotoxin induces alterations in the reproductive tract, weight changes in some endocrine glands (adrenal, thyroid, and pituitary), and changes in serum levels of ovary hormones with decreased fertility rates. However, no teratogenic effects have been described. In pigs, ZEN causes severe signs of hyperestrogenism [54]. The main effect observed in long-term rodent studies is low body weight gain. In mice, ZEN induces pituitary adenomas and produces liver lesions which evolve to hepatocellular adenomas. Moreover, alterations in the mammary gland and fibrosis of the uterus and cystic ducts have been reported [67].

8.2.5 Fumonisins

Fumonisins are a family of mycotoxins produced by several field fungi belonging to the genus *Fusarium*, primarily *F. verticillioides* and *F. proliferatum*, which frequently infect maize and other crops [68, 69]. The most prevalent and toxic analogs are fumonisin B_1 (FB$_1$) and fumonisin B_2 (FB$_2$) [60].

Fumonisins are poorly absorbed as less than 5% of the ingested fumonisins reaches the blood stream [70]. There is no in vivo evidence of significant FB$_1$ metabolism [70]. Toxic effects of fumonisins are mediated by the inhibition of the enzyme ceramide synthase due to the structural similarity that this family of mycotoxins has with the sphinganine. Toxin ingestion leads to an increase in the amount of sphinganine in serum, tissues, and urine. In addition, depletion of ceramide, sphingomyelin, and glycosphingolipids may occur [71–73].

Fumonisin B_1 causes a neurotoxic disease of horses known as equine leukoencephalomalacia. This is a fatal illness characterized by liquefactive necrosis of the cerebral white matter involving frontal and parietal lobes [74]. In swine, fumonisins induce pulmonary edema probably because of an acute left-sided heart failure [75]. Other toxic effects, such as nephrotoxicity and hepatotoxicity, have been observed in laboratory animals [76]. In humans, epidemiological studies relate these mycotoxins with neural tube defects and esophageal cancer [73, 77].

8.2.6 Patulin

PAT is a heterocyclic lactone that can be produced by various species of *Penicillium*, *Aspergillus*, and *Byssochlamys* [78], but mainly by *P. expansum* when it invades apples and other fruits during storage. This compound was isolated in 1943 as an antibiotic effective against Gram-positive and Gram-negative bacteria [79]. However, later studies showed that it was toxic to humans and animals. Nowadays, PAT is one of the best characterized mycotoxins found in agricultural products [80].

There is still little data about the absorption and metabolism of PAT. This toxin is mainly accumulated in the red blood cells [81]. PAT inhibits many enzymes since it is a highly reactive molecule with affinity for sulfhydryl-containing compounds, such as cysteine or glutathione [82]. One important aspect of PAT toxicity is its ability for suppressing catalase activity, which results in the increase of ROS generation leading to oxidative stress [83].

Acute toxic effects of PAT are associated with inflammatory alterations of the gastrointestinal tract, producing ulceration and inflammation of the mucosa of the stomach [84]. This may be caused by the destruction of the tight junctions in the epithelial cell layer [28]. Other clinicopathological alterations may include agitation, convulsions, metabolic alkalosis, pulmonary congestion, edema, oliguria, reduced plasma protein, and neutrophilia [82, 84]. Chronic exposure to PAT may involve neurotoxic and immunosuppressive effects [85].

8.2.7 Citrinin

Citrinin is produced by a number of *Monascus, Penicillium*, and *Aspergillus* species, including *M. purpureus, M. ruber, P. citrinum, P. verrucosum*, and *A. terreus* [86]. This mycotoxin is predominantly produced under storage conditions and can be found in grains, beans, fruits, and spices. Specifically, citrinin is commonly present in red mold rice, a fermented rice product derived from some strains of *Monascus* spp. such as *M. purpureus* [87, 88]. Its isolation dates back to the 1930s, when it was first obtained as an antibiotic, from a culture of *P. citrinum*, but interest on citrinin waned once its toxicity was detected. Although the mechanism of action of citrinin is not completely understood, it encompasses various pathways, including inhibition of DNA and RNA synthesis, elevation of ROS production, and activation of the caspase-cascade system, ultimately leading to apoptotic cell death [89, 90].

Acute exposure to citrinin results in kidney necrosis and hepatic impairment, while prolonged exposure leads to a gradual progression of histopathological alterations in the kidneys, leading to the development of interstitial nephritis and cell adenomas. Additionally, citrinin has been demonstrated to possess genotoxic properties across various test systems and has also been linked to reproductive toxicity [91, 92].

8.2.8 Ergot alkaloids

EAs belong to a class of mycotoxins produced by several genera of Ascomycota, such as *Claviceps, Aspergillus, Penicillium*, and *Epichloë*. However, the main producers of these alkaloids are found mainly within the genus *Claviceps*, being *C. purpurea* the most prominent species [93]. *Claviceps purpurea* is a parasitic fungus of cereals and grasses that invades immature ovaries at the flowering stage, typically unfertilized ovaries, and grows in place of the kernel, replacing the seeds with dense dark mycelial masses called sclerotia (known as ergots). EAs are subsequently produced and accumulated inside the sclerotia. While ergot infestation is significantly curtailed by the self-pollination nature of wheat, other cereals, such as rye, barley, and pearl millet, are more susceptible to hosting *Claviceps* ergots on a frequent basis [94].

Over 50 distinct EAs have been identified, and they are commonly categorized into three classes: clavine-type alkaloids (ergoclavines), ergoamides (including D-lysergic acid amide, D-lysergic acid α-hydroxyethylamide, and ergometrine), and ergopeptines, which constitute the largest and most diverse class of EAs. Notable analogs within this class include ergotamine, ergocryptine, and ergocristine [93].

In the Middle Ages, the consumption of food (mainly rye bread) contaminated with EAs led to numerous outbreaks of human poisoning, a condition called St. Anthony's fire. Today, improved agricultural practices and food processing techniques have significantly reduced the occurrence of severe outbreaks of this toxicosis, now known as ergotism. However, complete elimination of EAs from cereals remains challenging, and there is still a food safety hazard [95].

EAs exhibit a diverse array of activities as they function as agonists or antagonists targeting noradrenaline, dopamine, and serotonin receptors. These molecules can interfere at more than one receptor site and their inherent structural differences lead to different activities across alkaloids. Consequently, a wide range of pharmacological effects can arise from the numerous possible interactions. In fact, some EAs have already been extensively used as medicines over the centuries [96]. In general, acute exposure to EAs elicits signs of neurotoxicity such as muscular weakness, tremors, or rigidity. Repeated exposure can result in vasoconstriction-induced ischemia in specific body regions. Furthermore, EAs have detrimental effects on the reproductive system, including abortion and the suppression of lactation [97].

8.2.9 Emerging and modified toxins

Exposure to fungal toxins is not restricted to regulated mycotoxins. In this sense, the term "emerging mycotoxins" gained popularity in 2008 to address the presence of toxic secondary metabolites that are not currently regulated or routinely analyzed, but their documented occurrence was rapidly increasing. Initially, this designation encompassed *Fusarium* toxins such as ENNs, BEA, FUS, and MON. However, some metabolites produced by *Aspergillus* and *Alternaria* species also fit into this category. Currently, some of the most extensively researched emerging mycotoxins are STE, produced by *Aspergillus* species, and the *Alternaria* metabolites alternariol, alternariol methyl ether, and TeA [20–23, 98].

On the other hand, mycotoxins produced by fungi can undergo modifications through fungal, plant, or animal metabolism, resulting in potentially toxic compounds that are not considered in current legislation. These transformed products have often been referred to as masked, bound, conjugated, or hidden mycotoxins, although the usage of these terms has been inconsistent. The term "modified mycotoxins" was introduced to describe all types of mycotoxin modifications and was adopted by the EFSA in a scientific opinion about the modified forms of certain mycotoxins [99]. To prevent misidentifications, a systematic classification of mycotoxins into four levels has been proposed (Figure 8.2) [100], providing a comprehensive understanding of their different forms. The first level encompasses three groups: free mycotoxins, matrix-associated mycotoxins, and modified mycotoxins.

Free mycotoxins refer to the unmodified primary compounds that are produced as secondary metabolites by fungi. Matrix-associated mycotoxins are compounds that are either trapped or dissolved or covalently bound to the surrounding matrix. On the other hand, modified mycotoxins include both biologically and chemically modified compounds. Biologically modified compounds are the result of metabolic reactions occurring in animals, plants, or fungi, while chemically modified compounds are gener-

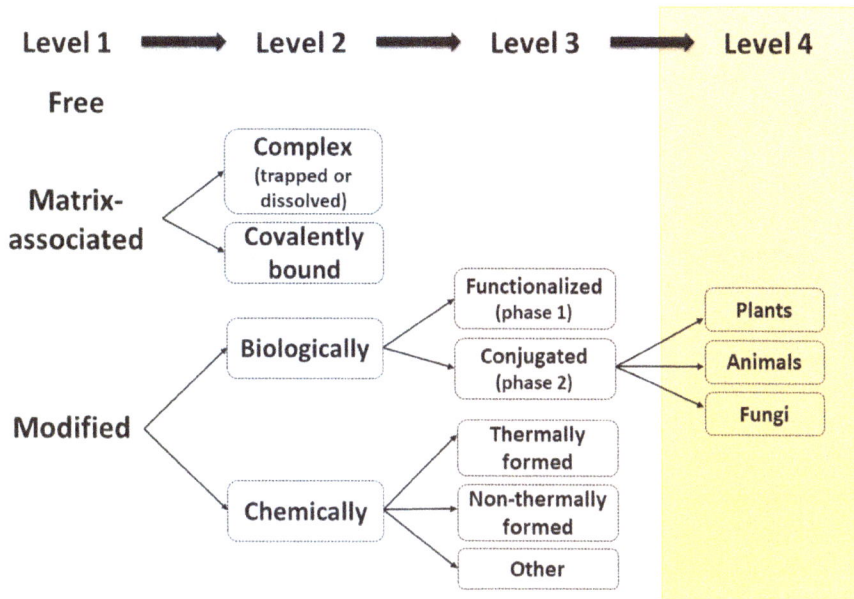

Figure 8.2: Systematic definition of mycotoxins (adapted from [100]).

ated through thermal, chemical, or light-induced processes. Although this classification provides clarity, some compounds can fall into multiple categories. In any case, the term "modified mycotoxins" serves as an overarching designation encompassing masked or conjugated compounds, whereas the term "masked mycotoxins" specifically refers to metabolites produced in plants.

Many modified mycotoxins have been recently discovered, but others remain unknown. Notably, acetylated derivatives of DON, namely 3-ADON and 15-acetyl-DON (15-ADON), phase I metabolites of ZEN such as α-zearalenol (α-ZEN) and β-zearalenol (β-ZEN), as well as glucose conjugates like DON-3-glucoside (DON-3G) and ZEN-14-glucoside (ZEN-14G), are among the most significant ones. The lack of sufficient toxicological studies for these compounds, coupled with the potential hydrolysis they may undergo during the digestion process, poses a genuine risk to public health [100, 101].

8.2.9.1 Beauvericin and enniatins

BEA and ENNs are cyclic hexadepsipeptide toxins produced by species of the genus *Fusarium.*

BEA was first isolated from the culture of the entomopathogenic fungus *Beauveria bassiana.* Next, its production by many *Fusarium* species was reported, including strains of *F. oxysporum, F. poae, F. proliferatum, F. sambucinum, F. semitectum* (currently, *F. incarnatum*), and *F. subglutinans*. This toxin is produced under moist and cool conditions and is mainly found in maize and maize-based products [102].

In the case of ENNs, 29 naturally occurring analogs have been identified so far [102, 103], of which seven (ENNs A, A1, B, B1, B2, B3, and B4) have been reported in cereals, being ENNs A, A1, B, and B1 most frequently found in food and feeds [104; 22]. ENNs, originally discovered from cultures of *F. oxysporum*, are also produced by other *Fusarium* species, including *F. avenaceum, F. poae, F. sambucinum, F. sporotrichioides,* and *F. tricinctum*, which infect cereals and other commodities [105].

Due to their similar structures, the biological activities of BEA and ENNs have traditionally been associated. The toxicity of these compounds has been linked to ionophoric properties since 1973, when it was suggested that ENN B formed complexes within biological membranes, facilitating the transport of ions. However, recent observations have revealed distinctions in the mechanism of action between ENNA1 and B1 despite both lead to apoptotic cell death in SH-SY5Y human neuroblastoma cells by altering calcium homeostasis. Specifically, ENNA1 exhibits more pronounced effects on calcium fluxes and increased release of ROS. On the other hand, BEA induces apoptosis in various cell lines, and its cytotoxicity could potentially be attributed to the disruption of calcium homeostasis. This hypothesis was supported by the finding that the introduction of a calcium chelator prevented BEA-induced cell death in human leukemia cells [106].

8.2.9.2 Fusaproliferin

FUS is a bicyclic sesterterpene that was first identified from maize kernels inoculated with *Fusarium proliferatum* [107]. In the following years, other *Fusarium* species, such as *F. subglutinans, F. temperatum,* and *F. verticillioides*, were also reported as FUS producers [4, 22, 108]. FUS has been found naturally occurring in cereals, particularly maize [107].

There are limited information on FUS toxicity and mode of action. FUS has been shown to be toxic on brine shrimp larvae and human B lymphocytes, and teratogenic on chicken embryos [109, 110], which suggest a potential hazard to humans and animals that needs to be investigated [22].

8.2.9.3 Moniliformin

MON is a low-molecular weight mycotoxin first reported from *Fusarium fujikuroi* (formerly *F. moniliforme*); since then, its production by several other *Fusarium* species, including *F. avenaceum, F. proliferatum, F. subglutinans,* and *F. verticillioides*, was reported. In addition, *Penicillium melanoconidium* has been reported to produce MON [111].

MON has low acute toxicity, but it can still induce adverse effects on the liver, leading to inflammation and changes in liver function. In addition, MON causes respiratory diseases and myocardial degeneration, probably as a consequence of the inhibition of the oxidation of intermediates in the tricarboxylic acid cycle. It has also been proposed that MON disrupts the synthesis of enzymes [112].

8.2.9.4 Sterigmatocystin

STE is a polyketide produced by fungi belonging to the genera *Aspergillus*, *Bipolaris*, *Botryotrichum*, *Humicola*, and *Penicillium*. However, the main producers are *A. flavus*, *A. parasiticus*, *A. nidulans*, and especially *A. versicolor* [113, 114]. These species of *Aspergillus* have been found to contaminate cereal grains, such as wheat, rye, maize, and barley, with STE. Interestingly, rice and oats grains exhibit a higher susceptibility to contamination by this toxin compared to other grains [115].

In aflatoxigenic species, this compound undergoes metabolism to produce AFB1 and AFG1, so it is rarely accumulated. However, it seems that other species, such as *A. nidulans* and *A. versicolor*, lack the necessary genes for converting STE into AFs; as a consequence, cereals or other substrates invaded by them can present high amounts of STE [114]. As an intermediate in the biosynthesis of AFs, STE exhibits several biological activities that closely resemble those of AFB1. In this sense, STE induces oxidative stress, mitochondrial dysfunction, apoptosis, cell cycle arrest, and immune system alterations. In vivo studies have demonstrated that STE has the ability to elicit hepatotoxic and nephrotoxic effects in various animal species. Additionally, STE has been found to possess genotoxic properties, forming DNA adducts and causing DNA damage. Notably, STE has been associated with the development of gastric cancer in humans [114].

8.2.9.5 *Alternaria* toxins

Alternaria species are widely distributed in nature and are responsible for causing diseases pre- and postharvest in various economically significant plants, including cereals like wheat, sorghum, and barley [116]. In addition, these fungi can contaminate oilseeds, such as sunflower and rapeseed, and affect several fruits and vegetables, including tomatoes, apples, cucumber, peppers, and olives, among others. The ability of *Alternaria* species to grow even at low temperatures makes them particularly problematic during the storage and transportation of food under unfavorable conditions, leading to food spoilage.

These species produce over 70 metabolites, which are characterized by their phytotoxicity, meaning they have harmful effects on plants. Among these metabolites, AOH, AME, TeA, altenuene (ALT), and altertoxins (ATX-I, ATX-II, and ATX-III) have been found as contaminants of food and have been identified to induce toxicity in animals; thus they are also classified as mycotoxins [117].

Alternaria toxins have been reported to exhibit carcinogenicity, mutagenicity, genotoxicity, cytotoxicity, and reproductive and developmental toxicity, among other effects. TeA has proved to be more toxic than AOH, AME, and ALT. In studies involving chickens, dogs, and mice, TeA was observed to be acutely toxic, without causing genotoxic effects. Both in dogs and chickens, it was shown to cause internal hemorrhages. Indications of precancerous changes caused by TeA have been reported in the oesophageal mucosa of mice. Interestingly, a high incidence of human oesophageal cancer in a Chinese area might be associated with the consumption of cereal grains contaminated with AOH and AME. Altertoxins have been found to possess mutagenic properties, being more potent mutagens to mice than AOH and AME [117–120].

Recent findings suggest that AOH and related chemicals have estrogenic and androgenic potential. Conversely, naturally occurring mixtures of *Alternaria* toxins exhibit antiestrogenic effects leading to their classification as endocrine disruptors [117–120]. The limited toxicological information available regarding altenuene indicates that it has a low acute toxicity [120].

Alternaria toxins are food safety hazards of growing concern, as evidenced by the fact that the European Commission has recently issued a recommendation on monitoring the presence of *Alternaria* toxins, providing indicative levels of alternariol, AME, and TeA in certain foods (Table 8.4), above which research should be conducted on the factors leading to their presence or on the effect of food processing.

Table 8.4: Main *Alternaria* toxins in foodstuffs, mycotoxigenic species that produce them [120], and EU indicative levels.

Mycotoxin	Food	Main producing fungi	EU indicative levels* (µg/kg)
Alternariol (AOH)	Cereal-based foods for infants and young children	*Alternaria alternata*	2
	Processed tomato products and sunflower oil	*Alternaria arborescens*	10
	Sesame and sunflower seeds	*Alternaria tenuissima*	30
Alternariol monomethyl ether (AME)	Cereal-based foods for infants and young children	*Alternaria alternata*	2
	Processed tomato products	*Alternaria arborescens*	5
	Sunflower oil	*Alternaria tenuissima*	10
	Sesame and sunflower seeds		30

Mycotoxin	Food	Main producing fungi	EU indicative levels[*] (µg/kg)
Tenuazonic acid (TeA)	Sesame seeds, sunflower oil, and tree nuts	*Alternaria arborescens Alternaria tenuissima*	100
	Cereal-based foods for infants and young children, and processed tomato products		500
	Dried figs and sunflower seeds		1,000
	Paprika powder		10,000

[*]Commission Recommendation of 5 April 2022 on monitoring the presence of *Alternaria* toxins in food.

8.3 Main mycotoxin-producing fungi

Although more than 50 fungal genera are known to include mycotoxigenic species, the vast majority and most important mycotoxigenic species are found in three fungal genera belonging to two orders in the Ascomycota: the genus *Fusarium* in the order Hypocreales and the genera *Aspergillus* and *Penicillium* in the order Eurotiales.

Species of *Fusarium* colonize forage and grain crops in the field, some of them causing severe diseases and other living endophytically (growing in a plant without affecting it), and produce mycotoxins before harvest. *Aspergillus* and *Penicillium* species generally grow on foodstuff and feed under inadequate conditions of drying, transport, storage, and processing. An exception is *Aspergillus flavus*, which can be a pathogen or an endophyte in the field, and a storage fungus, and can produce mycotoxins in the three cases.

The production of mycotoxins in the field by species of *Fusarium* is practically unavoidable because they are common components of the epiphytic and endophytic microflora in many crops. The importance of mycotoxin contamination by *Fusarium* will depend on environmental conditions during the crop growing season and the subsequent food storage. However colonization of storage agricultural commodities by *Penicillium* and *Aspergillus* species can be prevented by drying crops at harvest time and/or by controlling environmental conditions of storage [15].

The identification and classification of mycotoxigenic species requires some knowledge on the life cycle of filamentous ascomycetes, and especially on the morphological characters of sexual and asexual reproductive structures, together with the application of molecular methods and phylogenetic analysis developed in the 2000s. In recent years, the analysis of the mycotoxin profile produced by specific groups or species, particularly some belonging to the genus *Fusarium*, has received increasing attention as an additional tool for fungal classification and identification and to obtain information of the potential toxicity of contaminated foods.

8.3.1 Fungi: an overview

The kingdom Fungi, placed together with the animals in the eukaryotic supergroup Opisthokonta [121], is one of the largest and ubiquitous groups of living organisms on Earth. The number of extant fungal species has been estimated to be from ~1.5 to 5.1 million, although only about 135,000 have been formally classified and named by taxonomists [122–124]. Fungi are unicellular or multicellular nonchlorophyllic organisms, most of them obligate aerobes. The cell wall consists of various layers mainly composed of chitin and glucan [125]. Unicellular fungi are commonly referred to as yeasts. The vast majority of fungi is multicellular and exists as filamentous forms which undergo a vegetative and a reproductive stage. The vegetative body of a fungus is the thallus, which may be unicellular (in yeasts) or filamentous and multinucleated (in most fungi). Some fungi are dimorphic as they can exist as either in yeast or filamentous form.

All fungal species are heterotrophic. Considering the nutrient source, they can be classified as biotrophs, when nutrients are from a living host (plant or animal); saprotrophs, when nutrients are obtained from dead or decomposing organic matter, mainly dead plants but also dead animals; and necrotrophs, when the fungus infect a living host and kill host cells to get the nutrients [126].

Appropriate conditions of temperature, water availability, and pH are essential for fungal growth. Even though the temperature range is quite wide, most species grow well at temperatures around 25 °C. Most fungi are acidophilic, growing well between pH 4 and 6, and require a high water activity (a_w), with a minimum a_w of around 0.65 [127].

Fungi are currently classified according to macro- and microscopic morphological characteristics of their reproductive structures, and molecular, phylogenetic, and phylogenomic analysis in the following divisions (phyla): Ascomycota, Basidiomycota, Blastocladiomycota, Chytridiomycota, Cryptomicota, Microsporidia, Mucoromycota, Neocallimastigomycota, and Zoopagomycota [128–131]. The majority of the named fungal species are Ascomycota (64,163 species), followed by Basidiomycota (31,515 species) [99, 132], both phyla grouped in the subkingdom Dikarya as they have dikaryotic hyphae.

Within the Ascomycota, most known species (roughly 90%) belong to the subphylum Pezizomycotina (ca. 59,000), which likely comprises the largest fraction of unknown fungal diversity [133].

Known species of ascomycetes are mainly terrestrial although more than 3,000 species occur in freshwater habitats and there are more than 1,500 marine species [122].

Most species are decomposers that enzymatically digest organic compounds from the dead substrate where they live on and absorb the released nutrients through the chitinous cell walls of their hyphae. Their role in breaking down organic compounds made them important contributors to nutrient recycling in ecosystems. Some members of the Ascomycota are biotrophs that may colonize roots of most terrestrial plants and form mutualistic symbiosis called mycorrhizas (from the Greek *mykes* = fungus and *rhiza* = root). Both symbionts benefit from the association: the plant provides a carbon source to the fungus and the fungus absorbs phosphate ions and other mineral nutrients from the soil and transfers them to the plant. Also many species of Ascomycota live in a symbiotic relationship with an alga or cyanobacteria forming lichens. Other ascomycetes are used in the food industry (for making bread, cheeses, alcoholic beverages) and for the production of medicinally important compounds, as antibiotics.

The phylum Ascomycota also includes species that cause disease in animals, humans, and plants. Many agriculturally important plant pathogens are ascomycetes.

As the great majority of Basidiomycota, some Ascomycota (like morels and truffles) form macroscopic sexual fruiting bodies, commonly known as mushrooms. However most ascomycetes develop microscopic fruiting bodies and are referred to as microfungi [132].

8.3.2 Life cycle of Ascomycota

Although some Ascomycota, such as yeasts, are unicellular organisms, the great majority grows as filamentous forms. In the vegetative stage, spores of filamentous microfungi germinate and form long thread-like walled filaments termed hyphae which extend, branch, and intertwine within the supporting substrate as a network, denominated mycelium (the thallus). Hyphae are made from tubular cells attached to one another. Growth takes place at the tips of hyphae.

If nutrients are available from the substrate and environmental conditions are favorable, the hyphae of individual fungi extend endlessly via apical growth outward in all directions from the center, showing a radially expanding colonial growth, until nutrient sources are exhausted.

Hyphae are divided into individual cells by internal cross walls called septa (sing., septum), which are perforated by a single, central pore. Fungal cells contain haploid nuclei.

Filamentous microfungi reproduce by means of spores either sexually, with homothallic or heterothallic mycelia depending on the species, or asexually (Figure 8.3).

The sexual reproductive state in the Ascomycota is referred to as the teleomorph, typically a fruiting body, while the asexual state is called the anamorph. Some species produce more than one morphologically distinct anamorphs, which are then called synanamorphs. Fungal species that present a sexual state and one or more asexual states are pleomorphic.

Another frequent term is holomorph which refers to the whole life cycle of a fungus, including both the anamorph and the teleomorph.

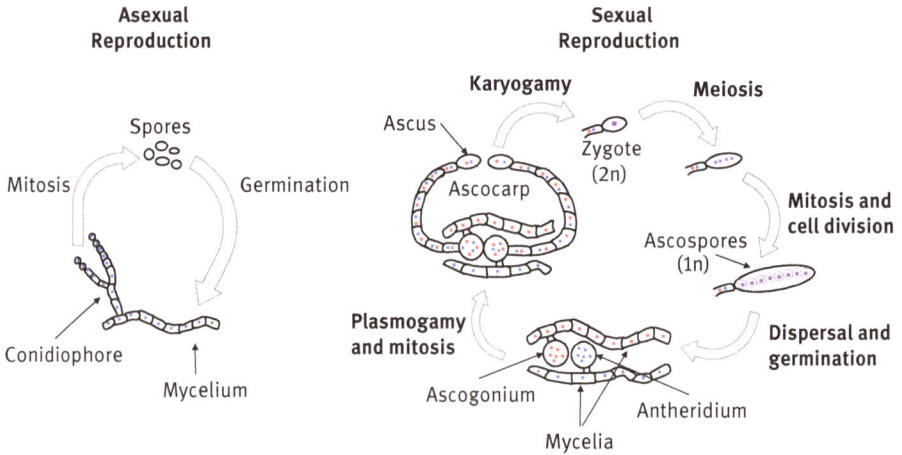

Figure 8.3: Life cycle of filamentous Ascomycota (adapted from: Zeeshan 93 – Own work, CC BY-SA 4.0, https://commons.wikimedia.org/w/index.php?curid=46636026).

8.3.2.1 Sexual reproduction

For sexual reproduction, the haploid hyphae of two mating fungal strains lie side by side, and then each of them develop a projection toward the other. The projections, which are sexual reproductive structures called gametangia (antheridium – ♂, and ascogonium - ♀), fuse together, and then cell walls break down leading to the production of dikaryotic hyphal strands. The tips of these strands eventually form sac-like structures called asci (sing. ascus). The sac-like structures are typical of Ascomycota which are often referred to as sac fungi. In the ascus, the two haploid nuclei from the parent strains fuse and form a diploid nucleus, which undergoes meiosis and forms four haploid nuclei. These haploid nuclei divide by mitosis to form eight haploid ascospores.

Asci may be protected by differentiated haploid hyphae, tightly interwoven, which form fruiting bodies called ascocarps (also known as ascomata). Ascocarps may be spherical (cleistothecia, sing. cleistothecium), globular or flask-like with an apical narrow opening (perithecia, sing. perithecium), or commonly stalked and broadly open like a disk or a cup (apothecia, sing. apothecium). When ascospores are released from ascocarps and germinate, they will develop into new haploid hyphae.

Homothallic fungi have both male and female nuclei derived from the same individual (the same thallus/mycelium) for sexual reproduction. Thus, two types of haploid uclei are produced from the same mycelium to form a diploid nucleus. This means that they reproduce sexually by self-fertilization. They can also reproduce sexually by mating a partner (i.e., another fungal individual of the same species) of compatible mating type.

Heterothallic fungi rely on outcrossing. This is because each individual produces only one type of mating haploid nuclei and needs a mating partner with compatible mycelium, for sexual reproduction. In this case two different mycelia (thalli) are needed to form a zygote. Some heterothallic fungi can also present self-fertilization under certain environmental conditions.

8.3.2.2 Asexual reproduction

Asexual reproduction is the main form of propagation in the Ascomycota. The fungal spore is a haploid cell produced by mitosis from a haploid parent cell, being therefore genetically identical to the parent cell. Asexually produced spores are usually called conidia. They are formed exogenously by fragmentation of the tips of specialized hyphae called conidiophores. Conidiophores may be from short hyphal-like structures to complex branched structures differing notably in appearance from the hyphae. Moreover, conidiophores may branch off from the mycelia, being arranged singly, or may be organized into four types of asexual structures: sporodochia (sing. sporodochium), which are cushion-shaped containers composed of a mass of hyphae covered with conidiophores; pycnidia (sing. pycnidium), hollow round to flask-shaped structure lined with conidiophores; acervuli (sing. acervulus), pile of hyphae bearing a compact layer of conidiophores; and synnemata (sing. synnema), columnar structure of united conidiophores bearing conidia principally at the apex [134].

In conidiophores arranged singly, the hyphal tip can be very similar to a normal hyphal tip or, commonly, be differentiated into a bottle-shaped cell called phialide, from which spores are produced.

Spores are dispersed by wind, water, and animals, and when they germinate they form a hypha which will develop into mycelium.

8.3.3 Naming of filamentous Ascomycota

Mycotoxigenic fungi have been traditionally identified and classified according to the morphological characteristics of their sexual reproductive structures, observed in the teleomorphic state. However, although the anamorphic and the teleomorphic states are known for most ascomycetes, many species have only been found as anamorphs (asexual fungi) and they were identified attending at the asexual reproductive characteristics.

Until recently, pleomorphic fungal species might have two scientific names (termed dual nomenclature): one for the asexual morph (anamorph) and another for the sexual morph (teleomorph), which might cause confusion. In July 2011, experts attending the International Botanical Congress in Melbourne determined to abandon the dual nomenclature for pleomorphic fungi. Consequently, the International Code of Nomenclature for Algae, Fungi, and Plants (ICN) adopted the single name nomenclature. On 1 January 2013, dual naming of fungi ended. Since then, one fungus can only have one name [135]. This change was possible, thanks to methods of DNA sequencing and phylogenetic analysis for fungal identification: the anamorph and the teleomorph of the same fungal species have the same DNA sequence data. The "One Fungus = One Name" nomenclature was applied to all fungal genera, including of course those harboring major mycotoxigenic fungi: *Fusarium*, *Aspergillus*, and *Penicillium* are anamorph names that were chosen over the teleomorph names.

8.3.4 *Fusarium*

The genus *Fusarium* belongs to the family Nectriaceae, order Hypocreales, in the phylum Ascomycota. *Fusarium* is one of the largest genera of fungi. It comprises nearly 1,500 species, subspecies, varieties, and *formae speciales* [136]. Some species of *Fusarium* are among the most economically important plant pathogens affecting agricultural crops worldwide, causing a number of diseases which not only reduce crop quality, since they produced mycotoxins, and yield, but also can lead to the death of plants [137]. This is the case of head blight of wheat, Fusarium ear rot of maize, and Fusarium stalk rot of maize.

8.3.4.1 Morphological characteristics of *Fusarium* species

From a morphological point of view, *Fusarium* isolates are identified and classified based on the characteristics of the asexual reproductive structures. The size, number of septa, general shape, and shape of apical and basal cells of macroconidia are important features for the identification of *Fusarium* species (Figure 8.4). When looking at a macroconidium, the more curved portion of the cell is to the top, the apical cell is to the left, and the basal cell is to the right [138].

Other useful characters for morphological identification of *Fusarium* isolates are the size, number of septa, and shape of microconidia, if present since not all *Fusarium* species produce them. Common shapes of microconidia are oval, kidney-shaped, obovoid with a truncate base, pyriform, napiform, globose, spherical, and fusiform (Figure 8.4).

In Figure 8.5, images of asexual structures of some important mycotoxigenic *Fusarium* species are shown.

Apart from microconidia, the conidiogenous cell on which they are borne, and the arrangement of microconidia on and around the conidiogenous cell are also important characters for species identification [138].

Two basic types of conidiogenous cells are distinguished: monophialides, which have a single opening per cell through which conidia are produced, and polyphialides, which have more than one opening. In the phialides, microconidia may be arranged singly, like in polyphialides, in chains, or in false heads, like in monophialides. The term false head is due to the fact that, despite they superficially resemble spore

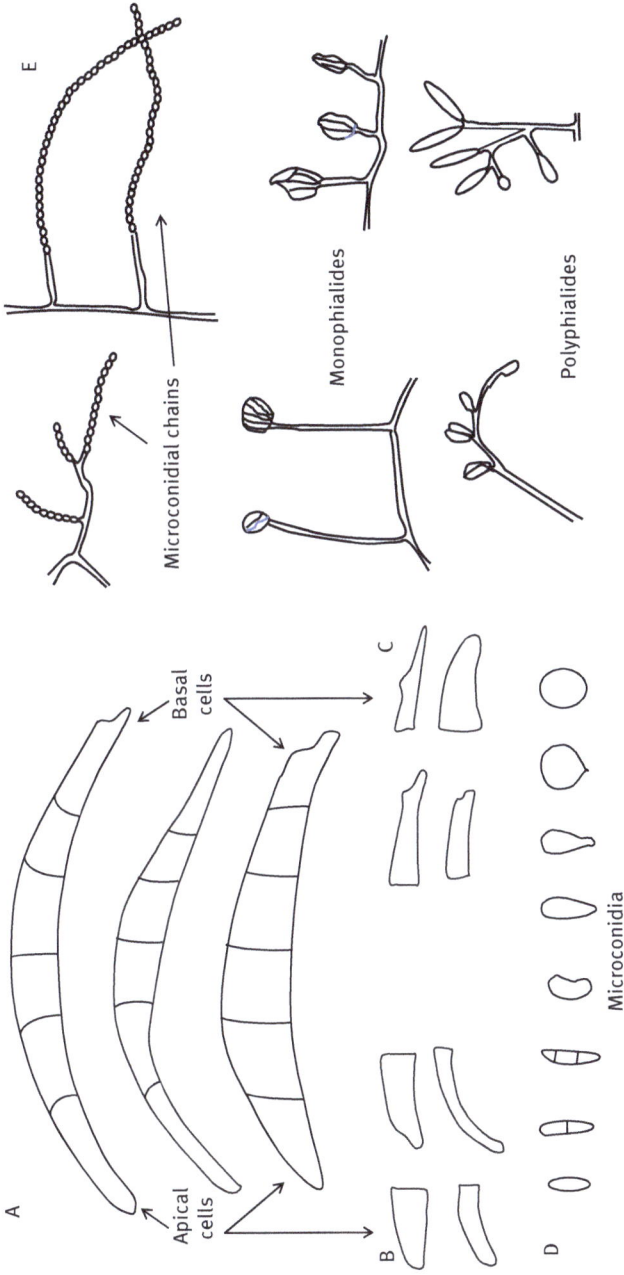

Figure 8.4: Morphological characteristics of macroconidia, apical and basal cells of macroconidia, micronidia, and phialides in the genus *Fusarium* (adapted from [98]). Drawings not to scale.

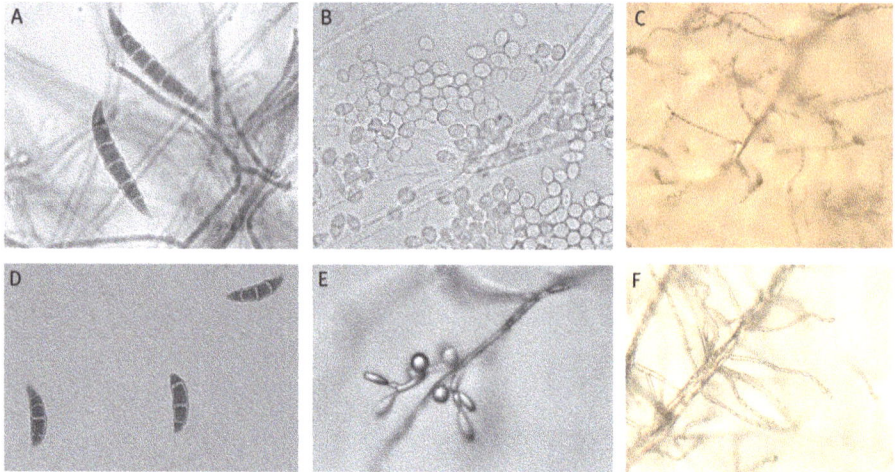

Figure 8.5: Morphological characteristics of asexual structures in the genus *Fusarium*. (A) macroconidia of *F. graminearum*; (B) microconidia of *F. poae*; (C, F) microconidial chains of *F. verticillioides*; (D) macroconidia of *F. culmorum*; and (E) polyphialides of *F. sporotrichioides*.

heads, *Fusarium* species do not form true heads of conidia as happens in *Aspergillus*, for instance, but clumps of spores at the end of the phialides [138].

Some *Fusarium* species also produce chlamydospores, which are enlarged thick-walled vegetative cells that form within hyphae (intercalary) or at hyphal tip (terminally). They separate from the parent hypha and behave as resting spores, surviving in unfavorable conditions. Chlamydospores may be born singly, doubly, in clumps and in chains [138].

In some mycotoxigenic *Fusarium* species, such as *F. avenaceum*, *F. sporotrichoides*, and *F. subglutinans*, fusoid conidia with up to three to four septa, called mesoconidia, are formed from polyphialides in the aerial mycelium, but not in sporodochia. They are larger than microconidia and often lack a notched basal cell.

Pigmentation of colonies growing on agar media (Figure 8.6), either on plates or slants, and growth rate in potato dextrose agar (PDA) medium at either 25 or 30 °C, are also used as significant characters in the identification of *Fusarium* species.

Many *Fusarium* species produce several mycotoxins. Also some mycotoxins are produced by several *Fusarium* species. However, the mycotoxigenic profile can be used as a secondary character for identification of some species.

Up to the introduction of the single nomenclature for fungal species names, several teleomorph genera were associated with species of *Fusarium*, the most common being the genus *Gibberella*. Some *Gibberella* species, which currently are named with the anamorph name, are important pathogens on cereals: *Gibberella zeae* = *Fusarium graminearum* and *G. moniliformis* = *F. verticillioides* [137].

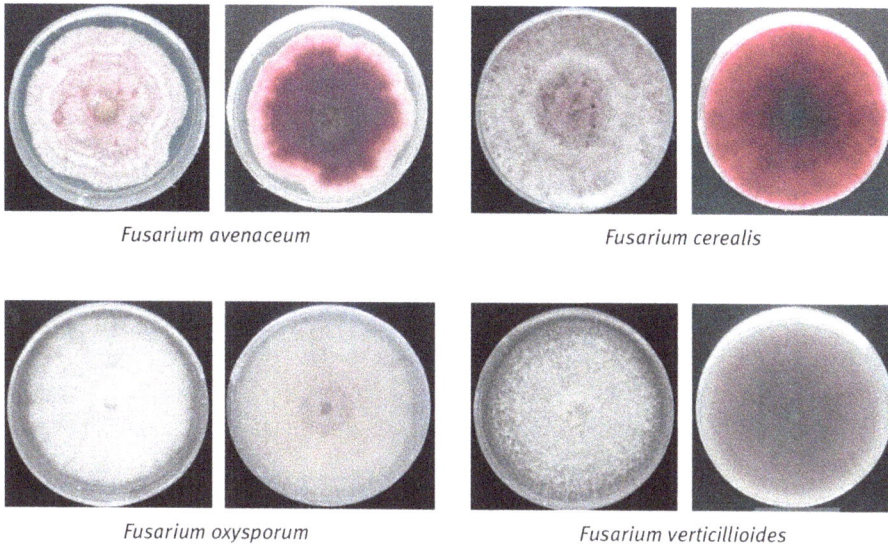

Fusarium avenaceum

Fusarium cerealis

Fusarium oxysporum

Fusarium verticillioides

Figure 8.6: Colony surface (left) and reverse (right) of isolates of *Fusarium avenaceum*, *F. cerealis*, *F. oxysporum*, and *F. verticillioides* grown on PDA at 25 °C.

8.3.5 *Aspergillus* and *Penicillium*

The genera *Aspergillus* and *Penicillium* belong to the family Aspergillaceae, order Eurotiales, in the phylum Ascomycota. They are characterized by the formation of flask-shaped or cylindrical phialides, by the production of asci inside cleistothecia or surrounded by Hülle cells, and by ascospores mainly having a furrow or slit [139].

8.3.5.1 *Aspergillus*

The genus *Aspergillus* comprises approximately 350 accepted species which share an asexual spore-forming structure called aspergillum [140, 141]. The name of the genus was decided by an Italian priest and biologist, Pier Antonio Micheli, in 1729, after viewing the fungal asexual structures of an isolate under a microscope, since they reminded him of the shape of an *aspergillum*, a Latin work to design a holy water sprinkler that derives from the verb *aspergere* (to sprinkle) [142].

    Aspergillus species are currently classified into four subgenera (namely *Aspergillus*, *Circumdati*, *Fumigati*, and *Nidulantes*) and 19 sections, each including related species [139], based on morphological characters but that largely correspond with the current published phylogenies [140].

Of particular interest is the section *Flavi* because it includes the two species most commonly implicated as causal agents of aflatoxin contamination: *Aspergillus flavus* and *A. parasiticus* [143]. *Aspergillus flavus* produces aflatoxins B_1 and B_2, and cyclopiazonic acid (CPA), which is a mycotoxin that has neurotoxic effects [144]. Two types of *A. flavus* strains have been found: the S-type, which produces numerous small sclerotia (<400 μm in diameter), relatively few conidia and high levels of aflatoxins (B_1 and B_2), and the L-type, characterized by producing fewer, larger sclerotia (>400 μm in diameter), more conidia and less aflatoxins than the S-type [145]. *Aspergillus parasiticus* produces both aflatoxins B_1 and B_2 and aflatoxins G_1 and G_2, but no CPA.

The genus *Aspergillus* is widely distributed worldwide, but most species, particularly those of the section *Flavi*, are mainly found in temperate and subtropical climates between latitudes 26° and 35° north and south of Ecuador [146], and therefore the risk of aflatoxin contamination in food and feed commodities is greater in those regions.

8.3.5.2 Main morphological characteristics of *Aspergillus* species

The morphology of the conidiophore is the predominant microscopic character used for the characterization and identification of *Aspergillus* species (Figure 8.7).

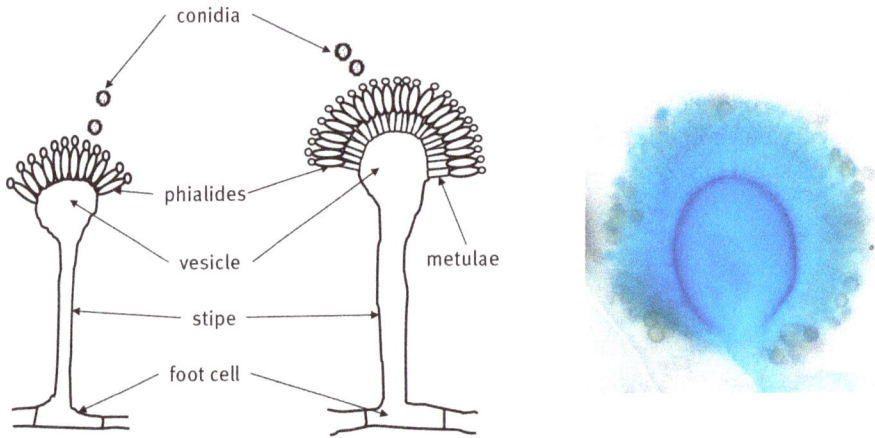

Figure 8.7: On the left, morphological characteristics of the conidiophores in the genus *Aspergillus* (drawings not to scale). On the right, photo of a conidiophore of *Aspergillus flavus* is stained with cotton blue.

In most aspergilli, the conidiophore has a foot cell at the base and a long stipe (usually nonseptate), which is a thick-walled hyphal branch which arises perpendicularly from the foot cell. The tip of the stipe is swollen forming a structure called vesicle. Conidiogenous cells, termed phialides, develop on the vesicle surface. In some species, the phialides are the only layer of supporting cells on the surface of the vesicle, and then it is said that the conidiophore has an uniseriate head. In other species, there is a layer of supporting cells, called metulae, on the surface of the vesicle and, over them, a second layer with the phialides. When metulae and phialides, which are borne simultaneously, are present, the conidiophore has a biseriate head. Conidia form by budding of the cytoplasm from the phialide cells. Additional phenotypic characters used for species identification are conidial color and ornamentation, growth rate on agar media, and growth rate at different temperatures and water activities [139, 147]. As an example, *Aspergillus* section *Flavi* includes species with conidial heads in shades of yellow-green (Figure 8.8).

Figure 8.8: Colony surface of isolates of *Aspergillus* spp. belonging to section *Flavi* grown on PDA (top left and right, bottom left) and malt extract agar (MEA; bottom right).

In addition to the conidial state (anamorph) characteristic of the genus, approximately one-third of *Aspergillus* species also have a sexual state (i.e., a teleomorph), all but five of which are homothallic [148]. For sexual reproduction, *Aspergillus* form cleistothecia, which are ascocarps (fruiting bodies) containing numerous asci. Meiospores called ascospores are formed within the asci. The ascocarps produced by the sexual states associated with *Aspergillus* are so different in morphology that the teleomorphs were until recently classified in 11 different genera, what reflects an enormous degree of phylogenetic and biological diversity [148, 149].

Traditionally nine teleomorph genera were linked to *Aspergillus* anamorph: *Chaetosartorya*, *Emericella*, *Eurotium*, *Fennellia*, *Hemicarpenteles* (now considered to belong to *Penicillium*), *Neosartorya*, *Petromyces*, *Sclerocleista*, and *Stilbothamnium*, and two more were added at the beginning of the 2000s: *Neopetromyces* and *Neocarpenteles*. When the single-name system came into use all the teleomorph genera were synonymized with *Aspergillus* [140].

Some aspergilli form masses of Hülle cells associated with cleisothecia. Hülle cells are thick-walled specialized multinucleate cells that originate from a nest-like aggregation of hyphae during sexual development. They serve as nurse cells to the developing cleistothecium. Other structures found in *Aspergillus* are sclerotia, which are asexual hardened masses of hyphae, yellow to black in color, capable of survive dormant in soil in harsh environments for long periods until conditions are favorable for growth. Species of *Aspergillus* section *Flavi* form dark sclerotia (Figure 8.9). In some species cleistothecia form is embedded within the sclerotia [156].

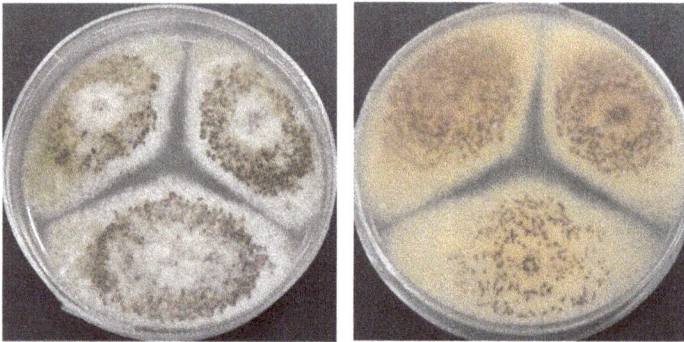

Figure 8.9: Formation of sclerotia by an isolate of *Aspergillus* sp. of section *Flavi* grown on CYA.

8.3.5.3 *Penicillium*

The genus *Penicillium* includes over 360 species [139]. Most decompose organic materials and cause devastating rots as pre- and postharvest pathogens on food crops, producing a wide range of mycotoxins, while other are common indoor air allergens [150]. *Penicillium* species contaminate a wide variety of foods and are capable of growing at refrigeration temperatures, thus frequently spoiling refrigerated food products.

The name *Penicillium* is derived from the Latin word *penicillus* which means little brush. It was introduced by Link in 1809 due to the similarity of the asexual structures of these fungi with a brush.

8.3.5.4 Main morphological characteristics of *Penicllium* species

The conidiophores of *Penicillum* branch near the apex, forming a brush-like structure (Figure 8.10). Enlarged cells, called metulae, are formed at the apex of the conidiophore.

The conidiophores exhibit a well-defined cluster of phialides that are either directly attached to a stipe, or through one or more stages of branching, depending on the species [151].

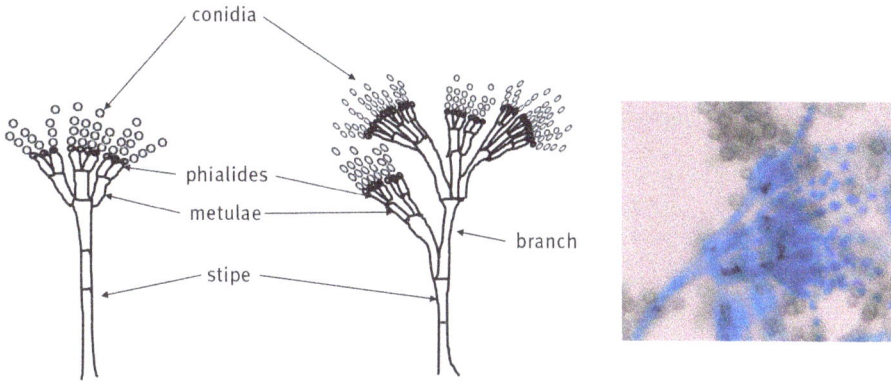

Figure 8.10: On the left, morphological characteristics of the conidiophores in the genus *Penicillium* (drawings not to scale). On the right, photo of a conidiophore of *Talaromyces purpureogenus* (formerly *Penicillium purpureogenum*) stained with cotton blue.

The conidia of the penicillia are colored, mostly in shades of gray to blue to blue-green (Figure 8.11).

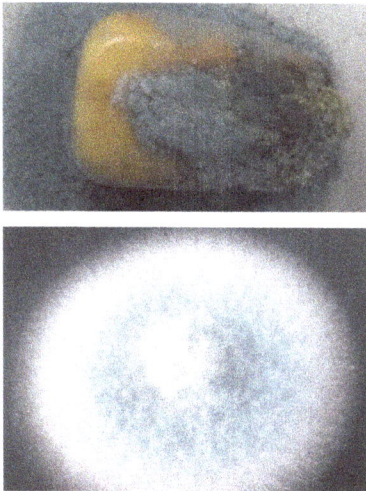

Figure 8.11: Above: blue-gray colonies of *Penicillium* spp. growing from a maize kernel on PDA (yellowish green colonies on the kernel tip correspond to *Aspergillus* sp. of section *Flavi*). Below: colony of a monosporic isolate of *Penicillium* sp. grown on MEA at 25 °C.

Some species of *Penicillium* have a sexual state, forming ascospores in cleistothecia. *Eupenicillium*, *Chromocleista*, and *Hemicarpenteles* are teleomorph genera currently included in the one name *Penicillium*. However the teleomorph genus *Talaromyces*, characterized by acerose phialides and usually symmetrical branched conidiophores, was not included and now it is acknowledged as a separate fungal genus [139].

Aspergillus and *Penicillium* are phylogenetically closely related [140], but can be easily differentiated by the morphological features of their conidiophores: the conidiophore of *Aspergillus* has a well-defined foot cell and an unbranched often nonseptate stipe, while the conidiophore of *Penicillium* is branched, has a septated stipe, and lacks a distinct foot cell. In addition, the phialides in *Penicillium* are born successively and not simultaneously as in *Aspergillus* [139].

8.4 Methods for isolation, identification, and characterization of mycotoxigenic fungi

Taxonomic identification of fungi is essential when working with isolates that contaminate food and feed products with mycotoxins not only for assessing food quality but also for the development of control strategies for ensuring food safety.

Identification of mycotoxigenic species is initially based on the observation of morphological characteristics of isolates, specifically macroscopic characters of colonies grown on culture media and microscopic features of the sexual and/or asexual structures. However, although classic identification methods can be useful to identify some fungi, they frequently provide poor identification sensitivity, have long turn-around times, and can lack specificity for species-level identification [152].

For these reasons, the traditional phenotypic identification is currently completed with molecular techniques, which include extraction of fungal DNA, polymerase chain reaction (PCR) amplification and sequencing of certain regions of the fungal genome, and phylogenetic analysis. In many occasions molecular techniques are essential for species identification. In biology, phylogenetics is the study of evolutionary relatedness among groups of organisms (e.g., species and populations), which are discovered through molecular sequencing data and morphological data matrices.

The classification and identification of *Fusarium*, *Aspergillus*, and *Penicillium* species were based on phenotypic characters until the early 1990s, but, since then, a polyphasic approach has been implemented in the taxonomy of the three genera. The polyphasic taxonomy integrates several kinds of data and information on the fungal isolates, as phenotypic (macromorphology of colonies and micromorphology of fungal structures) and physiological (e.g., growth on different cultivation media at different temperatures and water activity) data, extrolite profiles (in the case of *Aspergillus* and *Penicillium*), and multigene phylogenetic analysis [139, 153]. Extrolites are secondary metabolites, including mycotoxins.

The extrolite profiles are acquiring more and more importance in the polyphasic taxonomy since they are species-specific [151]. Nevertheless, current identification of

Aspergillus and *Penicillium* is strongly supported by molecular and phylogenetical analysis but still to a lesser extent by chemotaxonomy [140].

For phylogenetic species recognition of isolates from the three genera, fungal DNA is first extracted and then usually two to four loci are amplified and sequenced. The genes most often used are ITS, beta-tubulin (*BenA*), calmodulin (*CaM*), translation elongation factor 1 alpha (*TEF1α*), RNA polymerase II largest subunit (*RPB1*), RNA polymerase II second largest subunit (*RPB2*), and actin (*Act*). Sequences from the amplified genes are then used to construct gene genealogies and compare them. For a given fungal isolate, the concordance of two or more gene genealogies allows species recognition. The phylogenetic approach that recognize fungal species based on concordance of multiple gene genealogies is known as Genealogical Concordance Phylogenetic Species Recognition and was first endorsed by [154].

There has been a steep increase in the number of accepted *Fusarium*, *Aspergillus*, and *Penicillium* species in the last 10 years and will probably continue into the next [139]. This has been related to the introduction of molecular techniques and phylogenetic analysis in many laboratories worldwide.

8.4.1 Detection and isolation of mycotoxigenic fungi

Mycotoxigenic fungi must be isolated from representative food and feed samples properly collected and handled. At the time of collection, it is important to identify clearly each sample with a label. In case samples are perishable, they must be transported to the laboratory in portable coolers.

The detection of mycotoxigenic fungi in plant samples from the field and in food and feed products requires the following preliminary procedures:

1. Use a laminar flow hood. Prepare a safe and sterile workspace. Keep all surfaces dry and clean. Keep in mind that surfaces at the back of the hood and objects close to the table surface are more likely to remain sterile. When handling material potentially contaminated with mycotoxigenic fungi, wearing of a laboratory coat, disposable gloves, and safety glasses is recommended.
2. Prepare agar media for fungal cultivation (Figure 8.12). All ingredients must be mixed up and then heat to dissolve. Afterward media must be sterilized in an autoclave at 121 °C for 15 min. Verify for correct pH before autoclaving.

    The agar medium should be tempered to 45–50 °C before dispensing it aseptically into the base of Petri dishes of 9 mm diameter. This will minimize condensation in the lid of the dish. Also it is an essential step if a heat-sensitive solution (e.g., an antibiotic) is to be added [138]. Pass the neck of agar bottle through flame, open Petri dish lid as little as possible and dispense approximately 20 mL of agar medium per Petri dish of 90 mm diameter. Wait until solidify. Petri plates can be stored in an inverted position in a refrigerator until use in order to prevent drying of the agar.

Weigh components

Add components into a flask

Mix well and pour in a bottle

Sterilize medium

Temper the medium

Pour the melted medium into Petri dishes

Allow the medium to cool and harden

Wrap dishes with aluminum foil

Keep dishes in a refrigerator

Figure 8.12: Procedure for preparing agar media for fungal cultivation.

Although agar media in Petri dishes are usually employed, the same media can also be used in agar slants (agar slope) in test tubes. A slant is a tube placed at an angle during cooling to give a large slanted surface for fungal inoculation. In this case, once prepared, the medium is dispensed into the slants with a syringe or some other repeating dispenser before autoclaving. Test tubes of 10 × 75, 13 × 100, and 16 × 150 mm are most commonly used, receiving 1.25, 2.5, and 6.0 mL of medium per slant, respectively [138]. After adding the medium, tubes are stoppered with plastic foam plugs or aluminum foil, placed in a rack, and autoclaved. Afterward the medium in the test tube is allowed to harden in a slanted position (some test tubes are specifically set up for this), laying on their sides using a pipet to keep them tilted up. Slants should then be stored in a vertical position in a refrigerator until use.

3. Sterilize not only solutions and media but also all instruments before use. Autoclave all material at 121 °C for 15 min.

8.4.1.1 Media for isolation of *Fusarium*, *Aspergillus*, and *Penicillium* species

The standard media used for isolation of *Fusarium* are PDA, which is a general purpose media for growing a wide range of fungi, and Komada, which is a *Fusarium*-selective medium. Also malt extract agar (MEA) and water agar (WA) can be used. Spezieller Nährstoffarmer agar (SNA) and carnation leaf-piece agar (CLA) are also employed, together with PDA, for species identification [138].

PDA, MEA, and more complex agar media such as Czapek's agar (CZ) and Czapek yeast autolysate agar (CYA) are commonly used for the isolation of *Aspergillus* and *Penicillium* species.

Antibiotics, such as chloramphenicol and streptomycin, may be added to molten agar media (after autoclaving) to inhibit bacterial growth on plates. Also dichloran, Rose Bengal, or NaCl are routinely added to agar media to inhibit fast-growing molds, such as *Rhizopus* and *Mucor* species [143].

In Table 8.5, the composition of media recommended for isolation and identification of species of *Fusarium*, *Aspergillus*, and *Penicillium* is shown.

8.4.1.2 Plating methods

For the detection of mycotoxigenic fungi in food and feed products, dilution plating and direct plating in general and specific cultivation media are the most widely used methods [132].

In the dilution plating method, a food homogenate is prepared and then serial dilutions of the homogenate are made followed by plating. For this 1 g (dry weight) of the sample is ground up (if needed) and dispersed in 9 mL of sterile water. One millili-

Table 8.5: Media for fungal cultivation (adapted in part from [100, 110]).

Potato dextrose agar (PDA)

White potatoes	250 g
Dextrose	20 g
Agar	20 g
Distilled water	100 mL

Czapek stock solution (concentrate)

$NaNO_3$	30 g
KCl	5 g
$MgSO_4 \cdot 7H_2O$	5 g
$FeSO_4 \cdot 7H_2O$	0.1 g
Distilled water	100 mL

Trace elements' stock solution

$CuSO_4 \cdot 5H_2O$	0.5 g
$ZnSO_4 \cdot 7H_2O$	0.1 g
Distilled water	100 mL

Czapek's agar (CZ)

Czapek concentrate	10 mL
Sucrose	30 g
$CuSO_4 \cdot 5H_2O$	0.005 g
$ZnSO_4 \cdot 7H_2O$	0.001 g
Agar	20 g
Distilled water	1,000 mL

Czapek yeast autolysate agar (CYA)

Czapek concentrate	10 mL
Sucrose	30 g
Yeast extract	5 g
$CuSO_4 \cdot 5H_2O$	0.005 g
$ZnSO_4 \cdot 7H_2O$	0.01 g
Agar	20 g
Distilled water	1,000 mL
pH 6.2 ± 0.2	

Malt extract agar (MEA)

Malt extract	50 g
$CuSO_4 \cdot 5H_2O$	0.005 g
$ZnSO_4 \cdot 7H_2O$	0.01 g
Agar	20 g
Distilled water	1,000 mL

Malt extract 20% sucrose agar (M20S)

Malt extract	50 g
Sucrose	200 g
$CuSO_4 \cdot 5H_2O$	0.005 g
$ZnSO_4 \cdot 7H_2O$	0.01 g
Agar	20 g
Distilled water	1,000 mL
pH 5.4 ± 0.2	

Yeast extract sucrose agar (YES)

Yeast extract	20 g
Sucrose	150 g
$MgSO_4 \cdot 7H_2O$	0.5 g
$CuSO_4 \cdot 5H_2O$	0.005 g
$ZnSO_4 \cdot 7H_2O$	0.001 g
Agar	20 g
Distilled water	885 mL
pH 6.5 ± 0.2	

Spezieller Nährstoffarmer agar (SNA)

Sucrose	0.2 g
Glucose	0.2 g
KH_2PO_4	1 g
KNO_3	1 g
$MgSO_4 \cdot 7H_2O$	0.5 g
KCl	0.5 g
Agar	20 g
Distilled water	1,000 mL

Table 8.5 (continued)

Creatine sucrose agar (CREA)

Sucrose	30 g
Creatine·5H$_2$O	3 g
K$_3$PO$_4$·7H$_2$O	1.6 g
MgSO$_4$·7H$_2$O	0.5 g
KCl	0.5 g
FeSO$_4$·7H$_2$O	0.01 g
CuSO$_4$·5H$_2$O	0.005 g
ZnSO$_4$·7H$_2$O	0.01 g
Bromocresol purple	0.05 g
Agar	20 g
Distilled water	1,000 mL
pH 8 ± 0.2	1,000 mL

Oatmeal agar (OA)

Oatmeal flakes	30 g
CuSO$_4$·5H$_2$O	0.005 g
ZnSO$_4$·7H$_2$O	0.01 g
Agar	20 g
Distilled water	1,000 mL
pH 6.5 ± 0.2	

Autoclave flakes (121 °C for 15 min) in 1,000 mL of dH$_2$O. Squeeze mixture through cheese cloth and use flow through, topping up to 1,000 mL with distilled water with 20 agar.

Komada-Fusarium medium

Part 1	
Na$_2$B$_4$O$_7$·10H$_2$O	1 g
K$_2$HPO$_4$	1 g
KCl	0.5 g
MgSO$_4$·7H$_2$O	0.5 g
Fe-Na-EDTA	0.001 g
D-Galactose	20 g
L-Asparagine	2 g
Agar	15 g
Distilled water	1,000 mL

Komada-Fusarium medium

Part 2	
PCNB (pentachloronitrobenzene) (Terraclor 75%)	1 g
Oxgall (bile bovine)	0.5 g
Streptomycin sulfate	0.3 g
pH 3.9 ± 0.1	

Mix components of part 1 and boil to melt agar. Then, cool the medium until 50–55 °C and add components of part 2.

Dichloran glycerol agar (DG18)

Peptone	5 g
Glucose	10 g
KH$_2$PO$_4$	1 g
MgSO$_4$·7H$_2$O	0.5 g
Dichloran	0.002 g
Chloramphenicol	0.1 g
Agar	15 g
Distilled water	1,000 mL
pH 5.6 ± 0.2	

ter of this solution is transferred to a second tube containing 9 mL of sterile water, resulting in a 0.01 dilution of the spore mass in the original sample. This process is repeated to yield dilutions of 0.001, 0.0001, and so on. For plating, either the pour-plate method or the spread-plate method can be used. In the pour-plate method, the diluted sample is pipetted into a sterile Petri dish, and then a sterile melted medium is added and mixed with the sample. In the spread-plate method, the diluted sample is pipetted and spread evenly onto the surface of the agar plate with the help of a sterile glass spreader.

In the direct plating method, the sample is usually surface-disinfected with 5% sodium hypochlorite or 70% ethanol–water solution before plating. Working in a laminar airflow hood, forceps, sterilized by passing them through a flame and allowed to cool, should be used to remove pieces of the sample from the disinfecting solution. Then the sample pieces should be blotted on a sterile paper towel. In case samples are not surface-disinfested, it is advisable to keep them at –20 °C to kill mites and insects that might interfere with fungal isolation [143]. Plating consists of placing pieces of the sample on the surface of the agar in Petri dishes (Figure 8.13). Whatever the plating method used, agar plates are then incubated at 25 °C in the dark during 7 days or longer.

8.4.2 Obtaining monosporic fungal isolates

Fungal colonies grown from the sample pieces in agar plates will have to be subcultured until obtaining pure cultures. For species identification, it is essential to obtain pure cultures of the mycotoxigenic fungi that may contaminate food and feed commodities.

Small pieces of agar containing hyphal tips should be cut from the growing edge of every distinct colony visible on the plate and transferred to another agar plate and incubated again until obtaining pure cultures, with no contamination of other fungal isolates or bacteria. This can be done with the help of a dissecting microscope placed inside the laminar airflow hood.

Another way to obtain a pure culture of a fungal isolate is to try to remove single-germinated conidia with the help of a sterile needle, previously dipped in sterile distilled water and placed them onto agar medium (in a Petri dish or a slant). When dealing with *Fusarium* isolates, monosporic cultures are obtained, as a standard, from an agar piece carrying a single germinated conidium that is placed in the center of an agar plate or the centre of the surface in an agar slant (Figure 8.14). Monosporic isolates can be stored at 4 °C on agar plates.

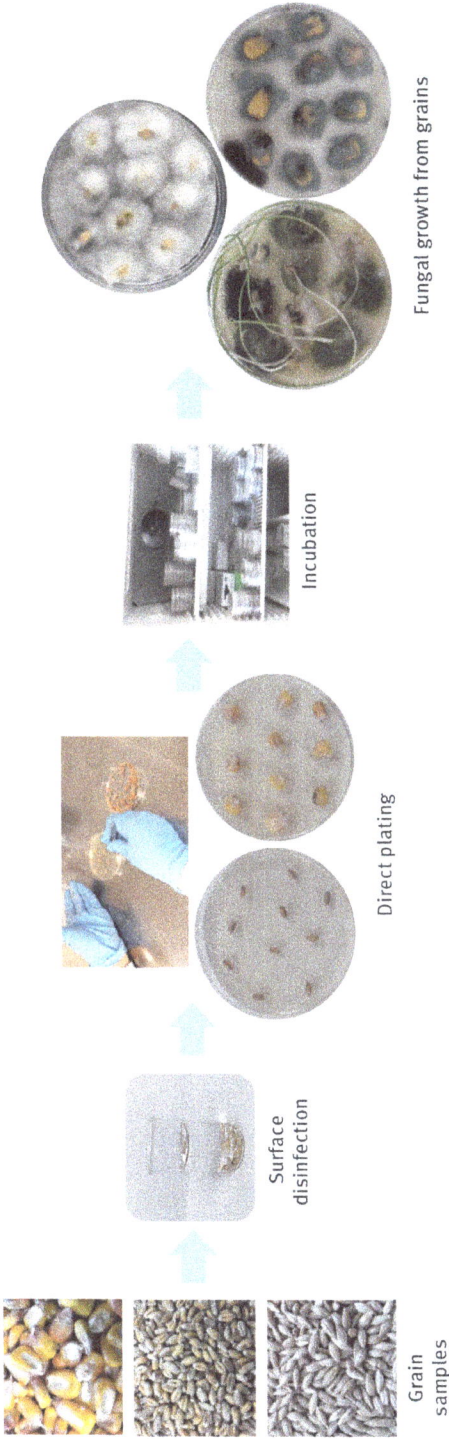

Figure 8.13: Isolation of mycotoxigenic fungi by direct plating method. On the left (from top to bottom): grains of maize, wheat, and barley. On the right: above, colonies of Fusarium spp. growing from barley grains; below, colonies of Aspergillus spp. (green and black mycelia) and Fusarium spp. (pinkish white mycelia) growing from wheat grains (left), and colonies of Penicillium spp. (blue grayish mycelia) and Aspergillus spp. (black mycelia) growing from maize kernels (right).

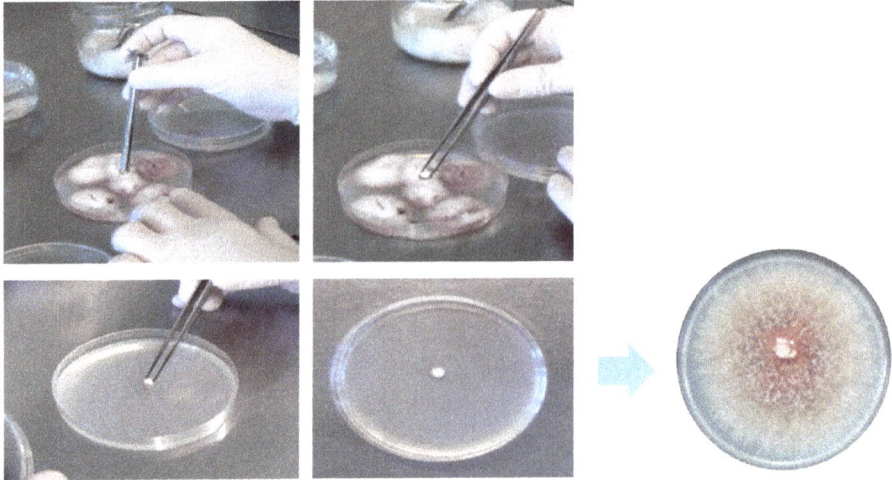

Figure 8.14: Subcultivation of fungal isolates grown from wheat grains for obtaining monosporic cultures. On the right, a monosporic isolate of *Fusarium culmorum* grown on PDA.

8.4.3 Morphological characterization

Once monosporic cultures of *Fusarium*, *Aspergillus*, and *Penicillium* are obtained, the next step is made subcultures on specific media for morphological characterization.

Morphological identification of *Fusarium* species generally is based on characters observed in pure cultures of isolates grown on PDA, SNA, or CLA. Colony morphology, pigmentation, and growth rates must be observed in PDA plates. Shape and size of macroconidia, shape, size and formation of microconidia, and production of chlamydospores can be recorded in colonies growing in PDA but especially in SNA and CLA [138]. Some *Fusarium* species have particular morphological features that allowed their identification, but many do not. In any case, morphological characterization of *Fusarium* species is currently complemented with molecular and phylogenetic analysis.

Samson et al. [140] and Visagie et al. [150] recently recommended the use of the following standardized methods for the subculture of *Aspergillus* and *Penicillium* isolates for their identification and characterization based on morphological characters:

1. Prepare spore suspensions in a 30% glycerol SS-buffer (0.5 g/L agar, 0.5 g/L Tween 80) solution, which can be stored at −80 °C. Spore suspensions can be also made in a 0.2% agar and 0.05% Tween 80 solution, and stored at 4 °C.

2. Prepare cultivation media in 90 mm Petri dishes (preferentially vented) with a volume of 20 mL of medium per plate. Media recommended as standard for *Aspergillus* and *Penicillium* are CYA and MEA. Additional media, such as CZ, yeast extract sucrose agar (YES), dichloran 18% glycerol agar (DG18), oatmeal agar (OA) and creatine sucrose agar (CREA), among others, can be used to observe a wider range of mor-

phological characters. Some morphological characters of *Aspergillus* isolates may vary depending on the cultivation media and incubation conditions.

3. Inoculate the corresponding spore suspension in CYA and MEA plates in three-point pattern (Figure 8.15) using a micropipette (0.5–1 μL per spot). Do not wrap plates with Parafilm® since air exchange restriction often inhibits growth and sporulation [155].

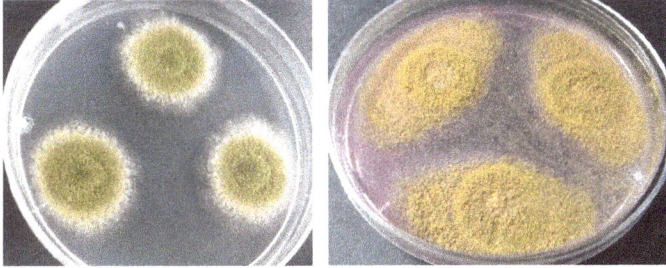

Figure 8.15: Colonies of two isolates of *Aspergillus* spp. of section *Flavi* grown from spore suspensions inoculated on CYA (left) and DG18 (right) plates in three-point pattern.

4. After inoculation, incubate CYA and MEA plates reverse side up at 25 °C during 7 days in the dark, setting additional CYA plates incubated at 30 and 37 °C that are useful to distinguish between species. For *Aspergillus* isolates of the section *Circumdati* additional CYA plates incubated at 30 °C are recommended, while at 45–50 °C for those of the section *Fumigati*.

After 7 days growth, macroscopic and microscopic features of the isolates must be recorded. Macroscopic characteristics of *Aspergillus* are colony growth rates, degree of sporulation, production of sclerotia or cleistothecia, colors of mycelia, formation of soluble pigments, exudates, colony reverses, Hülle cells, and cleistothecia [140].

In *Penicillium* isolates, important macroscopic characters are colony texture, degree of sporulation, the color of conidia, the abundance, texture and color of mycelia, the presence and colors of soluble pigments and exudates, colony reverse colors, and degree of growth and acid production on CREA (creatine sucrose agar). As in the case of *Aspergillus*, some species may produce cleistothecia or sclerotia, but mostly after longer incubation times, especially on OA. Thus OA plates should be incubated for prolonged periods [150].

Microscopic characteristics of asexual and sexual reproductive structures of *Aspergillus* and *Penicillium* are important diagnostic characters. In *Aspergillus* isolates, the shape of conidial heads, the presence or absence of metulae between vesicle and phialides (i.e., uniseriate or biseriate), color of stipes, and the dimension, shape and texture of stipes, vesicles, metulae (when present), phialides, conidia, and Hülle cells (when present) should be recorded. The same must be done for cleistothecia, asci, and ascospores. Also important for identifying species are the size, morphology, and, especially, ornamentation of ascospores.

The same applies for *Penicillium* isolates. Microscopic characteristics of conidiophores and cleistothecia (when produced) should be observed. It is important to accurately describe the branching patterns of conidiophores, to record the wall texture/ornamentation of stipes and conidia, the dimension, shape, and texture of stipes, vesicles, metulae/branches (when present), phialides, conidia, cleistothecia, asci, and ascospores (when present).

For observing microscopic characteristics, preparations are usually made from fungal cultures on PDA or SNA in the case of *Fusarium*, and on MEA in that of *Aspergillus* and *Penicillium*, after 7–10 days growth, although other media can also be used. In fact, OA is a good medium for observing cleistothecia. As a mounting fluid for preparations, it is recommended the use of lactic acid (60–70%) or Shear's solution or lactic acid with cotton blue. Lactofuchsin can be used for *Fusarium* cultures. The composition of mounting fluids is presented in Table 8.6.

Table 8.6: Mounting media.

Shear's mounting medium	
Potassium acetate	6 g
Glycerine	120 mL
Ethanol 95%	180 mL
Distilled water	300 mL
Ink blue	1 g
Lactofuchsin	
Acid fuchsin	0.1 g
Lactic acid (85%)	100 mL
Cotton blue in lactic acid	
Cotton blue	0.01 g
Lactic acid (85%)	100 mL

For staining fungal structures of *Fusarium*, *Aspergillus*, and *Penicillium* isolates, a small drop of mounting medium can be first placed on a microscope slide (Figure 8.16). Then, using a sterilized dissecting or inoculating needle, a small portion of the colony must be removed and placed in the mounting medium, and, with a second needle, tease it out. Mounts can be washed with drops of 70% ethanol to prevent air-bubbles and wash away excess conidia in the case of *Aspergillus* isolates. Then a cover-slip should be carefully placed over the mount, lowering one edge to the slide before the other.

When possible, the use of differential interference contrast light microscopy (= Nomarski) is highly recommended for best observation of conidial ornamentation and conidiophore characters [140, 150] since this technique produce high-contrast op-

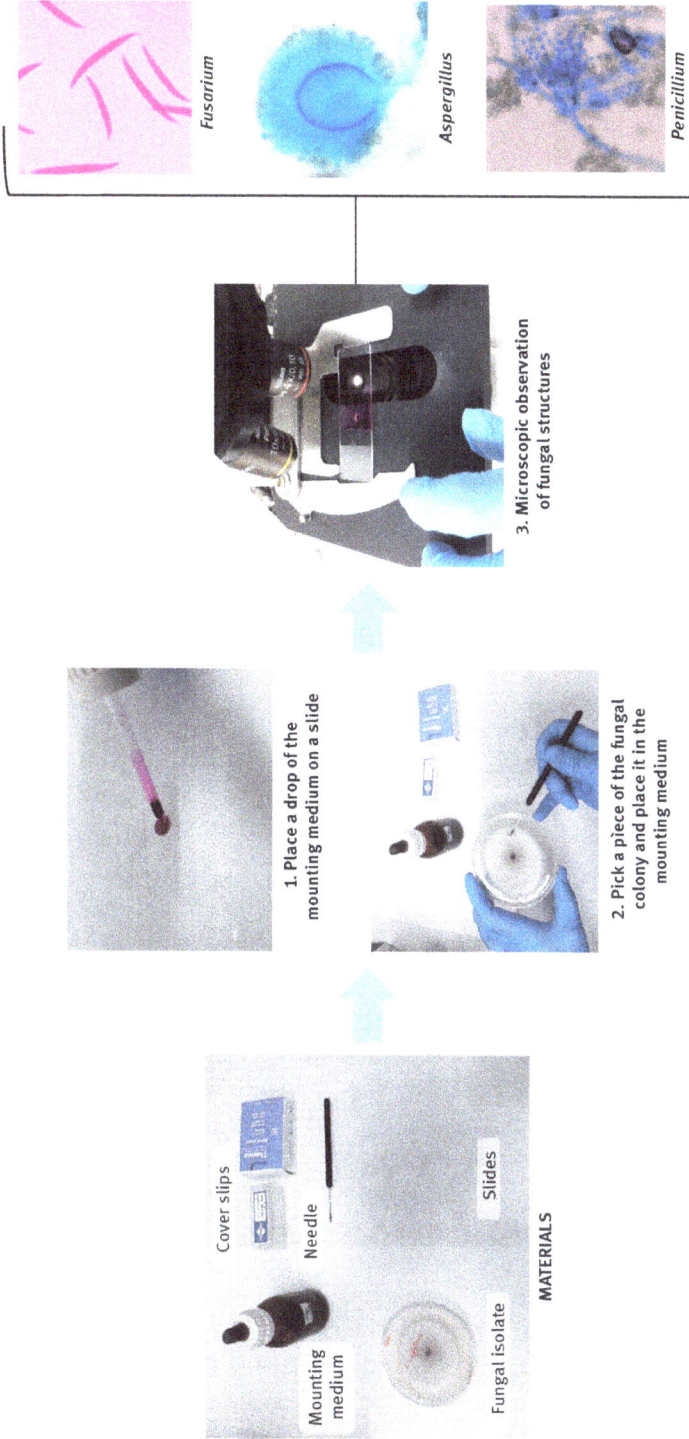

Figure 8.16: Procedure for staining and mounting slides for microscopic observation of fungal isolates. On the right (from top to bottom), macroconidia of *Fusarium graminearum*, a conidiophore of *Aspergillus flavus*, and a conidiophore of *Talaromyces purpureogenus* (formerly *Penicillium purpureogenum*) are shown.

tical images of the edges of objects and fine structural detail within transparent specimens (Figure 8.17).

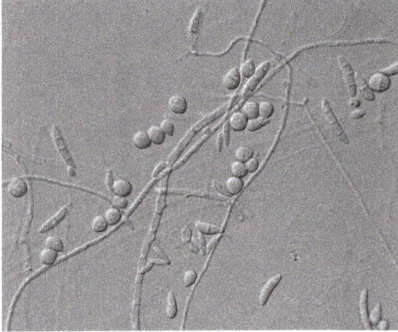

Figure 8.17: Microconidia of *Fusarium sporotrichioides* observed in a microscope by Nomarski interference contrast.

8.4.4 Molecular and phylogenetic identification

For phylogenetic identification of fungal species, the following molecular techniques have to be carried out (Figure 8.18): extraction of fungal DNA, PCR amplification, and sequencing of certain regions of the fungal genome that are associated with certain locations/functions and are generally referred to as molecular markers.

8.4.4.1 Extraction of genomic DNA

Accurate identification of mycotoxigenic fungi using a sequence-based approach requires an extraction method that yields template DNA pure enough for PCR or other types of amplification [156]. There are commercial kits that enable rapid extraction of fungal DNA extraction. Quantity, quality, and integrity of template DNA must be subsequently determined. The most common technique to determine DNA concentration and purity is measurement of absorbance (optical density) using an UV spectrophotometer. The DNA concentration is determined from the absorbance at 260 nm. The most common purity calculation is the ratio of the absorbance at 260 nm divided by the reading at 280 nm, which gives an indicator of purity form protein contaminants (protein absorbs at 280 nm while nucleic acids absorb at 260 nm). Good-quality DNA will have an A_{260}/A_{280} ratio of 1.7–2.0. Lower values may indicate protein contamination. Another indicator of sample purity is the 260/230 ratio that should be around 2.0–2.2. Low A_{260}/A_{230} ratios are related to contamination with buffer salts or organic compounds. DNA quantity and integrity can be determined by agarose (0.5% wt/vol) gel electrophoresis at 60 V using an appropriate DNA ladder as molecular weight markers, followed by gel staining and visualization under UV light.

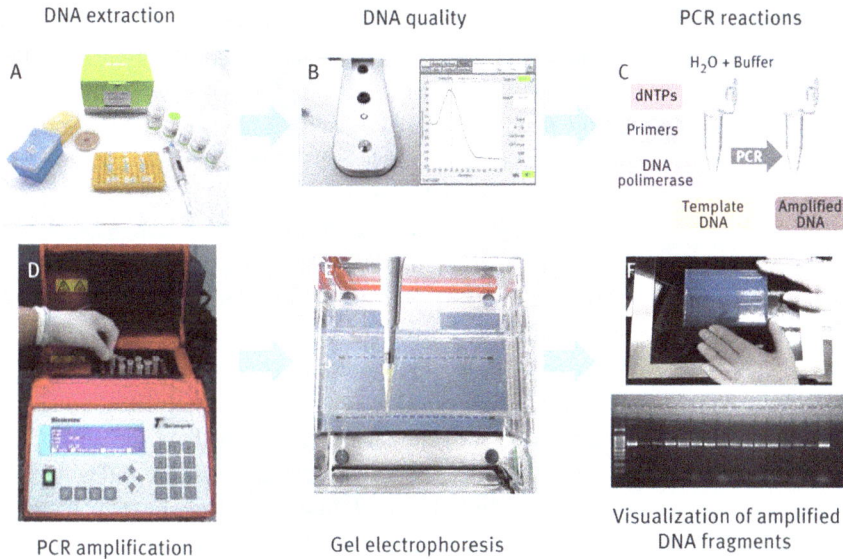

DNA extraction DNA quality PCR reactions

PCR amplification Gel electrophoresis Visualization of amplified DNA fragments

Figure 8.18: Methods for molecular identification of mycotoxigenic fungi: extraction of fungal DNA, determination of quality and quantity of template DNA, PCR amplification of certain regions of the fungal genome, electrophoresis of PCR products in agarose gels, and examination of amplified DNA products under a UV transilluminator.

8.4.4.2 PCR amplification and sequencing

Once good-quality fungal DNA is extracted, two to four loci are usually amplified and sequenced (Figures 8.18 and 8.19). The internal transcribed spacer region of the nuclear rDNA, which encompasses the noncoding transcribed spacers ITS1 and ITS2 and the 5.8S rDNA gene (ITS1-5.8S-ITS2), is the official barcode for fungi because it has primers that work universally [157]. However in *Fusarium*, *Aspergillus*, *Penicillium*, and many other genera of ascomycetes, the ITS region is not variable enough for distinguishing closely related species. In addition the ITS region is characterized by frequent size variation and the presence of repeated sequences, which hinders sequence alignment above the genus level [158].

As secondary marker, the more commonly sequenced gene for *Fusarium* is the *TEF1α*, although other genomic DNA genes, such as beta-tubulin and histone H3, and mitochondrial DNA genes have also been used. *TEF1α*, which encodes an essential part of the protein translation machinery, shows a high level of sequence polymorphism among closely related species, even in comparison to the intron-rich portions of protein-coding genes such as calmodulin, beta-tubulin, and histone H3. Thus, *TEF1α* has high phylogenetic utility because it is highly informative [159, 160].

Samson et al. [140] proposed the use of the *CaM* as a temporary secondary identification marker in *Aspergillus*, whereas Visagie et al. [150] proposed the beta-tubulin (*BenA*) for routine identification of isolates in the case of *Penicillium*. ITS1-ITS4 are primers used for amplification of the ITS region [161]), CMD5-CMD6 [162]), and CF1-CF4 [163] for the *CaM* gene and Bt$_2$a-Bt$_2$b for the *BenA* gene [134]. For amplification of ITS, *CaM*, and *BenA* genes, the following conditions are used: one cycle at 94 °C for 5 min; 35 cycles at 94 °C for 45 s, 55 °C for 45 s, 72 °C for 1 min; and a final cycle at 72 °C for 7 min.

The PCR products are electrophoresed in agarose gels that are subsequently stained and examined under a UV transilluminator (Figure 8.18). Products of PCR are then purified and sequenced. Sequences are usually compared with related fungal ITS, *TEF1a*, *CaM*, and *BenA* sequences using the BLAST search on NCBI's GenBank sequence database or other curated specialized databases containing fungal sequences (e.g., FUSARIUM-ID and RefSeq Target Loci (RTL)). Phylogenetic analyses of nucleotide sequences of ITS, *TEF1a*, *CaM*, and *BenA* genes are finally carried out for isolate identification. Sequences are used to construct and study phylogenetic trees (Figure 8.19). A phylogeny (also called a dendogram) is a graph-like structure whose topology describes the inferred evolutionary history among a set of biological entities, such as species or DNA sequences.

Figure 8.19: Methods for molecular phylogenetic identification of mycotoxigenic fungi: sequencing of amplified DNA regions, comparison of sequences by BLAST search and alignment, and construction and study of phylogenetic trees for species identification. At the bottom right, example of a phylogenetic tree (dendrogram) for identification of *Fusarium* isolates.

8.4.5 Chemotyping

As revised by Niessen [164], molecular studies on mycotoxigenic fungi are not only focused on species identification, what is important, but many are addressed to study mycotoxin pathway gene sequences that could allow to identify fungal species and their toxigenic potential, that is, their potential to produce mycotoxins.

Early studies on the toxigenic potential of *Fusarium cerealis* (once considered synonymous with *F. crookwellense*), *F. culmorum*, and *F. graminearum*, analyzing extracts from cultures by chromatographic methods, showed variability in type B trichothecene production by different isolates of the same species. This led to the definition of two main chemotypes associated with these *Fusarium* species and to considering chemotaxonomy as complementary to classical taxonomy. Chemotype I strains produce DON, DON producers being further differentiated as chemotypes IA and IB depending on whether they can produce 3-ADON or 15-ADON, respectively, and chemotype II strains produce NIV and/or 4-acetyl NIV [165, 166]. Now it is known that, considering type B trichothecenes, *F. culmorum*, and *F. graminearum*, depending on the isolate, have the potential to produce NIV, DON, and several acetylated derivatives of NIV and DON, but *F. cerealis* strains only produce NIV.

The genetics and regulation of the trichothecene biosynthesis can explain the differences in type B trichothecene production among *Fusarium* species and among isolates of the same *Fusarium* species. Trichothecene genes, like genes for other mycotoxins, are expressed as biosynthetic gene clusters. The thrichothecene biosynthesis gene (*TRI*) clusters and biosynthesis pathways in the *F. graminearum* species complex are the most studied [167]. There are sixteen *TRI* genes (*TRI1, TRI3, TRI4, TRI5, TRI6, TRI7, TRI8, TRI9, TRI10, TRI11, TRI12, TRI13, TRI14, TRI15, TRI16*, and *TRI101*), located at four different loci on different chromosomes: a two-gene *TRI1-TRI16* locus on chromosome 1, a 12-gene core *TRI* cluster on chromosome 2 (*TRI3-TRI14*), a single-gene *TRI15* locus on chromosome 3, and a single-gene *TRI101* locus on chromosome 4 [168–170].

In recent years, PCR-based genotyping of the TRI gene cluster has received much attention as a reliable and fast molecular tool to characterize the chemotype of isolates of different *Fusarium* species. Especially knowledge about the nucleotide sequences of the *TRI3, TRI7*, and *TRI13* genes has allowed the design of specific PCR primers for chemotyping. For example, some primers have been designed taking into account that *F. graminearum* isolates that produce NIV have functional *Tri7* and *Tri13* genes, but not DON-producing isolates [171–173]. Also, specific primers designed for the *TRI3* gene allow to identify 3-ADON and 15-ADOn chemotypes [174]. In Figure 8.20, the different chemotypes of *F. graminearum*, characterized by PCR amplification of target regions of the *TRI3* and *TRI13* genes, are shown.

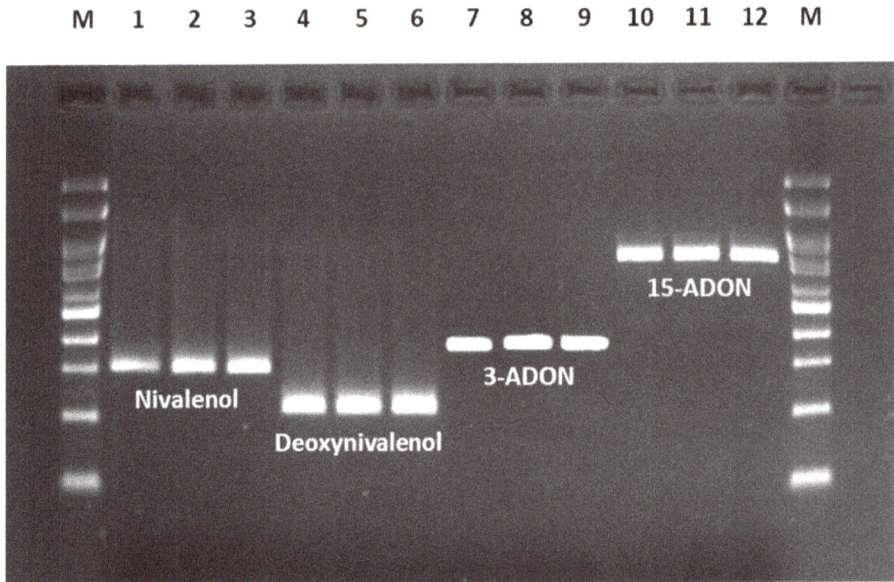

Figure 8.20: Agarose gel (2%) showing PCR chemotypes of 12 isolates of *Fusarium graminearum*. Size of the amplification of DNA products (primers [173, 174] indicated in brackets): 312 bp (Tri13NIVF/Tri13R) for NIV chemotypes (lanes 1–3); 282 bp (Tri13F/Tri13DONR) for DON chemotypes (lanes 4–6); 354 bp (Tri3F1325/Tri3R1679) for 3-ADON chemotypes (lanes 7–9); 708 bp (Tri3F971/Tri3R1679) for 15-ADON chemotypes (lanes 10–12). M: 100 bp DNA ladder marker.

Similar research has been conducted on fumonisin-producing *Fusarium* species and mycotoxigenic species of other fungal genera. For instance, it has been shown that genes *fum1* (= *fum5*), *fum6,* and *fum8* are only present in *F. verticillioides, F. prolifera-tum, F. fujikuroi,* and *F. nygamai,* which represent the principal producers of fumoni-sins within the *Fusarium fujikuroi* complex [175].

Despite the characterization of chemotypes through PCR-based *TRI* genotyping has become widespread, it should be noted that this constitutes a misusage of the term chemotype since the term chemotype refers to a chemical phenotype [176].

Other studies have aimed to combine qualitative and quantitative methods for detecting the toxigenic potential of fungal species colonizing food crops. One of the approaches is based on multiplex real-time PCR which have been used to detect and quantify fungal species in cereal grains by using markers targeting the trichothecene synthase (*tri5*) gene in trichothecene-producing *Fusarium* sp. isolates, the rRNA gene in *Penicillium verrucosum,* and the polyketide synthase gene (*Pks*) in *Aspergillus ochraceus* [177].

Molecular detection and characterization of mycotoxigenic fungi are thus valu-able tools that can help to optimize food and feed production processes for minimized risk of mycotoxin production [177].

Acknowledgments: The authors are grateful for the support provided from the following FEDER cofunded grants: Collaboration Agreement between the Xunta de Galicia and the University of Santiago de Compostela regulating the Specialization Campus "Campus Terra," CLIMIGAL (2022-PU016) and BreveRiesgo (2022-PU011); Conselleria de Cultura, Educacion e Ordenación Universitaria, Xunta de Galicia, GRC (ED431C 2021/01); Ministerio de Ciencia e Innovación IISCIII/PI19/001248, PID 2020-11262RB-C21, Grant CPP2021-008447 funded by MCIN/AEI/10.13039/501100011033 and by the European Union NextGenerationEU/PRT; and European Union Interreg Agritox EAPA-998-2018, H2020 778069-EMERTOX, and HORIZON-MSCA-2022-DN-01-MSCA Doctoral Networks 2022-101119901-BIOTOXDo.

References

[1]   Steyn PS. Mycotoxins, general view, chemistry and structure. Toxicol Lett. 1995;82(83):843–51.

[2]   Bennett JW, Klich M. Mycotoxins. Clin Microbiol Rev. 2003;16:497–516.

[3]   Sainz MJ, Alfonso A, Botana LM. Considerations about international mycotoxin legislation, food security, and climate change. In: Botana LM, Sainz MJ, eds. Climate change and mycotoxins. Berlin, Germany: De Gruyter, 2015;153–80.

[4]   Moss MO. Mycotoxins. Mycol Res. 1996;100:513–23.

[5]   Peraica M, Radić B, Lucić A, Pavlović M. Toxic effect of mycotoxins in humans. B World Health Organ. 1999;77:754–66.

[6]   Medina A, Akbar A, Baazeem A, Rodriguez A, Magan N. Climate change, food security and mycotoxins: Do we know enough? Fungal Biol Rev. 2017;31:143–54.

[7]   Alonso VA, Pereyra CM, Keller LA, Dalcero AM, Rosa CA, Chiacchiera SM, Cavaglieri LR. Fungi and mcyotoxins in silage: An overview. J Appl Microbiol. 2013;115:637–43.

[8]   Turner NW, Subrahmanyam S, Piletsky SA. Analytical methods for determination of mycotoxins: A review. Anal Chim Acta. 2009;632:168–80.

[9]   Alshannaq A, Yu JH. Occurrence, toxicity, and analysis of major mycotoxins in food. Int J Environ Res Public Health. 2017;14(6):632. doi:10.3390/ijerph14060632.

[10]   McKevith B. Nutritional aspects of cereals. Nutr Bull. 2005;30:13–26.

[11]   Medina A, González-Jartín JM, Sainz MJ. Impact of global warming on mycotoxins. Curr Opin Food Sci. 2017;18:76–81.

[12]   Battilani P, Toscano P, Van der Fels-Klerx HJ, Moretti A, Camardo Leggieri M, Brera C, Rortais A, Goumperis T, Robinson T. Aflatoxin B1 contamination in maize in Europe increases due to climate change. Sci Rep. 2016;6:24328. doi:10.1038/srep24328.

[13]   Medina A, Akbar A, Baazeem A, Rodriguez A, Magan N. Climate change, food security and mycotoxins: Do we know enough? Fungal Biol Rev. 2017;31:143–54.

[14]   Ostry V, Malir F, Toman J, Grosse Y. Mycotoxins as human carcinogens – The *IARC Monographs* classification. Mycotoxin Res. 2017;33:65–73.

[15]   Stoev SD. Foodborne mycotoxicoses, risk assessment and underestimated hazard of masked mycotoxins and joint mycotoxin effects or interaction. Environ Toxicol Pharmacol. 2015;39:794–809.

[16]   Nielsen KF. Mycotoxin production by indoor molds. Fungal Genet Biol. 2003;39:103–17.

[17]   Smith JE, Lewis CW, Anderson JG, Solomon GL Mycotoxins in Human Nutrition and Health. Directorate General XII Science, Research and Development. EUR 16048 EN, 1994.

[18]   Prandini A, Tansini G, Sigolo S, Filippi L, Laporta M, Piva G. On the occurrence of aflatoxin M1 in milk and dairy products. Food Chem Toxicol. 2009;47:984–91.

[19] Freire L, Sant'Ana AS. Modified mycotoxins: An updated review on their formation, detection, occurrence, and toxic effects. Food Chem Toxicol. 2018;111:189–205.

[20] Abbas HK, Shier WT. Chapter 3. Mycotoxin contamination of agricultural products in the Southern United States and approaches to reducing it from pre-harvest to final food products. In: Appell M, Kendra D, Trucksess MW, eds. Mycotoxin prevention and control in agriculture. American Chemical Society Symposium Series 1031. Oxford, UK: Oxford University Press, 2009;37–58.

[21] Aichinger G, Favero G, Del Warth B, Marko D. Alternaria toxins – Still emerging? Compr Rev Food Sci Food Saf. 2021;20:4390–406.

[22] Gruber-Dorninger C, Novak B, Nagl V, Berthiller F. Emerging mycotoxins: Beyond traditionally determined food contaminants. J Agric Food Chem. 2017;65(33):7052–70.

[23] Coppock RW, Dziwenka MM. Mycotoxins. In: Gupta RC, ed. Biomarkers in toxicology. 2nd ed. Amsterdam: Academic Press/Elsevier, 2019;615–26.

[24] Fraeyman S, Croubels S, Devreese M, Antonissen G. Emerging *Fusarium* and *Alternaria* mycotoxins: Occurrence, toxicity and toxicokinetics. Toxins. 2017;9:228.

[25] Gratz SW. Do plant-bound masked mycotoxins contribute to toxicity? Toxins. 2017;9:85. doi:10.3390/toxins9030085.

[26] González-Jartín JM, Alfonso A, Sainz MJ, Vieytes MR, Botana LM. Detection of new emerging type-A trichothecenes by untargeted mass spectrometry. Talanta. 2018;178:37–42.

[27] IARC. Chemical agents and related occupations. Vol 100F. A review of human carcinogens. IARC Monog Eval Carc. Lyon, France: International Agency for Research on Cancer, 2012.

[28] IARC. Some traditional herbal medicines, some mycotoxins, naphthalene and styrene. Vol 82. IARC Monog Eval Carc. Lyon, France: International Agency for Research on Cancer, 2002.

[29] Van der Fels-Klerx HJ, Liu C, Battilani P. Modelling climate change impacts on mycotoxin contamination. World Mycotoxin J. 2016;9:717–26.

[30] D'Mello J, Placinta C, Macdonald A. Fusarium mycotoxins: A review of global implications for animal health, welfare and productivity. Anim Feed Sci Tech. 1999;80:183–205.

[31] Mishra HN, Das C. A review on biological control and metabolism of aflatoxin. Crit Rev Food Sci Nutr. 2003;43:245–64.

[32] Bbosa GS, Kitya D, Lubega A, Ogwal-Okeng J, Anokbonggo WW, Kyegombe DB. Review of the biological and health effects of aflatoxins on body organs and body systems. In: Razzaghi-Abyaneh M, ed. Aflatoxins – Recent advances and future prospects. Rijeka, Croatia: InTech, 2013;239–265.

[33] EFSA. Scientific Opinion of the Panel on Contaminants in the Food Chain on a request from the European Commission related to the potential increase of consumer health risk by a possible increase of the existing maximum levels for aflatoxins in almonds, hazelnuts and pistachios and derived products. EFSA J. 2007;446:1–127.

[34] Bbosa GS, Kitya D, Odda J, Ogwal-Okeng J. Aflatoxins metabolism, effects on epigenetic mechanisms and their role in carcinogenesis. Health. 2013;5:14–34.

[35] Hamid AS, Tesfamariam IG, Zhang Y, Zhang ZG. Aflatoxin B1-induced hepatocellular carcinoma in developing countries: Geographical distribution, mechanism of action and prevention. Oncol Lett. 2013;5:1087–92.

[36] Williams JH, Phillips TD, Jolly PE, Stiles JK, Jolly CM, Aggarwal D. Human aflatoxicosis in developing countries: A review of toxicology, exposure, potential health consequences, and interventions. Am J Clin Nutr. 2004;80:1106–22.

[37] Lewis L, Onsongo M, Njapau H, Schurz-Rogers H, Luber G, et al. Aflatoxin contamination of commercial maize products during an outbreak of acute aflatoxicosis in eastern and central Kenya. Environ Health Perspect. 2005;113:1763–67.

[38] Bui-Klimke T, Wu F. Ochratoxin A and human health risk: A review of the evidence. Crit Rev Food Sci Nutr. 2014;55(13):1860–69.

[39] Gallo A, Bruno KS, Solfrizzo M, Perrone G, Mule G, Visconti A, Baker SE. New insight into the ochratoxin A biosynthetic pathway through deletion of a nonribosomal peptide synthetase gene in *Aspergillus carbonarius*. Appl Environ Microbiol. 2012;78:8208–18.

[40] EFSA. Opinion of the scientific panel on contaminants in the food on a request from the commission related to ochratoxin A in food. EFSA J. 2006;365:1–56.

[41] Walker R, Larsen JC. Ochratoxin A: Previous risk assessments and issues arising. Food Addit Contam. 2005;22:6–9.

[42] Di Paolo N, Guarnieri A, Loi F, Sacchi G, Mangiarotti AM, Di Paolo M. Acute renal failure from inhalation of mycotoxins. Nephron. 1993;64:621–25.

[43] Al-Anati L, Petzinger E. Immunotoxic activity of ochratoxin A. J Vet Pharmacol Ther. 2006;29:79–90.

[44] Gayathri L, Dhivya R, Dhanasekaran D, Periasamy VS, Alshatwi AA, Akbarsha MA. Hepatotoxic effect of ochratoxin A and citrinin, alone and in combination, and protective effect of vitamin E: In vitro study in HepG2 cell. Food Chem Toxicol. 2015;83:151–63.

[45] Patil RD, Dwivedi P, Sharma AK. Critical period and minimum single oral dose of ochratoxin A for inducing developmental toxicity in pregnant Wistar rats. Reprod Toxicol. 2006;22:679–87.

[46] Ekwomadu TI, Akinola SA, Mwanza M. Fusarium mycotoxins, their metabolites (free, emerging, and masked), food safety concerns, and health impacts. Int J Environ Res Public Health. 2021;18:11741.

[47] Thrane U, Adler A, Clasen PE, Galvano F, Langseth W, et al. Diversity in metabolite production by *Fusarium langsethiae*, *Fusarium poae*, and *Fusarium sporotrichioides*. Int J Food Microbiol. 2004;95:257–66.

[48] Piec J, Pallez M, Beyer M, Vogelgsang S, Hoffmann L, Pasquali M. The Luxembourg database of trichothecene type B *F. graminearum* and *F. culmorum* producers. Bioinformation. 2016;12:1–3.

[49] Dohnal V, Jezkova A, Jun D, Kuca K. Metabolic pathways of T-2 toxin. Curr Drug Metab. 2008;9:77–82.

[50] Wu Q, Dohnal V, Huang L, Kuca K, Yuan Z. Metabolic pathways of trichothecenes. Drug Metab Rev. 2010;42:250–67.

[51] Janik E, Niemcewicz M, Podogrocki M, Ceremuga M, Stela M, Bijak M. T-2 toxin – The most toxic trichothecene mycotoxin: Metabolism, toxicity, and decontamination strategies. Molecules. 2021;26(22):6868.

[52] Rocha O, Ansari K, Doohan FM. Effects of trichothecene mycotoxins on eukaryotic cells: A review. Food Addit Contam. 2005;22:369–78.

[53] Creppy EE. Update of survey, regulation and toxic effects of mycotoxins in Europe. Toxicol Lett. 2002;127:19–28.

[54] Pestka JJ, Smolinski AT. Deoxynivalenol: Toxicology and potential effects on humans. J Toxicol Environ Health B Crit Rev. 2005;8:39–69.

[55] Holanda DM, Kim SW. Mycotoxin occurrence, toxicity, and detoxifying agents in pig production with an emphasis on deoxynivalenol. Toxins. 2021;13(2):171.

[56] Kumar P, Mahato DK, Gupta A, Pandey S, Paul V, Saurabh V, et al. Nivalenol mycotoxin concerns in foods: An overview on occurrence, impact on human and animal health and its detection and management strategies. Toxins. 2022;14(8):527.

[57] Turner P. Deoxynivalenol and nivalenol occurrence and exposure assessment. World Mycotox J. 2010;3(4):315–21.

[58] Zingales V, Fernández-Franzón M, Ruiz MJ. Occurrence, mitigation and in vitro cytotoxicity of nivalenol, a type B trichothecene mycotoxin – Updates from the last decade (2010–2020). Food Chem Toxicol. 2021;152:112182.

[59] Jimenez M, Manez M, Hernandez E. Influence of water activity and temperature on the production of zearalenone in corn by three *Fusarium* species. Int J Food Microbiol. 1996;29:417–21.

[60] Camean Fernández AM, Repetto Jiménez M. Toxicología alimentaria. Madrid, Spain: Ediciones Díaz de Santos, 2012.

[61] Shin BS, Hong SH, Bulitta JB, Hwang SW, Kim HJ, et al. Disposition, oral bioavailability, and tissue distribution of zearalenone in rats at various dose levels. J Toxicol Environ Health A. 2009;72:1406–11.

[62] Malekinejad H, Colenbrander B, Fink-Gremmels J. Hydroxysteroid dehydrogenases in bovine and porcine granulosa cells convert zearalenone into its hydroxylated metabolites alpha-zearalenol and beta-zearalenol. Vet Res Commun. 2006;30:445–53.

[63] Mirocha CJ, Pathre SV, Robison TS. Comparative metabolism of zearalenone and transmission into bovine milk. Food Cosmet Toxicol. 1981;19:25–30.

[64] Pistol GC, Braicu C, Motiu M, Gras MA, Marin DE, et al. Zearalenone mycotoxin affects immune mediators, MAPK signalling molecules, nuclear receptors and genome-wide gene expression in pig spleen. PLoS One. 2015;10(5):e0127503.

[65] Fitzpatrick DW, Picken CA, Murphy LC, Buhr MM. Measurement of the relative binding affinity of zearalenone, alpha-zearalenol and beta-zearalenol for uterine and oviduct estrogen receptors in swine, rats and chickens: An indicator of estrogenic potencies. Comp Biochem Physiol C. 1989;94:691–94.

[66] Malekinejad H, Maas-Bakker R, Fink-Gremmels J. Species differences in the hepatic biotransformation of zearalenone. Vet J. 2006;172:96–102.

[67] Program NT. Carcinogenesis bioassay of zearalenone (CAS No. 17924-92-4) in F344/N rats and B6C3F1 mice (feed study). Natl Toxicol Program Tech Rep Ser. 1982;235:1–155.

[68] Marín S, Magan N, Ramos AJ, Sanchis V. Fumonisin-producing strins of *Fusarium*: A review of their ecophysiology. J Food Protect. 2004;67:1792–805.

[69] Bryla M, Roszko M, Szymczyk K, Jedrzejczak R, Obiedzinski MW, Sekul J. Fumonisins in plant-origin food and fodder – A review. Food Addit Contam Part A. 2013;30:1626–40.

[70] Shier WT. The fumonisin paradox: A review of research on oral bioavailability of fumonisin B_1, a mycotoxin produced by *Fusarium moniliforme*. J Toxicol: Toxin Rev. 2000;19:161–87.

[71] Wang E, Ross PF, Wilson TM, Riley RT, Merrill AH Jr. Increases in serum sphingosine and sphinganine and decreases in complex sphingolipids in ponies given feed containing fumonisins, mycotoxins produced by *Fusarium moniliforme*. J Nutr. 1992;122:1706–16.

[72] Merril AH Jr, Sullards MC, Wang E, Voss KA, Riley RT. Sphingolipid metabolism: Roles in signal transduction and disruption by fumonisins. Environ Health Perspect. 2001;109(Suppl 2):283–89.

[73] Marasas WF, Riley RT, Hendricks KA, Stevens VL, Sadler TW, et al. Fumonisins disrupt sphingolipid metabolism, folate transport, and neural tube development in embryo culture and in vivo: A potential risk factor for human neural tube defects among populations consuming fumonisin-contaminated maize. J Nutr. 2004;134:711–16.

[74] Marasas WF, Kellerma TS, Gelderblom WC, Coetzer JA, Thiel PG, van der Lugt JJ. Leukoencephalomalacia in a horse induced by fumonisin B1 isolated from *Fusarium moniliforme*. Onderstepoort J Vet Res. 1988;55:197–203.

[75] Smith GW, Constable PD, Eppley RM, Tumbleson ME, Gumprecht LA, Haschek-Hock WM. Purified fumonisin B_1 decreases cardiovascular function but does not alter pulmonary capillary permeability in swine. Toxicol Sci. 2000;56:240–49.

[76] Stockmann-Juvala H, Savolainen K. A review of the toxic effects and mechanisms of action of fumonisin B_1. Hum Exp Toxicol. 2008;27:799–809.

[77] Chu FS, Li GY. Simultaneous occurrence of fumonisin B1 and other mycotoxins in moldy corn collected from the People's Republic of China in regions with high incidences of esophageal cancer. Appl Environ Microbiol. 1994;60:847–52.

[78] Dombrink-Kurtzman MA, Blackburn JA. Evaluation of several culture media for production of patulin by *Penicillium* species. Int J Food Microbiol. 2005;98:241–48.

[79] Birkinshaw JH, Bracken A, Michael SE, Raistrick H. Biochemistry and chemistry. The Lancet. 1943;242:625–30.

[80]   Rychlik M, Kircher F, Schusdziarra V, Lippl F. Absorption of the mycotoxin patulin from the rat stomach. Food Chem Toxicol. 2004;42:729–35.

[81]   Dailey RE, Blaschka AM, Brouwer EA. Absorption, distribution, and excretion of [^{14}C] patulin by rats. J Toxicol Environ Health. 1977;3:479–89.

[82]   Puel O, Galtier P, Oswald IP. Biosynthesis and toxicological effects of patulin. Toxins. 2010;2:613–31.

[83]   Jin H, Yin S, Song X, Zhang E, Fan L, Hu H. p53 activation contributes to patulin-induced nephrotoxicity via modulation of reactive oxygen species generation. Sci Rep. 2016;6. doi:10.1038/srep24455.

[84]   McKinle ER, Carlton WW, Boon GD. Patulin mycotoxicosis in the rat: Toxicology, pathology and clinical pathology. Food Chem Toxicol. 1982;20:289–300.

[85]   Moake MM, Padilla-Zakour OI, Worobo RW. Comprehensive review of patulin control methods in foods. Compr Rev Food Sci F. 2005;4:8–21.

[86]   Ostry V, Malir F, Ruprich J. Producers and important dietary sources of ochratoxin A and citrinin. Toxins. 2013;5(9):1574–86.

[87]   Liao CD, Chen YC, Lin HY, Chiueh LC, Shih DYC. Incidence of citrinin in red yeast rice and various commercial Monascus products in Taiwan from 2009 to 2012. Food Control. 2014;38:178–83.

[88]   Sankawa U, Ebizuka Y, Noguchi H, Isikawa Y, Kitaghawa S, Yamamoto Y, Kobayashi T, Iitak Y. Biosynthesis of citrinin in *Aspergillus terreus*: Incorporation studies with [2-^{13}C, 2-^{2}H$_3$], [1-^{13}C, ^{18}O$_2$] and [1-^{13}C, ^{17}O]-acetate. Tetrahedron. 1983;39:3583–91.

[89]   Kamle M, Mahato DK, Gupta A, Pandhi S, Sharma N, Sharma B, et al. Citrinin mycotoxin contamination in food and feed: Impact on agriculture, human health, and detection and management strategies. Toxins. 2022;14(2):85.

[90]   Salah A, Bouaziz C, Amara I, Abid-Essefi S, Bacha H. Eugenol protects against citrinin-induced cytotoxicity and oxidative damages in cultured human colorectal HCT116 cells. Environ Sci Pollut Res. 2019;26:31374–83.

[91]   Mariappan AK, Munusamy P, Latheef SK, Singh SD, Dhama K. Hepato nephropathology associated with inclusion body hepatitis complicated with citrinin mycotoxicosis in a broiler farm. Vet World. 2018;11(2):112–17.

[92]   de Oliveira Filho JWG, Islam MT, Ali ES, Uddin SJ, de Oliveira Santos JV, de Alencar MVOB, et al. A comprehensive review on biological properties of citrinin. Food Chem Toxicol. 2017;110:130–41.

[93]   Mukherjee J, Menge M. Progress and prospects of ergot alkaloid research. Adv Biochem Eng Biotechnol. 2000;68:1–20.

[94]   Schardl CL. Introduction to the toxins special issue on ergot alkaloids. Toxins. 2015;7(10):4232–37.

[95]   Tudzynski P, Correia T, Keller U. Biotechnology and genetics of ergot alkaloids. Appl Microbiol Biotechnol. 2001;57(5–6):593–605.

[96]   Klotz JL. Activities and effects of ergot alkaloids on livestock physiology and production. Toxins. 2015 27;7(8):2801–21.

[97]   Griffith RW, Grauwiler J, Hodel C, Leist KH, Matter B. Toxicologic considerations. In: Berde B, Schild HO, eds. Ergot alkaloids and related compounds. Handbook of experimental pharmacology. Berlin: Springer Verlag, 1978;805–38.

[98]   Jestoi M. Emerging *Fusarium*-mycotoxins fusaproliferin, beauvericin, enniatins, and moniliformin – A review. Crit Rev Food Sci Nutr. 2008;48(1):21–49.

[99]   EFSA CONTAM Panel. Scientific opinion on the risks for human and animal health related to the presence of modified forms of certain mycotoxins in food and feed. EFSA J. 2014;12(12):3916.

[100] Rychlik M, Humpf HU, Marko D, Dänicke S, Mally A, Berthiller F, Klaffke H, Lorenz N. Proposal of a comprehensive definition of modified and other forms of mycotoxins including "masked" mycotoxins. Mycotoxin Res. 2014;30(4):197–205.

[101] Berthiller F, Crews C, Dall'Asta C, Saeger SD, Haesaert G, Karlovsky P, Oswald IP, Seefelder W, Speijers G, Stroka J. Masked mycotoxins: A review. Mol Nutr Food Res. 2013;57(1):165–86.

[102] Logrieco A, Moretti A, Castella G, Kostecki M, Golinski P, Ritieni A, Chelkowski J. Beauvericin production by *Fusarium* species. Appl Environ Microbiol. 1998;64(8):3084–88.

[103] EFSA CONTAM Panel. Scientific opinion on the risks to human and animal health related to the presence of beauvericin and enniatins in food and feed. EFSA J. 2014;12(8):3802.

[104] Santini A, Meca G, Uhlig S, Ritieni A. Fusaproliferin, beauvericin and enniatins: Occurrence in food – A review. World Mycotoxin J. 2012;5:71–81.

[105] Altomare C, Logrieco AF, Gallo A. Mycotoxins and mycotoxigenic fungi: Risk and management. A challenge for future global food safety and security. In: Zaragoza O, Casadevall A, eds. Encyclopedia of mycology. Oxford, UK: Elsevier, 2021;64–93.

[106] Pérez-Fuentes N, Alvariño R, Alfonso A, González-Jartín J, Gegunde S, Vieytes MR, Botana LM. Enniatins A1 and B1 alter calcium homeostasis of neuronal cells leading to apoptotic death. Food Chem Toxicol. 2022;168:113361.

[107] Ritieni A, Fogliano V, Randazzo G, Scarallo A, Logrieco A, Moretti A, Mannina L, Bottalico A. Isolation and characterization of fusaproliferin, a new toxic metabolite from *Fusarium proliferatum*. Nat. Toxins. 1995;3:17–20.

[108] Ćeranić A, Svoboda T, Berthiller F, Sulyok M, Samson JM, Güldener U, Schuhmacher R, Adam G. Identification and functional characterization of the gene cluster responsible for fusaproliferin biosynthesis in *Fusarium proliferatum*. Toxins. 2021;13:468.

[109] Logrieco A, Moretti A, Fornelli F, Fogliano V, Ritieni A, Caiaffa MF, Randazzo G, Bottalico A, Macchia L. Fusaproliferin production by *Fusarium subglutinans* and its toxicity to *Artemia salina*, SF-9 insect cells, and IARC/LCL 171 human B lymphocytes. Appl Environ Microbiol. 1996;62(9):3378–84.

[110] Ritieni A, Monti SM, Randazzo G, Logrieco A, Moretti A, Peluso G, Ferracane R, Fogliano V. Teratogenic effects of fusaproliferin on chicken embryos. J Agric Food Chem. 1997;45:3039–43.

[111] Frisvad JC. A critical review of producers of small lactone mycotoxins: Patulin, penicillic acid and moniliformin. World Mycotoxin J. 2018;11(1):73–100.

[112] EFSA CONTAM Panel. Scientific opinion on the risks to human and animal health related to the presence of moniliformin in food and feed. EFSA J. 2018;16(3):5082.

[113] Rank C, Nielsen KF, Larsen TO, Varga J, Samson RA, Frisvad JC. Distribution of sterigmatocystin in filamentous fungi. Fungal Biol. 2011;115(4–5):406–20.

[114] Zingales V, Fernández-Franzón M, Ruiz M-J. Sterigmatocystin: Occurrence, toxicity and molecular mechanisms of action – A review. Food Chem Toxicol. 2020;146:111802.

[115] Mol HGJ, Pietri A, MacDonald SJ, Anagnostopoulos C, Spanjer M. Survey on sterigmatocystin in food. EFSA Supporting Publication. 2015;774:56.

[116] Tralamazza SM, Piacentini KC, Iwase CHT, Rocha Lde O. Toxigenic *Alternaria* species: Impact in cereals worldwide. Curr Opin Food Sci. 2018;23:57–63.

[117] Chen A, Mao X, Sun Q, Wei Z, Li J, You Y, et al. Alternaria mycotoxins: An overview of toxicity, metabolism, and analysis in food. J Agric Food Chem. 2021;69(28):7817–30.

[118] Aichinger G, Del Favero G, Warth B, Marko D. *Alternaria* toxins – Still emerging? Compr Rev Food Sci Food Saf. 2021;20(5):4390–406.

[119] EFSA on Contaminants in the Food Chain (CONTAM). Scientific opinion on the risks for animal and public health related to the presence of *Alternaria* toxins in feed and food. EFSA J. 2011;9(10):2407.

[120] Fernández Pinto VEF, Patriarca A. Alternaria species and their associated mycotoxins. In: Moretti A, Susca A, eds. Mycotoxigenic fungi: Methods and protocols, mMethods in mMolecular bBiology. Vol. 1542. Springer Science New York, NY, USA: Humana Press, 2017;13–32.

[121] Burki F. The eukaryotic tree of life from a global phylogenomic perspective. CSH Perspect Biol. 2014;6:a016147.

[122] Blackwell M. The fungi: 1.1.3 . . . 5.1 million species. Am J Bot. 2011;98:426–38.

[123] Hibbett DS, Ohman A, Glotzer D, Nuhn M, Kirk P, Nilsson RH. Progress in molecular and morphological taxon discovery in *Fungi* and options for formal classification of environmental sequences. Fungal Biol Rev. 2011;25:38–47.

[124] Hibbett D, Abarenkov K, Kõljalg U, Öpik M, Chai B, Cole J, et al. Sequence-based classification and identification of Fungi. Mycologia. 2016;108:1049–68.

[125] Bowman SM, Free SJ. The structure and synthesis of the fungal cell wall. BioEssays. 2006;28:799–808.

[126] Carris LM, Little CR, Stiles CM. Introduction to Fungi. Plant Health Instructor. 2012. doi:10.1094/PHI-I-2012-0426-01.

[127] Walker GM, White NA. Introduction to fungal physiology. In: Kavanagh K, ed. Fungi: Biology and applications. 3rd ed. Hoboken, NJ, USA: John Wiley & Sons, Inc, 2018;1–35.

[128] Hibbett DS, Binder M, Bischoff JF, Blackwell M, Cannon PE, et al. A higher-level phylogenetic classification of the Fungi. Mycol Res. 2007;111:509–54.

[129] Jones MDM, Forn I, Gadelha C, Egan MJ, Bass D, Massana R, Richards TA. Discovery of novel intermediate forms redefines the fungal tree of life. Nature. 2011;474:200–203.

[130] Adl SM, Simpson AGB, Lane CE, Lukes J, Bass D, et al. The revised classification of Eukaryotes. J Eukaryot Microbiol. 2012;59:429–93.

[131] Spatafora JW, Chang Y, Benny GL, Lazarus K, Smith ME, et al. A phylum-level phylogenetic classification of zygomycete fungi based on genome-scale data. Mycologia. 2016;108:1028–46.

[132] Kirk PM, Cannon PF, Minter DW, Stalpers JA. Dictionary of the fungi. 10th ed. Wallingford, UK: CABI Publishing, 2008.

[133] Lutzoni F, Miadlikowska J, Arnold AE. A global phylogeny and classification of the Pezizomycotina. In: Abstracts of the Second international workshop on ascomycete systematics, 2015;6.

[134] Barnes EH. The ascomycetes. In: Atlaas and manual of plant pathology. Boston, MA, USA: Springer, 1979;150–54.

[135] Gams W. Recent changes in fungal nomenclature and their impact of naming of microfungi. In: Li D-W, ed. Biology of microfungi, fungal biology. Springer Inernational Publishing Switzerland, 2016;7–23.

[136] Hibbett DS, Taylor JW. Fungal systematics: Is a new age of enlightenment at hand? Nat Rev Microbiol. 2013;11:129–33.

[137] Summerell BA, Laurence MH, Liew ECY, Leslie JF. Biogeography and phylogeography of *Fusarium*: A review. Fungal Divers. 2010;44:3–13.

[138] Leslie JF, Summerell BA. The *Fusarium* laboratory manual. Oxford, UK: Blackwell Publishing, 2006.

[139] Houbraken J, Samson RA, Yilmaz N. Taxonomy of *Aspergillus*, *Penicllium* and *Talaromyces* and its significance for biotechnology. In: de Vries RP, Gelber IB, Anderse MR, eds. *Aspergillus* and *Penicillium* in the post-genomic era. Norfolk, UK: Caister Academic Press, 2016;1–15.

[140] Samson RA, Visagie CM, Houbraken J, Hon S-B, Hubka V, et al. Phylogeny, identification and nomenclature of the genus *Aspergillus*. Stud Mycol. 2014;78:141–73.

[141] Kocsubé S, Perrone G, Magistà D, Houbraken J, Varga J, Szigeti G, Hubka V, Hong S-B, Frisvad JC, Samson RA. *Aspergillus* is monophyletic: Evidence from multiple gene phylogenies and extrolites profiles. Stud Mycol. 2016;85:199–213.

[142] Bennett JW. An overview of the genus *Aspergillus*. In: Machida M, Gomi K, eds. *Aspergillus*: Molecular biology and genomics. Norfolk, UK: Caister Academic Press, 2010;1–17.

[143] Klich MA. *Aspergillus flavus*: The major producer of aflatoxin. Mol Plant Pathol. 2007;8:713–22.

[144] Bryden WL. Mycotoxin contamination of the feed supply chain: Implications for animal productivity and feed security. Anim Feed Sci Technol. 2012;173:134–58.

[145] Probst C, Schulthness F, Cotty PJ. Impact of *Aspergillus* section *Flavi* community structure on the development of lethal levels of aflatoxins in Kenyan maize (*Zea mays*). J Appl Microbiol. 2010;108:600–10.

[146] Klich MA. Biogeography of *Aspergillus* species in soil and litter. Mycologia. 2002;94:21–27.

[147] Bhatnagar D, Ehrlich KC, Moore GG, Payne GA. *Aspergillus flavus*. In: Batt CA, Tortorello M-L, eds. Encyclopedia of food microbiology. 2nd ed. London: Academic Press, 2014;83–91.

[148] Geiser DM. Sexual structures in *Aspergillus*: Morphology, importance and genomics. Med Mycol. 2009;47:S21–S26.

[149] Chang P-K, Horn BW, Abe K, Gomi K. Aspergillus. In: Batt CA, Tortorello M-L, eds. Encyclopedia of food microbiology. 2nd ed. London: Academic Press, 2014;77–82.

[150] Visagie CM, Houbraken J, Frisvad JC, Hong S-B, Klaassen CHW, et al. Identification and nomenclature of the genus *Penicillium*. Stud Mycol. 2014;78:343–71.

[151] Pitt JI, Hocking AD. Fungi and food spoilage. 3rd ed. London, UK: Springer, 2009.

[152] Fredricks DN, Smith C, Meier A. Comparison of six DNA extraction methods for recovery of fungal DNA as assessed by quantitative PCR. J Clin Microbiol. 2005;43:5122–28.

[153] Frisvad JC, Filtenborg O. Terverticillate Penicillia: Chemotaxonomy and mycotoxin production. Mycologia. 1989;81:831–61.

[154] Taylor JW, Jacobosn DJ, Krokem S, Kasuga T, Geiser DM, Hibbett DS, Fisher MC. Phylogenetic species recognition and species concepts in fungi. Fungal Gen Biol. 2000;31:21–32.

[155] Okuda MA, Klich KA, Seifert KA. . . . Media and incubation effect on morphological characteristics of *Penicillium* and *Aspergillus*. In: Samosn RA, Pitt JI, eds. Integration of modern taxonomic methods for *Penicillium* and *Aspergillus* classification. Amsterdam, The Netherlands: Harwood Academic Publishers, 2000;83–99.

[156] Romanelli AM, Fu J, Herrera ML, Wickes BL. A universal DNA extraction and PCR amplification method for fungal rDNA sequence-based identification. Mycoses. 2014;57:612–22.

[157] Schoch CL, Seifert KA, Huhndorf S, Robert V, Spouge JL, Levesque CA, Chen W, Fungal Barcoding Consortium. Nuclear ribosomal internal transcribed spacer (ITS) region as a universal DNA barcode marker for *Fungi*. Proc Natl Acad Sci USA. 2012;109:6241–46.

[158] Mitchell JI, Zuccaro A. Sequences, the environment and fungi. Mycologist. 2006;20:62–74.

[159] Geiser DM, Jiménez-Gasco MM, Kang S, Makalowska I, Veeraraghavan N, et al. FUSARIUM-ID v.1.0: A DNA sequence database for identifying *Fusarium*. Eur J Plant Pathol. 2004;110:473–79.

[160] White TJ, Bruns T, Lee SB, Taylos JW. Amplification and direct sequencing of fungal ribosomal RNA genes for phylogenetics. In: Innis MA, Gelfand DH, Shinsky TJ, White TJ, eds. PCR protocols: A guide to methods and applications. New York, USA: Academic Press Inc, 1990;315–22.

[161] Hong SB, Go SJ, Shin HD, Frisvad JC, Samson RA. Polyphasic taxonomy of *Aspergillus fumigatus* and related species. Mycologia. 2005;97:1316–29.

[162] Peterson SW, Vega FE, Posada F, Nagai C. *Penicillium coffeae*, a new endophytic species isolated from a coffee plant and its phylogenetic relationship to *P. fellutanum*, *P. thiersii* and *P. brocae* based on parsimony analysis of multilocus DNA sequences. Mycologia. 2005;97:659–66.

[163] Glass NL, Donaldson GC. Development of primer sets designed for use with the PCR to amplify conserved genes from filamentous Ascomycetes. Appl Environ Microbiol. 1195;61:1323–30.

[164] Niessen L. PCR-based diagnosis and quantification of mycotoxin producing fungi. Int J Food Microbiol. 2007;119:38–46.

[165] Miller JD, Greenhalgh R, Wang YZ, Lu M. Trichothecene chemotypes of three *Fusarium* species. Mycologia. 1991;83:121–30.

[166] Sydenham EW, Shephard GS, Thiel PG, Marasas WF, Stockenstrom S. Fumonisin contamination of commercial corn-based human foodstuffs. J Agric Food Chem. 1991;39:2014–18.

[167] Chen Y, Kistler HC, Ma Z. *Fusarium graminearum* trichothecene mycotoxins: Biosynthesis, regulation, and management. Annu Rev Phytopathol. 2019;57:15–39.

[168] Gale LR, Bryant JD, Calvo S, Giese H, Katan T, O'Donnell K, et al. Chromosome complement of the fungal plant pathogen *Fusarium graminearum* based on genetic and physical mapping and cytological observations. Genetics. 2005;171:985–1001.

[169] Alexander NJ, Proctor RH, McCormick SP. Genes, gene clusters, and biosynthesis of trichothecenes and fumonisins in *Fusarium*. Toxin Rev. 2009;28:198–215.

[170] Merhej J, Richard-Forget F, Barreau C. Regulation of trichothecene biosynthesis in *Fusarium*: Recent advances and new insights. Appl Microbiol Biotechnol. 2011;91:519–28.

[171] Brown DW, McCormick SP, Alexander NJ, Proctor RH, Desjardins AE. A genetic and biochemical approach to study trichothecene diversity in *Fusarium sporotrichioides* and *Fusarium graminearum*. Fungal Genet Biol. 2001;32:121–33.

[172] Lee T, Han Y, Kim K-H, Yun S-W, Lee YW. *Tri13* and *Tri7* determine deoxynivalenol- and nivalenol-producing chemotypes of *Gibberella zeae*. Appl Environ Microbiol. 2002;68:2148–54.

[173] Chandler EA, Simpson DR, Thomsett MA, Nicholson P. Development of PCR assays to *Tri7* and *Tri13* trichothecene biosynthetic genes, and characterisation of chemotypes of *Fusarium graminearum*, *Fusarium culmorum* and *Fusarium cerealis*. Physiol Mol Plant Pathol. 2003;62:355–67.

[174] Quarta A, Mita G, Haidukowski M, Santino A, Mulè G, Visconti A. Further data on trichothecene chemotypes of European *Fusarium culmorum* isolates. Food Addict Contamin. 2005;22:309–15.

[175] González-Jaén MT, Mirete S, Patiño B, López-Errasquín E, Vázquez C. Genetic markers for the analysis of variability and for production of specific diagnostic sequences in fumonisin-producing strains of *Fusarium verticillioides*. Euro J Plant Pathol. 2004;110:525–32.

[176] Desjardins AE. Natural product chemistry meets genetics: When is a genotype a chemotype? J Agric Food Chem. 2008;56:7587–92.

[177] Vegi A, Wolf-Hall CE. Multiplex real-time PCR method for detection and quantification of mycotoxigenic fungi belonging to three different genera. J Food Sci. 2013;78:M70-6.

Ana M. Botana
9 Analysis of environmental toxicants

9.1 Introduction

The movement and fate of environmental contaminants are key aspects to determine their impact in the environment. As a general principle, the transportation and fate of contaminants are controlled by their physical transport and reactivity. It involves either chemical or biochemical reactions or other physical interactions with other phases; therefore, those chemicals released into the environment rarely remain in the form, or at the location, of release. Figure 9.1 shows the main ways involved in transport and chemical fate. Most often, those substances originate in the anthroposphere and they can pass to the air, earth, water (either surface or ground waters), sediments, and to the biota (plants and animals).

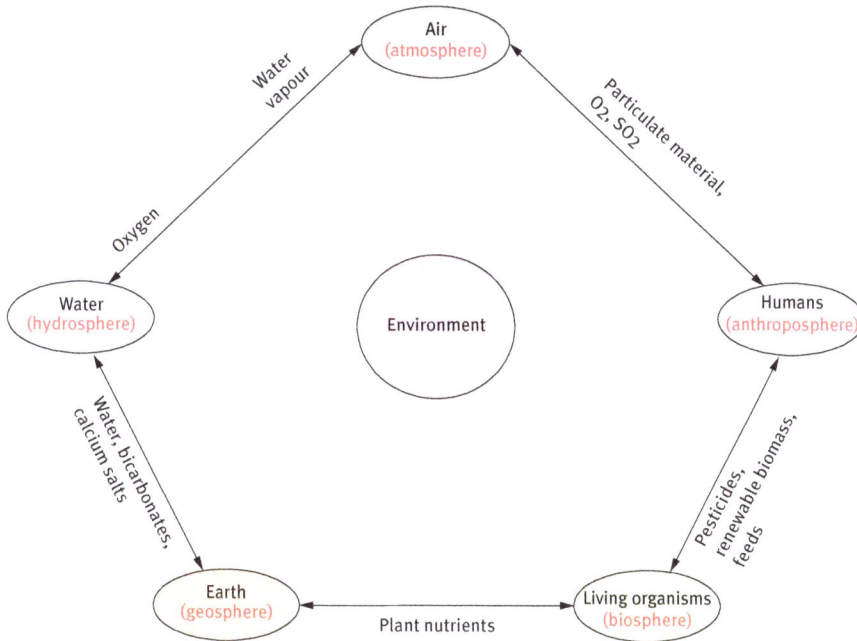

Figure 9.1: Interchange of some environmental toxicants delivered between anthroposphere and the other environmental sections that are involved in chemical fate.

The dilution of the toxicant in question as well as transfer among living creatures can happen and therefore concentration or bioaccumulation takes place, while at the same time most transport between environmental phases results in wider dissemina-

https://doi.org/10.1515/9783111014449-009

tion; for instance, lipid-soluble toxicants are readily taken up by organisms following exposure in air, water, or soil. They can persist in the tissues long enough to be transferred to the next trophic level, unless they are quickly metabolized. At this moment, if the organism is more susceptible than those at the previous level, the toxicant can become deleterious.

In order to study toxicants from the point of view of the analytical methodologies to identify them, it is convenient to classify them according to their chemical behavior. However, no single classification method is applicable for the entire spectrum of toxic agents present in the environment. In this chapter, the classification is based on the analytical behavior of toxicants due to the fact that toxicants can be grouped according to the analytical procedures involved to measure them. In the Stas–Otto scheme [1], toxicants have been divided into the following groups:

(a) Volatile toxicants, for example, hydrocyanic acid, alcohols, acetone, phenol, and chloral hydrate.
(b) Extractive toxicants:
    (i) Toxicants extractable by ether from acid solution, for example, organic acids and nitro compounds.
    (ii) Toxicants extractable by ether from alkaline solution, for example, alkaloids.
    (iii) Metals and metalloids, for example, copper, mercury, zinc, silver, and antimony.

9.2 General guidelines of chemical analysis in the environment

Chemical analysis of toxicants in the environment can be carried out in two different ways: whether weather toxicants are in high concentrations or in low concentrations. The second possibility is worth looking at in much detail, because the analytical procedures involved will be much more complex. In order to be able to determine very small amounts (very low concentrations) of chemicals in the environment, it is necessary to follow a series of operations (Figure 9.2):

1. Isolation (extraction and separation) of the chemicals of interest from sample matrix (air, water, sediment, living beings, etc.)
2. Separation and purification of the chemical of interest from other co-extracted chemicals (sample cleanup)
3. If necessary, sample concentration
4. Measurement by highly selective and sensitive analytical equipment

Occasionally, it is also necessary to derivatize (chemically modify) the chemicals of interest prior to their analysis.

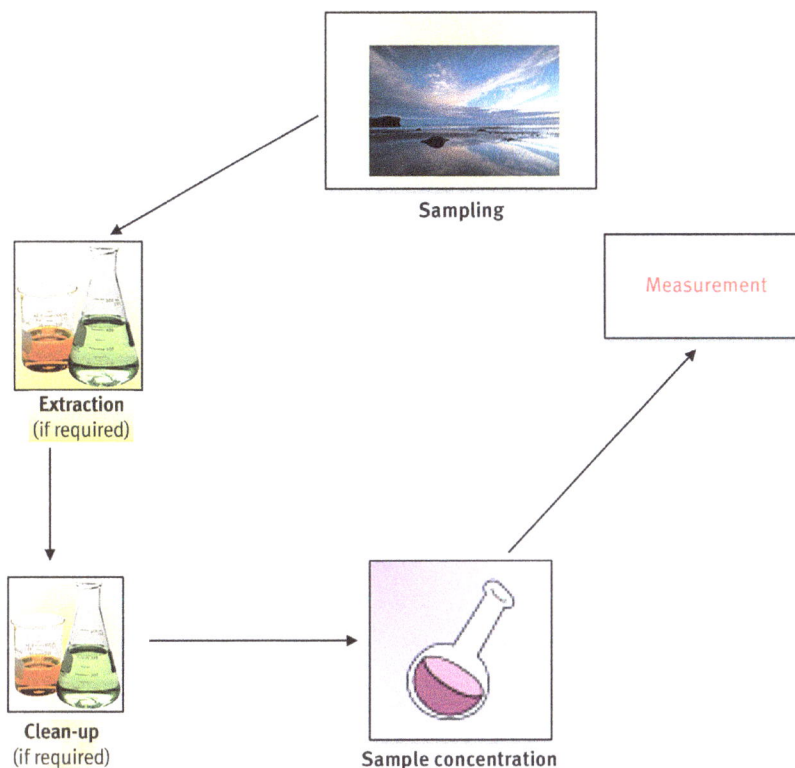

Figure 9.2: General guidelines in chemical analysis of environmental samples.

Within the broad range of instrumental methods used in environmental analysis, there are three methods which are the most commonly implemented ones, namely atomic spectrometry, gas chromatography (GC), and liquid chromatography (LC). The following section describes the series of operations in more detail, focusing on these three groups of instrumental methods.

9.2.1 Isolation (extraction and separation)

The methods for separation of toxicants depend on the type of sample, and in this section, samples are divided into three types: water samples; sediment, soil, and biological samples; and air samples. In Figure 9.3, the extraction and separation methods regarding the type of sample are shown.

Special attention is needed for sample preparation techniques based on microextraction: miniaturization is nowadays a matter of high interest in analytical methodologies [2]. In this way, it is possible to reduce the amount of reagents and solvents to be used, which implies decreasing waste generation and therefore less environmental

Extraction / Separation Methods

Type of sample	Method

Selection of extraction method: based on molecular weight, boiling point and polarity

Water

Volatile organic compounds
→ Purge & trap method
→ Headspace method

Semi-volatile chemicals
→ Liquid-liquid extraction
→ Solid-phase extraction (SPE)
→ Solid-phase microextraction (SPME)

Sediment, soil and biological samples

Purge & trap method

Organic solvent extraction method
→ Soxhlet extraction
→ Extraction after mechanical mixing
→ Ultrasonic extraction

Steam distillation

Supercritical fluid extraction (SFE)

Air samples

Solvent extraction from filter papers or adsorbents

Thermal desorption

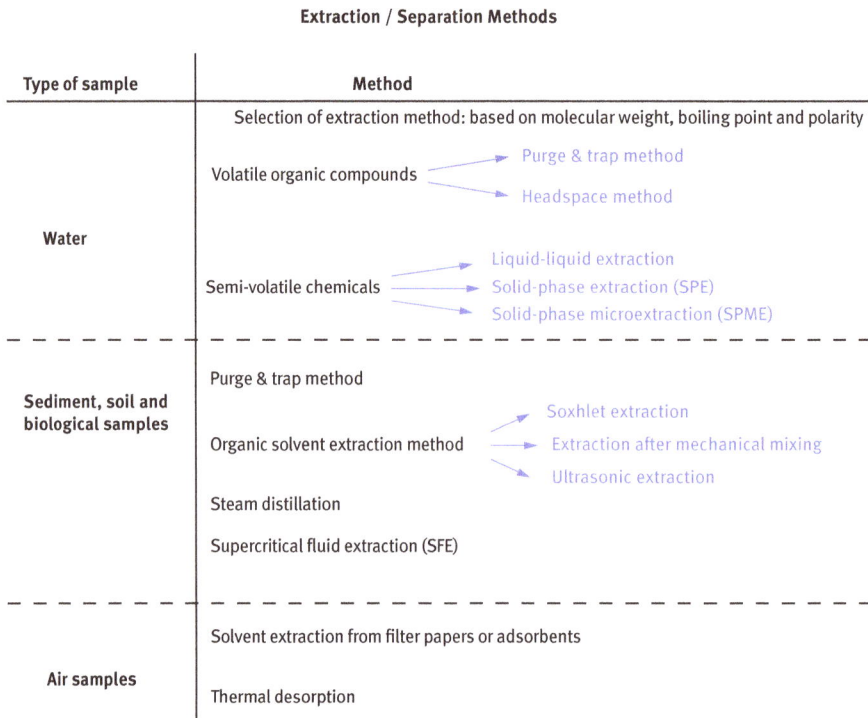

Figure 9.3: Extraction and separation methods regarding the type of sample.

pollution. A variety of sample preparation microextraction techniques have been developed [3, 4]. They can be classified as solid-phase-based microextraction (SPME) and liquid-phase-based microextraction techniques, although this classification is not based on the type of sample.

9.2.1.1 Solid-phase-based microextraction techniques

Direct immersion-SPME mode is often used for the analysis of pollutants present in clean matrices; a fiber is immersed into the sample solution and then the analytes are either eluted with suitable solvents or they are thermally desorbed from the fiber at the GC injector port. The headspace SPME technique has the advantage of selectively extracting volatile and semivolatile compounds and therefore can be directly coupled to GC, although it is possible to use this technique with some other instrumentation where liquid desorption is used. It has been used in the analysis of contaminants such as synthetic musk fragrances, UV filters, and trimethyl phosphate, which are all volatile or semivolatile contaminants of emerging concern [5–7].

9.2.1.2 Liquid-phase-based microextraction techniques

An alternative to conventional liquid–liquid extraction is the introduction of single-drop microextraction, where the extractant phase is a microdrop. It comprises both extraction and preconcentration of those analytes of interest, and the holder of the drop is a syringe, which at the same time is used in the injection for analysis.

Another technique to improve the stability and reliability of liquid–liquid extraction was the introduction of microporous hollow fiber membranes: the analytes are extracted from an aqueous sample and kept in a water-immiscible solvent which is immobilized in the pores of a hollow fiber. This new solution can then be analyzed by GC or LC, or even by capillary electrophoresis if the solvent is evaporated and the analytes reconstituted in an aqueous medium.

The drawback in terms of extraction has been the limited stability of the extractant phase. In this sense, the introduction of dispersive liquid–liquid microextraction in 2006 [8] represented an important accomplishment. In this technique, the extractant phase is an immiscible organic solvent; it is mixed with a dispersion solvent (methanol, acetone, acetonitrile, etc.) that shows high miscibility with both the extractant solvent and the sample. This mixture is quickly injected into the sample and thus a cloudy solution is formed. The sharp improvement of liquid–liquid mass transfer rates is directly related to the enormous improvement in interfacial area. Emulsion is then broken down, either by centrifugation or solvent-based de-emulsification, and the enriched extractant phase is taken for analysis [9]. There are many applications of dispersive liquid–liquid microextraction in environmental samples like the analysis of pesticides [10], nanomaterials [11], and disinfection by-products [12].

9.2.2 Separation and purification

Extracts from environmental samples can be complicated mixtures. Components of these mixtures can interfere with the analytical methods to be used, especially in the case of GC and LC analyses by giving poor separation because of over separation capacity of a column or by containing compounds that elute at the same time as the peaks of the compounds of interest. It is, therefore, necessary to clean up, or remove, those co-extracted compounds as much as possible. Cleanup methods include acid–alkaline partition, acetonitrile–hexane partition, and column chromatography. Column chromatography separation can involve adsorption, partition, gel permeation, ion exchange, and so on. Figure 9.4 summarizes all possibilities.

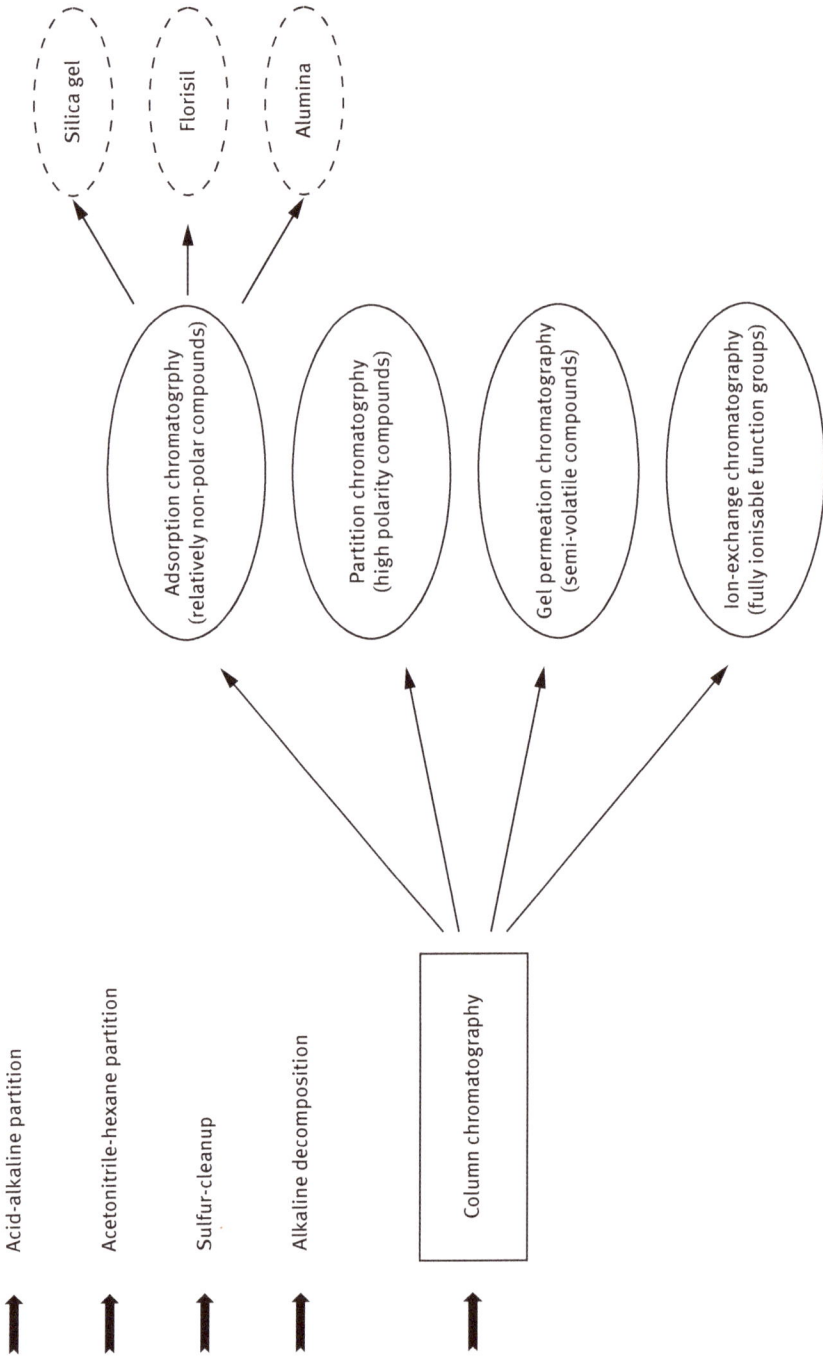

Figure 9.4: Cleanup methods to be used after extraction in environmental samples.

9.2.3 Sample concentration

When sample needs to be concentrated, either a concentrator (like a Kuderna–Danish concentrator) or rotary evaporator can be used: it depends on the boiling point of the compounds of interest, their sublimation character, timeframe for analysis, and so on.

The concentration setup takes longer than the evaporation setup, and is applicable to low-boiling point compounds as well as to high-boiling point compounds. The rotary evaporation can concentrate large volumes of samples in a relatively short period of time, although it causes big evaporative losses and is not suitable for low-boiling point compounds. However, for further concentration, one must use a microcolumn or evaporate under a stream of nitrogen.

9.2.4 Measurement: instrumental analytical methods

Modern instrumental techniques of analysis are essential for determining environmental toxicants. We are in a very exciting period in the evolution of analytical chemistry, where the development and optimization of new and improved analytical techniques are taking place. They allow the detection of much lower amounts of chemical compounds and so it is possible to determine contaminants that would not have been possible to detect otherwise. Although it is quite common in literature to find the instrumental method of analysis related to some specific type of samples, in this chapter, the analytical techniques are described as well as their properties: in each case, the nature of samples that may be analyzed is mentioned.

There are three important groups of analytical techniques to take into consideration: atomic spectrometry, GC, and LC. A better sensitivity is acquired in many cases with the use of hyphenated techniques; therefore, they will also be mentioned in each group.

9.3 Atomic spectrometry

Atomic spectrometry studies those elements that can be analyzed as atoms. Out of 118 identified elements, about 91 of them are called metals. These metallic elements are traditionally analyzed with this group of techniques and can be divided into two classes: those that are essential for survival, such as iron and calcium, and those that are nonessential or toxic, such as cadmium and lead. These toxic metals, unlike some organic substances, are not metabolically degradable and their accumulation in living tissues can cause death or serious health threats. Furthermore, these metals, dissolved in wastewaters and discharged into surface waters, will be concentrated as they travel up the food chain. Eventually, extremely poisonous levels of toxins can migrate to the immediate environment of the public. Metals that seep into groundwaters will contaminate drinking water wells and harm the consumers of that water. Pollution from manmade sources can easily create local conditions of elevated presence, which could lead to disastrous effects on animals and humans. Actually, man's exploitation of the world's mineral resources and the technological activities tend to unearth, dislodge, and disperse chemicals, particularly metallic elements, which have recently been brought into the environment in unprecedented quantities and concentrations and at extreme rates.

Heavy metals can be defined in several ways. One possible definition is the following: heavy metals form positive ions in solution and they have a density five times greater than that of water. They are of particular toxicological importance. Many metallic elements play an essential role in the function of living organisms. Humans receive their allocation of trace elements from food and water, an indispensable link in the food chain being plant life, which also supports animal life. It is well established that assimilation of metals takes place in the microbial world as well as in plants, and these elements tend to get concentrated as they progress through the food chain. Imbalances or excessive amounts of a metal species along this route lead to toxicity symptoms, disorders in the cellular functions, long-term debilitating disabilities in humans, and eventually death.

Due to the different routes of metals in the environment, the samples to be analyzed are also very different and so are the methods of analysis. Figure 9.5 gives a

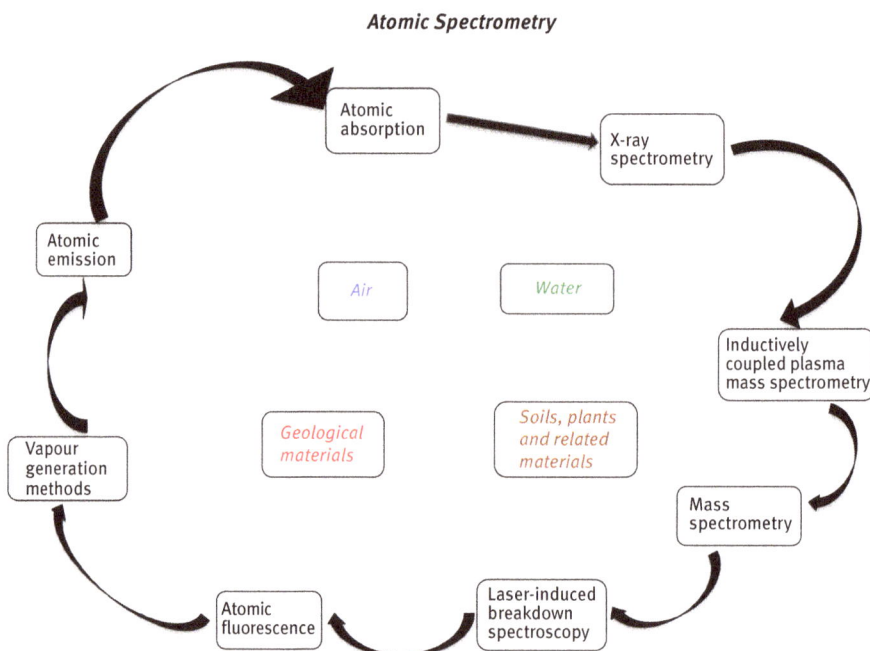

Figure 9.5: Different techniques that are in use nowadays for environmental samples.

coarse idea of the different techniques in use nowadays depending on the type of sample and Table 9.1 summarizes the samples related to specific techniques.

Table 9.1: Summary of the techniques in use nowadays regarding the type of sample.

Technique	Sample
Atomic absorption spectrometry	Air; water; soils, plants, and related materials; geological materials
Atomic emission spectrometry	Air; soils, plants, and related materials; geological materials
Vapor generation methods	Water
Atomic fluorescence spectrometry	Air; soils, plants, and related materials
Laser-based spectroscopy	Water
Laser-induced breakdown spectroscopy	Soils, plants, and related materials
Mass spectrometry	Air
Inductively coupled plasma mass spectrometry	Water; soils, plants, and related materials; geological materials
Other mass spectrometric techniques	Geological materials
X-ray spectrometry	Air; water; soils, plants, and related materials; geological materials

9.4 Gas chromatography

There are many research works on GC in different analytical and toxicological fields, and the methods are increasing in sophistication to identify and quantify environmental pollutants adequately. They can be distributed and dispersed by different means as we already know: it can be via water, soils, food, industrial activities, air, and so on. Due to environmental persistence, they remain in nature and are subject to bioaccumulation, which make them very hazardous compounds. This is more difficult because these compounds are present at trace concentrations [13]. Thus, the accurate determination of trace-level toxicants is a demanding analytical task. GC is one of the most frequently used techniques to analyze volatile and semivolatile organic compounds. In this sense, some major environmental pollutants are polychlorinated dioxins and dibenzofurans, polychlorinated biphenyls (PCBs), polycyclic aromatic hydrocarbons, polybrominated diphenyl ethers, toxaphenes, organochlorines, pesticides, and their major metabolites.

The toxicants have to be separated from co-extracted materials, which is difficult in case of complex matrices as is the case with most environmental samples: there are many closely related anthropogenic compounds present at orders of magnitude of higher con-

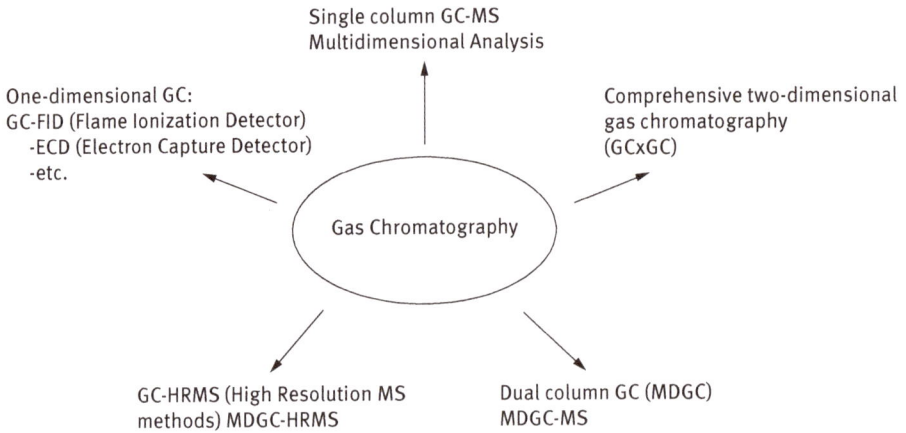

Figure 9.6: Gas chromatographic techniques that are used in the analysis of environmental toxicants.

centrations. Therefore, the analytical methods have to be very sensitive and selective and this is what chromatographic methods have been developed for. Figure 9.6 shows a selection of different GC techniques that have been applied in environmental toxicant analysis.

One-dimensional GC (1-D GC) was the main method for the determination of environmental toxicants, and in many laboratories, this is still the preferred or available method. However, for analyzing complex pollutant mixtures, this single-column GC technique has remarkable drawbacks such as lack of resolution, lack of robustness, and uncertainty in the identification. It is almost impossible to separate all components in a single chromatographic run, because the peak capacity of the column in the region where the components must be eluted is exceeded; so, to improve resolution several solutions need to be applied, like the use of a longer or narrower column, or a combination of both. To analyze dioxins, a very long column would be required in order to separate quite a big number of compounds, although several critical pairs still exist within this group [14, 15]. On the other hand, the determination, for instance, of individual congeners and atropisomers in PCBs is perhaps one of the most difficult applications in environmental analysis of PCBs. One way to improve identification is by using the hyphenated technique of gas chromatography-mass spectrometry (GC-MS): GC coupled to a mass spectrometric detector; but it still has problems as it is not possible to distinguish co-eluting congeners with the same chlorine number. They have the same or very similar mass spectra.

A different way to improve separation and therefore identification is to run the sample in two columns of different polarities to obtain a second set of retention times [16], although in this case it is not still possible to get a complete resolution when different polarities and efficiencies of both columns are optimized.

One alternative to improve separation is by fitting in parallel traps, columns, valves, flow switches, and so on, but all this will only lead to longer analysis times and technical difficulties. Another way is by using the corresponding analytical procedure to clean up the sample prior to its separation by GC, which will lead to a cleaner sample and therefore less compounds present to interfere. Although it must be made many times when the samples are complex, it is also convenient to reduce these steps to a minimum as it is time-consuming, and difficult to automate and reproduce. It is also difficult to get good recoveries with these steps and the samples are susceptible to be contaminated when working at trace levels.

A different alternative to reduce the sample handling steps and increase resolution is the use of multidimensional GC (MDGC). It is based on the separation that takes place in two or more independent separation steps/mechanisms [17], and the components remain separated until the overall analysis is completed; therefore, it becomes the only practical alternative to increase resolution. Although GC-MS might be considered a good alternative, there is still isomer co-elution because the separation takes place in one single column so that MDGC will be a valuable tool to improve or completely separate complex mixtures of toxicants [18].

A typical schematic diagram of a MDGC setup is shown in Figure 9.7. As can be seen, it consists of two independent ovens and detectors; a T-piece allows quantitative and reproducible transfer of small, unresolved selected fractions of the eluate from the first column to the second column, where separation takes place.

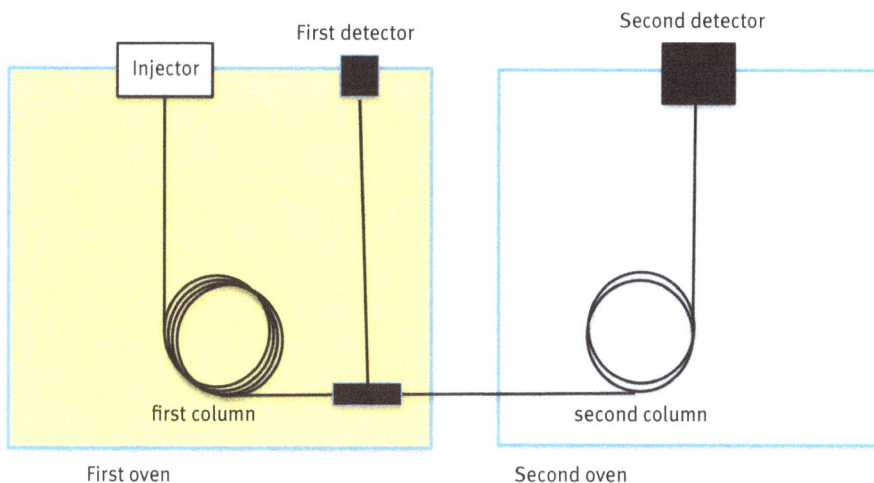

Figure 9.7: Typical schematic diagram of an MDGC setup.

All types of detectors can be used for two-dimensional GC, like electron capture detector and the MS, operated in the selected/single-ion monitoring mode because of their high sensitivity and specificity, for organohalogen analysis.

The resolution of an MDGC system is determined by the column(s)' dimensions and the difference in separation power between the two stationary phases. The longer the columns and the smaller internal diameters, the better the separation. A key aspect is the difference in selectivity for the two columns, which will remarkably affect the final separation; for instance, if polarity is quite different for both stationary phases, it may be expected to improve separation.

Some of the applications of MDGC analysis are tabulated in Table 9.2. There are hundreds of papers published on this topic when MDGC technique is used, so this list is just indicative of the need to use it to improve resolution in separation for volatile compounds in environmental analysis. A common goal in all the research is the resolution in specific congener analysis and the isolation of the components of interest for subsequent measurement of chiral ratios [19].

Table 9.2: List of analysis of some environmental toxicants employing multidimensional gas chromatography.

Environmental toxicants studied	Analysis goals
Organophosphorus pesticides	Identification of pesticides in food
Halogenated and organophosphorus pesticides	Identification of pesticide residues in food samples employing two columns of different polarities
PCBs	Analysis of PCB congeners using column switching
Non-*ortho*-chlorinated chlorobiphenyls (CBs)	Showed analysis of samples without preseparation will underestimate planar CB concentration
Tetrachlorodibenzodioxin (TCDD) isomers	Detection of complex isomeric mixtures of TCDDs
Polychlorinated dibenzo-*p*-dioxins and toxaphene	Study of gas-phase photodegradation of polychlorinated dibenzodioxins; MDGC interfaced to the photoreactor analysis of technical toxaphene mixture containing many congeners and biological samples
Polycyclic xenobiotics	Detection of the enantiomer ratios of polycyclic xenobiotics
Dioxins and PCBs	Determination of dioxins and PCBs

Although MDGC has many attractive points of interest, it is not the only technique used for analysis of volatile and semivolatile contaminants. The lack of robustness has restricted its use in routine laboratories where 1-D GC is still used. On the other hand, it has the relevant limitation that its increased separation power can only be applied to a few regions of the chromatogram, rather than to the whole sample. Thus, in many cases, analysis of environmental toxicants is based on GC followed by high-resolution mass spectrometric detection [20]. The mass spectrometer is often the detector of choice in MDGC as well because it can operate in a universal detection mode (similar to a flame ionization detector) and in a specific ion mode, the latter property being of particular importance for the analysis of a specific target compound.

Mass spectrometry is an analytical technique that is based on the mass-to-charge ratio (m/z) information: the identification takes place with the knowledge of m/z for the different analytes and their fragment ions under study. The biochemical analysis is one of the most important fields where this technique has been applied, and MDGC coupled with MS are routinely applied for the analysis of ultra-trace levels of organic pollutants [21].

Mass spectrometry has been enormously developed over the years: the single quadrupole mass spectrometers have a mass filter that allows one mass channel at a time to reach the detector when mass range is scanned. They have become much more sensitive. On the other hand, they do not have a high power of resolution, but are robust and easy to work with; they are mainly used for qualitative analysis.

In case of triple quadrupole mass spectrometry, it can be used either for quantification or for structural analysis: it has two quadrupoles, the first and the third one, that act as mass filters and the second one that causes fragmentation of the analyte when interacts with a quadrupole radiofrequency or collision gas. This technique has better selectivity, sensitivity, accuracy, and more comprehensive linear range.

The most widely used detectors for MDGC are low-resolution time-of-flight mass spectrometers (TOFMS). They can analyze complex pollutants from various environmental matrices at ultra-trace levels. It has the advantage compared with conventional GC that prevents interferences with a lower cost and is user-friendly. TOFMS allows separation due to its high spectral frequencies and deconvolution abilities and allows to obtain good sensitivity [22].

In case of high-resolution TOFMS (HR-TOFMS), it has the ability to generate spectrum data with high resolution and mass accuracy. For instance, HR-TOFMS equipment can detect pollutants at 0,03 pg/μL [23]. It can be used for targeted and untargeted analytes. In case of untargeted ones, latter identification is possible with the use of full spectrum data.

The analysis of environmental pollutants requires many times a tremendous effort to confirm separation conditions of multiple components, which is not realistic regarding the time and cost involved; this is the reason for mass spectrometry to become an important tool for the analysis of environmental pollutants. Just when it was believed that MS coupled to MDGC had matured [22] and with the development of a new data algorithm [24, 25] for better detection, new perspectives might be created because of the computational approaches that have been actively attempted [26–31].

9.5 Liquid chromatography (LC)

It is a fact that there is an overwhelming number of chemicals in use in the world; therefore, they are affecting all environmental areas and the ecosystems where they enter. As it has already been mentioned at the beginning of this chapter, the transformation products of these chemicals or their metabolites can alter even more the environmental balance. The number of organic micropollutants increases and their eco-toxicity can be comparable or even higher than that of the original compound [32, 33]. The problem at this stage is that, when this happens, their chemical structures and toxicological effects are not completely known; therefore, it is important to develop analytical strategies to monitor as many chemicals as possible and trace their fate in the environment.

The first list of priority pollutants created by the US Environmental Protection Agency contained analyzable compounds by GC-MS; however, since the late 1980s, the hyphenated technique LC-MS has grown very quickly to determine and control environmental toxicants. It offers a series of advantages with regard to the chemical nature of those compounds, because most of them are polar organic contaminants: compared to GC-MS, it is possible to avoid the step of derivatization for those nonvolatile compounds. It increases the number of pollutants that is possible to analyze and reduce the total analysis time [34].

A total of 700 substances have been categorized into 20 classes (NORMAN network) in the European surface waters. The most relevant classes are pesticides, pharmaceuticals, disinfection by-products, wood preservation, and industrial chemicals. In general, there are several groups of compounds that emerged as particularly relevant:
– Algal and cyanobacterial toxins
– Brominated flame retardants
– Disinfection by-products
– Hormones and other endocrine-disrupting compounds
– Drugs of abuse and their metabolites
– Organometallics
– Organophosphate flame retardants and plasticizers
– Nanomaterials (nanoparticles)
– Perfluorinated compounds
– Pharmaceuticals and personal care products
– Polar pesticides and their degradation/transformation products
– Surfactants and their metabolites

However, the different physicochemical properties of those compounds make impossible to develop one method to screen and determine all, as well as their degradation products. The scientific community needs to take up the challenge to develop more advanced instrumentation, to search for new chromatographic materials, and to make more effective analytical approaches. Since the impact on aquatic life and human health can be dramatic, a rigorous evaluation of analytical methodology is crucial for contaminants that can compromise flora, fauna, and public health integrity.

Simple and fast sample treatments have been developed to extract and analyze as many compounds as possible simultaneously, in the last years: the main objective has been to save time, expense, and labor. As different pollutants have different physicochemical properties, a balance must be found between handling "dirty" extracts and accepting low recoveries; on the other hand, in case of samples like aqueous ones containing pesticides and drugs, which are very polar and hydrophilic, it has been necessary to lyophilize and evaporate under reduced pressure to enrich them. However, conventional offline SPE (solid-phase extraction) on disks and cartridges has still been the most common and used technique due to its high simplicity and flexibility for many different samples. The classical materials for SPE include C18, graphitized carbon black, *N*-vinylpyrrolidone-divinylbenzene copolymers, mixed-mode cation-exchange cartridges, mixed-mode anion-exchange cartridges, and weak anion-exchange cartridges. Therefore, depending on the compound a different sorbent is chosen, which allows to obtain high recovery percentages and enrichment factors. They vary between 20 and 1,000.

In the last years, research has been focused on the development of nanomaterials (carbon nanostructured materials, metallic nanosized structures, and metal organic frameworks) because of their potential as sorbents in SPE operations either on conventional or miniaturized scales. Their properties of chemical stability; thermal, mechanical, and electronic properties [35–37], as well as their large surface area and durability make them suitable for a broad variety of environmental applications. Fullerenes, nanotubes, nanofibers, and graphene are the carbon nanostructured materials where more research is developed. They show a very good affinity for hydrophobic compounds, especially aromatic compounds that strongly interact with their graphitic portion. They are also very good for polar compounds, because a preliminary oxidizing treatment introduces polar functionality (hydroxyl, carboxyl, and carbonyl groups) and makes them suitable to interact with polar compounds and for chemical derivatizations [38, 39]. These materials, oxidized or not, can be used in generic sample treatments to extract a large number of organic micropollutants, and with an adequate functionalization, they can change their selectivity in a dramatic way, thus becoming specialized sorbents for specific methods of interest.

As an example of the goals accomplished, here is the description of the analysis of a group of marine toxins that are a serious problem in environmental analysis: saxitoxins and their analogues. They are commonly known as PSP (paralytic shellfish poisoning) toxins, and are rather common worldwide and the most lethal of marine toxin intoxication. They are a group of more than 21 tetrahydropurines, usually quantified by a semiquantitative mouse bioassay [40], which was the reference method internationally accepted in monitoring programs until 2019. Since then, the chemical methodology based on pre-column oxidation LC with fluorescence detection has been established as the EU reference method [41, 42]. The chemical methods used to determine PSP toxins are fluorimetric assays, high-pressure liquid chromatography (HPLC) with fluorimetric detection (either pre-column or post-column oxidation), LC-MS, and capillary electrophoresis methods.

The HPLC methods are widely used to quantify PSP toxins present in seafood samples, but they are also useful in providing the PSP profile because in chromatography it is possible to identify each toxin as well. These toxins have only a weak chromophore group and it should be modified before detection: when they are oxidized in an alkaline solution, a purine is formed that becomes fluorescent at acidic pH. This reaction can either be a pre-column or post-column one, and obtained purines are monitorized with a fluorescence detector. Figure 9.8 describes the general outline of these two procedures, where in the post-column method different types of columns have been used (1 and 2 options in the scheme) [43, 44].

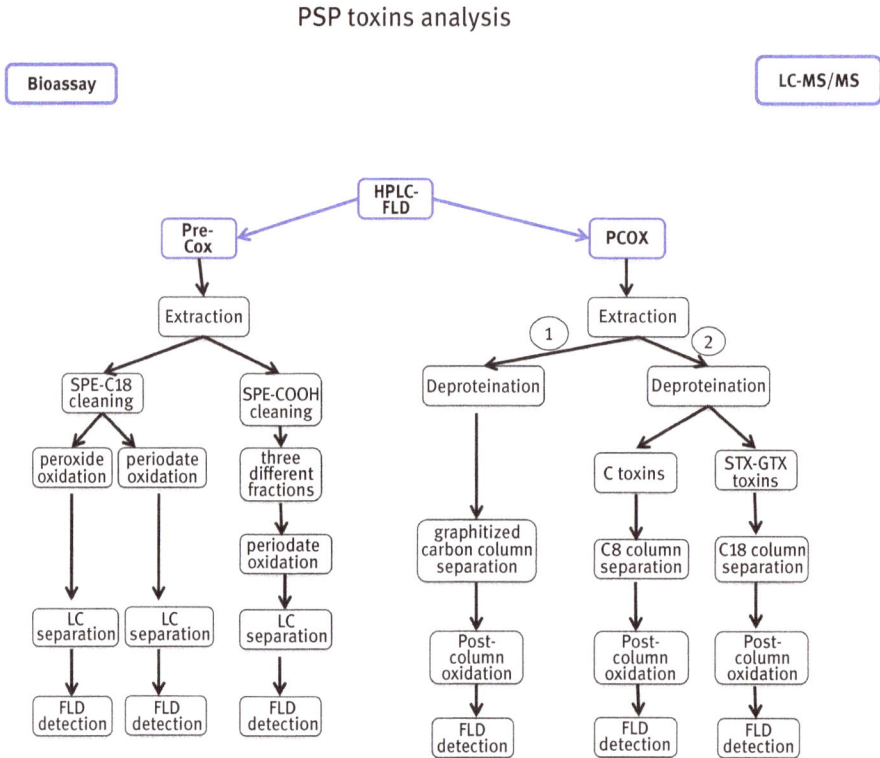

Figure 9.8: General outline of the pre-column and post-column oxidation procedures to determine PSP toxins by HPLC.

Pre- and post-column HPLC methods present as advantages a high sensitivity for low concentrations and low variability for results; but their drawbacks are also important. In the case of hydrophobic analogs, they are retained by C18 resins [45]; therefore, HPLC methods do not allow determining their presence in monitoring programs [46, 47]. LC-MS methods are actually being developed to get a good characterization of these compounds, hence it is recommended that the presence of PSP toxins is confirmed by MS. However, the use of reversed-phase conditions, which generally consist of some organic solvent and nonvolatile salts, is not suitable for LC-MS; mobile phases with phosphate content as well as ion-pair formers are a handicap for an efficient application of the LC-MS technique. Therefore, the application of ionic exchange chromatography with eluents containing only volatile compounds to quantify PSP toxins either with fluorimetric or MS detection has been proposed.

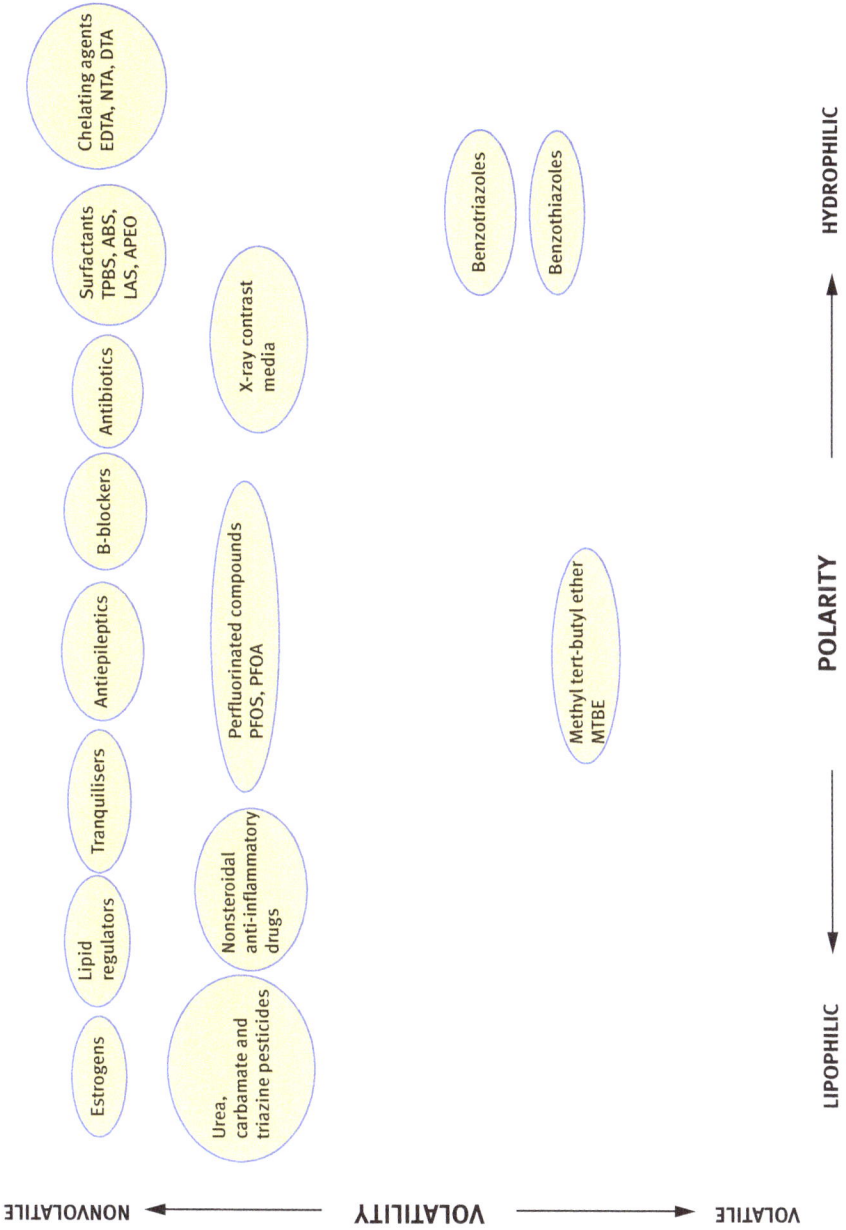

Figure 9.9: LC-MS and some of its applications.

Nowadays, environmental analysis is one of the most important application areas of LC-MS, mainly related to the study of occurrence and fate of organic micropollutants, that have been neither considered as a risk nor included in national monitoring plans so far. Figure 9.9 provides a coarse idea of the wide application range of LC-MS and how it is the technique of choice for analysis of most of these compounds.

In the last years, there has been a growing interest in the development of high-throughput, robust, and sensitive chromatographic methods, regardless of the specific research area. Ultra-high-performance LC (UPLC) technology is one of them, and it only dates back to 2004. It can deliver the mobile phase at pressures up to 1,000 bar, allowing columns packed with very small particles (17–18 μm) to reach their theoretical performance. Stationary phases based on sub-2 mm particles enable elution of analytes in much narrower and more concentrated bands, resulting in better chromatographic efficiency, resolution, and sensitivity with negligible intracolumn band dispersion. Compared to HPLC, this extra efficiency occurs at a higher flow rate and can be achieved in a shorter analysis time. UPLC columns with sub-2 μm porous particles have already successfully been employed to speed up the analysis of a large variety of organic micropollutants in environmental samples: drugs [48–52], personal care products [49], pesticides [33, 49, 53–55], perfluorinated compounds [56], and endocrine disruptors [57, 58].

Keywords: Atomic spectrometry, gas chromatography, liquid chromatography, metals, volatile compounds, nonvolatile compounds

Abbreviations: 1-D GC, one-dimensional gas chromatography; GC, gas chromatography; GC-MS, gas chromatography-mass spectrometry; HPLC, high-pressure liquid chromatography; HR-TOFMS, high-resolution time-of-flight mass spectrometry; LC, liquid chromatography; LC-MS, liquid chromatography-mass spectrometry; MDGC, multidimensional gas chromatography; MDGC-MS, multidimensional gas chromatography-mass spectrometry; PCBs, polychlorinated biphenyls; PSP, paralytic shellfish poisoning; SPE, solid-phase extraction; SPME, solid-phase microextraction; TOFMS, time-of-flight mass spectrometry; UPLC, ultra-high-performance liquid chromatography.

References

[1]   Hodgson E. A textbook of modern toxicology. 3rd ed. Hoboken, NJ: John Wiley & Sons, Inc, 2004.

[2]   Pena-Pereira F, Bendicho C, Mutavdzic Pavlovi´c D, Martín-Esteban A, Díaz Alvarez M, Pan Y, Cooper J, Yang Z, Safarik I, Pospiskova K, Segundo MA, Psillakis E. Miniaturized analytical methods for determination of environmental contaminants of emerging concern – A review. Anal Chim Acta. 2021;1158:238108.

[3]   Hansen F, Pedersen-Bjergaard S. Emerging extraction strategies in analytical chemistry. Anal Chem. 2020;92:2–15.

[4]   Reyes-Garcés N, Gionfriddo E, Gómez-Ríos GA, Alam MN, Boyaci E, Bojko B, Singh V, Grandy J, Pawliszyn J. Advances in solid phase microextraction and perspective on future directions. Anal Chem. 2018;90:302–60.

[5]   Godayol A, Gonzalez-Olmos R, Sánchez JM, Antico E. Assessment of the effect of UV and chlorination in the transformation of fragrances in aqueous samples. Chemosphere. 2015;125:25–32.

[6]   Castro O, Trabalón L, Schilling B, Borrull F, Pocurull E. Solid phase microextraction arrow for the determination of synthetic musk fragrances in fish samples. J Chromatogr A. 2019;1591:55–61.

[7]   Liu S, Pan G, Yang H, Cai Z, Zhu F, Ouyang G. Determination and elimination of hazardous pollutants by exploitation of a Prussian blue nanoparticles-graphene oxide composite. Anal Chim Acta. 2019;1054:17–25.

[8]   Rezaee M, Assadi Y, Milani Hosseini M-R, Aghaee E, Ahmadi F, Berijani S. Determination of organic conmpounds in water using dispersive liquid-liquid microextraction. J Chromatogr A. 2006;1116:1–9.

[9]   Zacharis CK, Tzanavaras PD, Roubos K, Dhima K. Solvent-based deemulsification dispersive liquid-liquid microextraction combined with gas chromatography-mass spectrometry for determination of trace organochlorine pesticides in environmental water samples. J Chromatogr A. 2010;1217:5896–900.

[10]   Hellín P, Pastor-Belda M, Garrido I, Campillo N, Viñas P, Flores P, Fenoll J. Dispersive liquid-liquid microextraction for the determination of new generation pesticides in solids by liquid chromatography and tandem mass spectrometry. J Chromatogr A. 2015;1394:1–8.

[11]   Wang Q, Li L, Long CL, Luo L, Yang Y, Yang Z-G, Zhou Y. Detection of C60 in environmental water using dispersive liquid-liquid micro-extraction followed by high-performance liquid chromatography. Environ Technol. 2020;41:1015–22.

[12]   On J, Pyo H, Myung S. Effective and sensitive determination of eleven disinfection byproducts in drinking water by DLLME and GC-MS. Sci Total Environ. 2018;639:208–16.

[13]   de Boer J, Udo A. TCDD equivalents of mono-ortho substituted chlorobiphenyls. Influence of analytical error and uncertainty of toxic equivalency factors. Anal Chim Acta. 1994;289:261–62.

[14]   Cochran JW, Frame, GM. Recent developments in the high-resolution gas chromatography of polychlorinated biphenyls. J Chromatogr A. 1999;843:323–68.

[15]   GM Frame. A collaborative study of 209 PCB congeners and 6 Arochlors on 20 different HRGC columns: Part I. Retention and coelution database. Fresenius' J Anal Chem. 1997;357:701–13.

[16]   Samuel C, Davis JM. The need for two-dimensional gas chromatography: Extent of overlap in one-dimensional gas chromatograms. J High Resolut Chromatogr. 2000;23:235–44.

[17]   Giddings J. Multidimensional gas chromatography. New York: Marcel Dekker, 1990.

[18]   Schomburg G, Weeke F, Schaefer RG. Direct determination of the phenanthrene and methyl-phenanthrene isomer distribution in crude oils by multi-dimensional capillary GC. J High Resolut Chromatogr Chromatogr Commun. 1985;8:388–90.

[19]   Marriott PJ, Haglund P, Ong RCY. A review of environmental toxicant analysis by using multidimensional gas chromatography and comprehensive GC. Clin Chim Acta. 2003;328:1–19.

[20]   Dimandja JMD, Grainger J, Patterson DG Jr. New fast single and multidimensional gas chromatographic separations coupled with high resolution mass spectrometry and time-of-flight mass spectrometry for assessing human exposure to environmental toxicants. Organohalog Compd. 1999;40:23–26.

[21]   Vaye O, Ngumbu RS, Xia D. A review of the application of comprehensive two-dimensional gas chromatography MS-based techniques for the analysis of persistent organic pollutants and ultra-trace level of organic pollutants in environmental samples. Rev Anal Chem. 2022;41:63–73.

[22]   Tranchida PQ, Franchina FA, Dugo P, Mondello L. Comprehensive two-dimensional gas chromatography-mass spectrometry: Recent evolution and current trends. Mass Spectrometry Rev. 2014;2008:1–11.

[23] Xia D, Gao L, Zheng M, Wang S, Liu G. Simultaneous analysis of polychlorinated biphenyls and polychlorinated naphthalenes by isotope dilution comprehensive two-dimensional gas chromatography high-resolution time-of-flight mass spectrometry. Anal Chim Acta. 2016;937:160–67.

[24] Wang Y, Xu X, Yin L, Cheng H, Mao T, Zhang K, et al. Coupling of comprehensive two-dimensional gas chromatography with quadrupole mass spectrometry: Application to the identification of atmospheric volatile organic compounds. J Chromatogr A. 2014;1361:229–39.

[25] Hoh E DN, Lehotay SI, Pangallo KC, Reddy CM, Maruya KA. Non-targeted analysis of electronics waste by comprehensive two-dimensional gas chromatography combined with high-resolution mass spectrometry: Using accurate mass information and mass defect analysis to explore the data. J Chromatogr A. 2015;1395:152–59.

[26] Hammer J, Matsukami H, Endo S. Congener-specific partition properties of chlorinated paraffins evaluated with COSMOtherm and gas chromatographic retention indices. Sci Rep. 2021;11:4426.

[27] Jaramillo R, Dorman FL. Retention time prediction of hydrocarbons in cryogenically modulated comprehensive two-dimensional gas chromatography: A method development and translation application. J Chromatogr A. 2020;1612:460696.

[28] Subraveti SG, Li Z, Prasad V, Rajendran A. Can a computer "learn" nonlinear chromatography?: Physics based deep neural networks for simulation and optimization of chromatographic processes. J Chromatogr A. 2022;1672:463037.

[29] Veenaas C, Linusson A, Haglund P. Retention-time prediction in comprehensive two-dimensional gas chromatography to aid identification of unknown contaminants. Anal Bioanal Chem. 2018;1626:461308.

[30] Poole CF. Evaluation of the solvation parameter model as a quantitative structure-retention relationship model for gas and liquid chromatography. J Chromatogr A. 2020;1626:461308.

[31] Aalizadeh R, Alygizakis NA, Schymanski EL, Krauss M, Schulze T, Ibáñez M, McEachran AD, Chao A, Williams AJ, Gago-Ferrero P, Covaci A, Moschet C, Young TM, Hollender J, Slobodnik J, Thomaidis NS. Development and application of liquid chromatographic retention time indices in HRMS-based suspect and nontarget screening. Anal Chem. 2021;93:11601–11.

[32] Bletsou AA, Jeon J, Hollender J, Archontaki E, Thomaidis NS. Targeted and non-targeted liquid chromatography-mass spectrometry workflows for identification of transformation products of emerging pollutants in the aquatic environment. TrAC-Trends Anal Chem. 2015;66:32–44.

[33] La Farré M, Picó Y, Barceló D. Application of ultra-high pressure liquid chromatography linear ion-trap Orbitrap to qualitative and quantitative assessment of pesticide residues. J Chromatogr A. 2014;1328:66–79.

[34] Pérez-Fernández V, Rocca LM, Tomai P, Fanali S, Gentili A. Recent advancements and future trends in environmental analysis: Sample preparation, liquid chromatography and mass spectrometry. Anal Chim Acta. 2017;983:9–41.

[35] Socas-Rodriguez B, Herrera-Herrera A, Asensio-Ramos M, Hernández-Borges J. Recent applications of carbon nanotube sorbents in analytical chemistry. J Chromatogr A. 2014;1357:110–46.

[36] Wen Y, Chen L, Li J, Liu D, Chen L. Recent advances in solid-phase sorbents for sample preparation prior to chromatographic analysis. TrAC-Trends Anal Chem. 2014;59:26–41.

[37] González-Sálamo J, Socas-Rodríguez B, Hernández-Borges J, Rodríguez-Delgado MA. Nanomaterials as sorbents for food sample analysis. TrAC-Trends Anal Chem. 2016;85:203–20.

[38] Liu Q, Zhou Q, Jiang GB. Nanomaterials for analysis and monitoring of emerging chemical pollutants. TrAC-Trends Anal Chem. 2014;58:10–22.

[39] El-Sheikh AH, Sweileh JA, Al-Degs YS, Insisi AA, Al-Rabady N. Critical evaluation and comparison of enrichment efficiency of multi-walled carbon nanotubes, C18 silica and activated carbon towards some pesticides from environmental waters. Talanta. 2008;74:1675–80.

[40] AOAC. Paralytic shellfish poisoning toxins in shellfish. Prechromatographic oxidation and liquid chromatography with fluorescence detection. Official Methods of Analysis of the Association of Official Analytical Chemists 2005:Method 200506: First Action, 2005.

[41] R E. Regulation (EC) No 853/2004 of the European Parliament and of the Council, laying down specific hygiene rules for food of animal origin. Official Journal of the European Union. 2004; L226:83–127.

[42] O'Neill A, Turner AD. Performance characteristics of AOAC method 2005.06 for the determination of paralytic shellfish toxins in manila clams, European otter clams, grooved carpet shell clams, surf clams, and processed king scallops. J AOAC Int. 2015;98(3):628–35.

[43] Rey V, Alfonso A, Botana LM, Botana AM. Influence of different shellfish matrices on the separation of PSP toxins using a postcolumn oxidation liquid chromatography method. Toxins. 2015;7:1324–40.

[44] Rey V, Botana AM, Alvarez M, Antelo A, Botana LM. Liquid chromatography with a fluorimetric detection method for analysis of paralytic shellfish toxins and tetrodotoxin based on a porous graphitic carbon column. Toxins. 2016;8:196–211.

[45] Negri A, Stirling D, Quilliam M, Blackburn S, Bolch C, Burton I, et al. Three novel hydroxybenzoate saxitoxin analogues isolated from the dinoflagellate Gymnodinium catenatum. Chen Res Toxicol. 2003;16:85–93.

[46] Vale P. Complex profiles of hydrophobic paralytic shellfish poisoning compounds in Gymnodinium catenatum identified by liquid chromatography with fluorescence detection and mass spectrometry. J Chromatogr A. 2008;1195:85–93.

[47] Vale P, Rangel I, Silva B, Coelho P, Vilar A. Atypical profiles of paralytic shellfish poisoning toxins in shellfish from Luanda and Mussulo bays, Angola. Toxicon. 2009;53:176–83.

[48] López-Serna R, Petrović M, Barceló D. Development of a fast instrumental method for the analysis of pharmaceuticals in environmental and wastewaters based on ultra-high performance liquid chromatography (UHPLC)-tandem mass spectrometry (MS/MS). Chemosphere. 2011;85:1390–99.

[49] Gracia-Lor E, Martínez M, Sancho JV, Peñuela G, Hernández F. Multi-class determination of personal care products and pharmaceuticals in environmental and wastewater samples by ultra-high performance liquid-chromatography-tandem mass spectrometry. Talanta. 2012;99:1011–23.

[50] Hernández F, Bijlsma L, Sancho JV, Díaz R, Ibañez M. Rapid wide-screening of drugs of abuse, prescription drugs with potential for abuse and their metabolites in influent and effluent urban wastewater by ultrahigh pressure liquid chromatography-quadrupole-time-of-flight-mass spectrometry. Anal Chim Acta. 2011;684:96–106.

[51] Masia A, Campo J, Blasco PY. Ultra-high performance liquid chromatography-quadrupole time-of-flight mass spectrometry to identify contaminants in water: An insight on environmental forensics. J Chromatogr A. 2014;1345:86–97.

[52] Bourdat-Deschamps M, Leang S, Bernet N, Daudin JJ, Nélieu S. Multi-residue analysis of pharmaceuticals in aqueous environmental samples by online solid-phase extraction-ultra-high-performance liquid chromatography-tandem mass spectrometry: Optimisation and matrix effects reduction by quick, easy, cheap, effective, rugged and safe extraction. J Chromatogr A. 2014;1349:11–23.

[53] Chen ZF, Ying GG, Lai HJ, Chen F, Su HC, Liu YS, Peng FQ, Zhao J-L. Determination of biocides in different environmental matrices by use of ultra-high-performance liquid chromatography-tandem mass spectrometry. Anal Bioanal Chem. 2012;404:3175–88.

[54] Wode F, Reilich C, Van Baar P, Dünnbier U, Jekel M, Reemstma T. Multi-residue analytical method for the simultaneous determination of 72 micropollutants in aqueous samples with ultra-high performance liquid chromatography-high resolution mass spectrometry. J Chromatogr A. 2012;1270:118–26.

[55] Köck-Schulmeyer M, Olmos M, López de Alda M, Barceló D. Development of a multiresidue method for analysis of pesticides in sediments based on isotope dilution and liquid chromatography-electrospray-tandem mass spectrometry. J Chromatogr A. 2013;1305:176–87.

[56] Onghena M, Moliner-Martinez Y, Picó Y, Campins-Falcó P, Barceló D. Analysis of 18 perfluorinated compounds in river waters: Comparison of high performance liquid chromatography-tandem mass spectrometry, ultra-high-performance liquid chromatography-tandem mass spectrometry and capillary liquid chromatography-mass spectrometry. J Chromatogr A. 2012;1244:88–97.

[57] Huerta B, Jakimska A, Llorca M, Ruhí A, Margoutidis G, Acuña V, Sabater S, Rodriguez-Mozaz S, Barceló D. Development of an extraction and purification method for the determination of multi-class pharmaceuticals and endocrine disruptors in freshwater invertebrates. Talanta. 2015;132:373–81.

[58] Gorga M, Petrovic M, Barceló D. Multi-residue analytical method for the determination of endocrine disruptors and related compounds in river and waste water using dual column liquid chromatography switching system coupled to mass spectrometry. J Chromatogr A. 2013;1295:57–66.

Jesús M. González-Jartín

10 Natural toxins in food: occurrence and detoxification strategies

10.1 Introduction

The food chain is susceptible to various hazards that pose risks to human and animal health. These hazards can be classified into four main types: biological, chemical, physical, and allergenic [1]. The outbreaks of foodborne illnesses occurred in the 1990s made consumers aware of the importance of food safety. During that decade, the establishment of the Single European Market brought the issue of food safety to the forefront of the political agenda, mainly due to the far-reaching consequences of various food scandals [2]. Following the bovine spongiform encephalopathy crisis, the European Commission (EC) established a comprehensive framework for Community food law. The primary objectives of this legislation were to guarantee the protection of public health, safety, and consumer interests. To effectively achieve these goals, the regulatory strategy adopted by the EC encompassed the entire food chain and prioritized legislation based on scientific evidence [3].

The increase in the agri-food trades is currently one of the most significant challenges to global food safety. In this sense, several nations have inadequate food safety standards and have yet to establish effective surveillance or reporting methods to identify and track foodborne illnesses. Therefore, standardization and equivalence of regulatory and standard frameworks is crucial [4]. On the other hand, climate change is related with future temperature, precipitation, and CO_2 concentration changes which are expected to increase the incidence of several hazards such as natural toxins, a category of contaminants produced by plants, animals, or microorganisms that can pose significant risks to human and animal health when ingested. For instance, in the case of mycotoxin contamination of cereal crops, it is expected that climate change will have an impact on the geographical distribution of particular cereals, mycotoxigenic fungus, and their mycotoxins [5]. Similarly, the increase of the pH and the temperature of ocean seawater may lead to variations in the phytoplankton abundance and the frequency of harmful algal blooms (HABs). In inland water bodies, cyanobacterial blooms are likely to last longer and be more severe due to a combination of factors including increased evaporation and nutrient concentration, water stratification, and higher phosphorus levels in deeper layers [6].

In order to protect public health, the EC has established in 2006 maximum tolerances for specific contaminants, including natural toxins, in different foods [7, 8]. Since their

Jesús M. González-Jartín, Departamento de Farmacología, Facultad de Farmacia, Universidade de Santiago de Compostela, 15705 Santiago de Compostela, Spain

https://doi.org/10.1515/9783111014449-010

publication, the document was amended substantially many times and replaced in 2023 by a new document [9]. From the moment food legislation was established in the European Union (EU), food operators and the European Food Safety Authority (EFSA) have made a continuous monitoring on the presence of the regulated toxins in several matrices. In this sense, the European Rapid Alert System for Food and Feed (RASFF) plays a vital role in ensuring food safety. This system facilitates the exchange of information among food safety authorities regarding potential health risks associated with food or feed, enabling immediate actions to mitigate these risks. Its purpose is to prevent harm and protect consumers. Furthermore, annual reports are published detailing the incidents identified, which is a valuable resource for monitoring contaminants and their trends over time [10–23]. However, in the last few decades the improvements in the field of food analysis allowed the detection of new compounds, generally known as emerging toxins, which are not legislatively regulated or routinely monitored, but new evidences about their extensive incidence have become an important food safety issue [24].

10.2 Mycotoxins

Mycotoxins are naturally occurring food toxins produced by certain fungi that can grow on various crops both before and after harvesting. While over 500 mycotoxins have been identified, only about 20 of them are of significant concern for food safety because they can reach harmful levels in food and feed. These toxins include aflatoxins (AFs), deoxynivalenol (DON), zearalenone (ZEN), T2/HT2 toxins, fumonisins (FBs), ochratoxin A (OTA), patulin (PAT), citrinin (CTN), and ergot alkaloids (EAs). These compounds are highly toxic, and their presence in food and feed is regulated in Europe and many other countries around the world to protect public health. On the other hand, the term "emerging mycotoxins" gained widespread use in 2008, and it was initially used for fusaproliferin, beauvericin (BEA), enniatins (ENNs), and moniliformin. However, other mycotoxins such as sterigmatocystin, mycophenolic acid, alternariol, and tenuazonic acid also belong to this category [25].

Mycotoxins are a significant concern and are consistently ranked as the third most frequently reported hazard category in RASFF. However, the frequency of mycotoxin notifications varies from year to year. For example, there were 1,847 notifications between 2010 and 2012, but this number decreased by 30% to 1,279 notifications between 2020 and 2022 (Figure 10.1). It is important to highlight that this decrease is primarily attributed to a reduction in the number of notifications related to AFs. Border rejections are responsible for approximately three-quarters of these notifications, indicating the need for rigorous control measures at the borders.

The presence of AFs accounted for a substantial majority of the notifications recorded between 2020 and 2022 comprising 89% of the total. Following, OTA emerged as the second most frequently detected mycotoxin, identified in 8.9% of the cases.

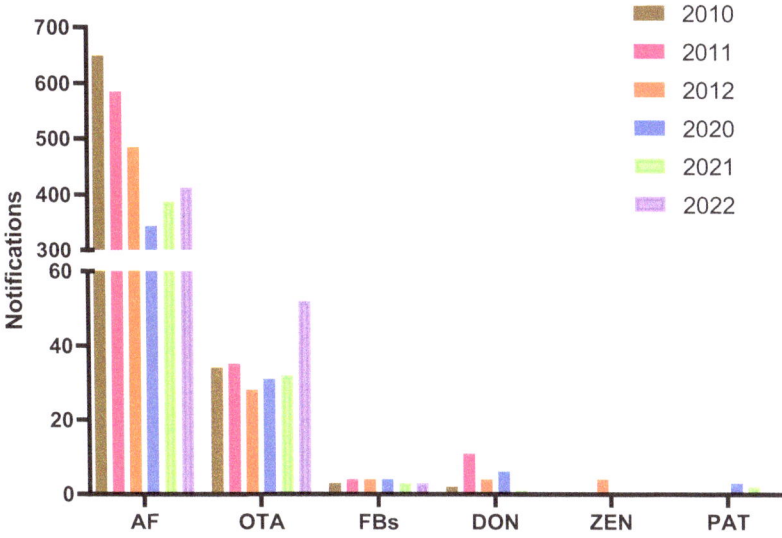

Figure 10.1: Number of notifications according to the toxins and the year.

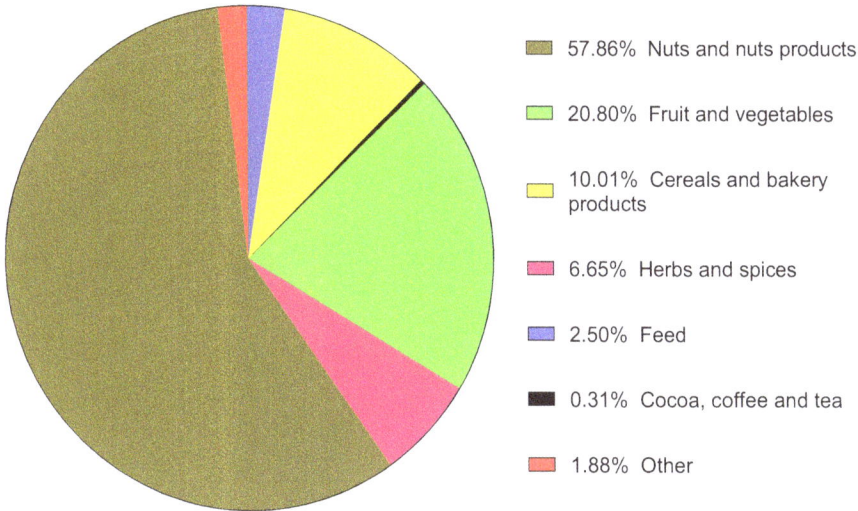

Figure 10.2: Distribution of notifications according to the matrix.

Other toxins are less frequently reported, with sporadic occurrences. When examining the type of product associated with these notifications (Figure 10.2), it is evident that a significant majority (57.86%) can be attributed to nuts and nut products. Following, fruits and vegetables represent the second most commonly reported category at 20.80%, while cereals and bakery products comprise 10% of the total notifications.

10.2.1 Aflatoxins

AFs are a group of approximately 18 related compounds primarily produced by *Aspergillus* species, with *A. flavus* and *A. parasiticus* being the most prominent sources. These toxins, including the four major naturally occurring compounds (AFB$_1$, AFB$_2$, AFG$_1$, and AFG$_2$), are commonly found worldwide in various foodstuffs such as nuts, spices, cereals, oils, fruits, vegetables, and meat. AFM$_1$ is a hydroxylated metabolite of AFB$_1$, which can be present in milk when animals ingest feed contaminated with AFB$_1$. Although there have been sporadic reports of acute human aflatoxicosis, such as the 2004 outbreak in Kenya that caused the death of 125 people, the primary concern is the long-term exposure to these toxins. Chronic exposure to AFs is associated with an increased risk of liver cancer and other health issues [26].

The majority of notifications regarding the presence of AFs in food are specifically linked to nuts (Figure 10.3). However, it is notable that in recent years, there has been a significant decline in the number of such notifications compared to the numbers recorded a decade ago [27]. However, the occurrence varies significantly across the world and also depends on the type of nut. A systematic review showed that the mean concentration of AFB$_1$ in peanuts was 37.82 µg/kg, but this varies depending on the product's origin. The highest levels were observed in Argentina (530 µg/kg), Congo (163.22 µg/kg), Nigeria, and Uganda (>103 µg/kg), while levels lower than 1 µg/kg were observed in Turkey, Spain, Thailand, and Morocco. The mean concentration in pistachios was 39.44 µg/kg, with the highest levels found in Taiwan (233.60 µg/kg), Morocco (158.00 µg/kg), and Iran (64.65 µg/kg). Cyprus, Pakistan, Saudi Arabia, and Turkey presented levels vary from 4.55 to 8.20 µg/kg, while low levels were detected in Korea,

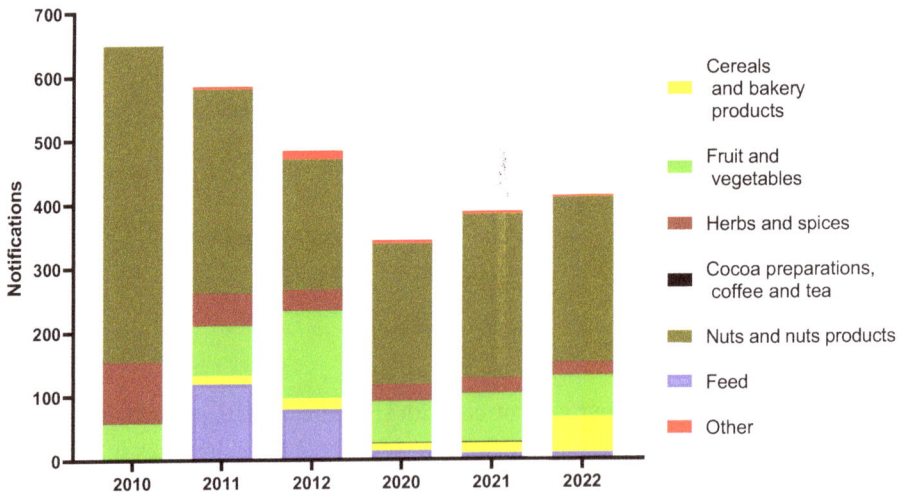

Figure 10.3: Number of AF notifications per year according to the matrix.

Italy, and Spain. In other nuts, levels are usually lower such as in walnuts (22.23 μg/kg) and in almonds (3.93 μg/kg) [28].

Since maximum levels of AFs are frequently exceeded, the EU adopted special conditions at the Union level for the import of certain foodstuffs from certain third countries including the increase of official controls at border control posts and control points. In this sense, the frequency of identity check and the sampling for analysis of AFs are carried out in up to 50% of the consignments of foodstuffs [29]. Consequently, most of the alerts (84%) that were produced from 2020 to 2022 were due to border rejection notifications. Similar data was collected previous years; for instance, in 2018, various types of nuts from different countries had recurrent issues summarized in Figure 10.4 [21].

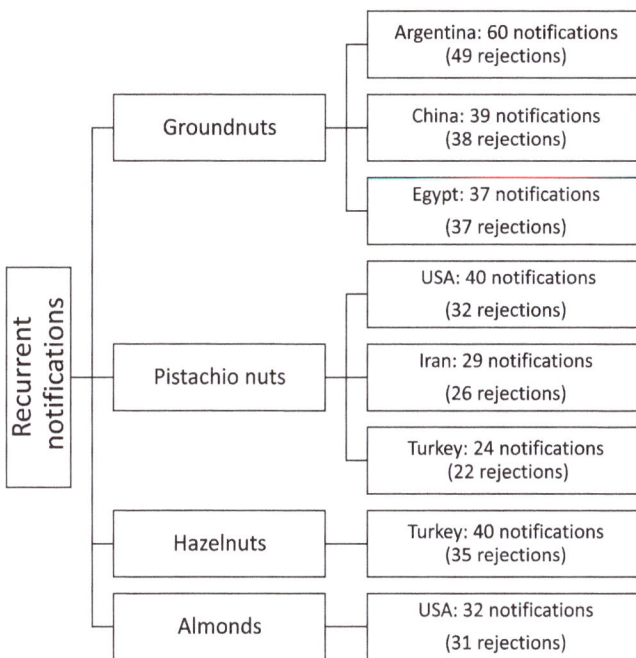

Figure 10.4: Recurrent notifications in 2018. Notifications per country and border rejections.

Dried figs from Turkey have also been a recurring issue consistently. In 2020, there were 54 notifications raised due to AFs in dried figs from Turkey. This number increased slightly to 56 alerts in 2021 and decreased to 42 notifications in 2022. However, due to the special import measures to which these fruits are subjected, most of the notifications correspond to rejections at the border and only six alerts have been produced in which the products had entered the market. Human exposure to AFs can also occur through the ingestion of oilseeds, such as sunflower and coconut, as well as their derived products, including oil. Extensive research and reports have highlighted the high incidences of AF contamination in plant-derived oils in various regions, notably China, Sudan, India, and Sri Lanka [30]. Furthermore, it is important to note that spices, particularly red pepper, have been found to exhibit high levels of AFs contamination. An example is the contamination of approximately 30% of species samples from Iran, where toxin levels ranged from 0.2 to 57.5 µg/kg [31]. These data are in agreement with the alerts identified in the RASFF portal from 2020 to 2022 (Figure 10.5). Out of the 62 notifications, 20 were associated with Chile, with 11 of them originating from India. Concerning Nutmeg, the majority of alerts (12 out of 15) were linked to Indonesia. Additionally, alerts related to seasoning products primarily involved Pakistan.

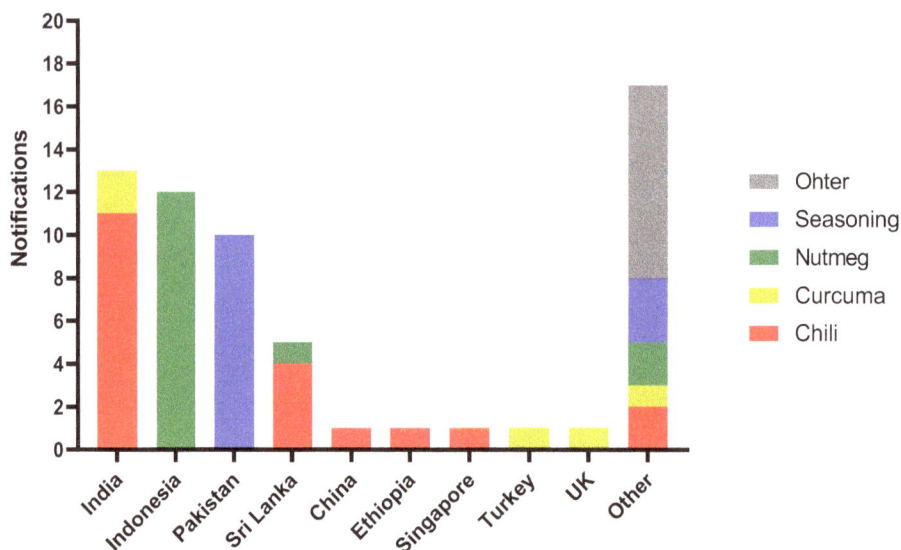

Figure 10.5: Number of AF notifications from 2020 to 2022 according to the spices and the origin of the product.

Although cereals generally exhibit lower contamination levels compared to other food items, they are still a significant source of human exposure to AFs due to their high consumption levels. Among cereals, maize (*Zea mays* L) is the most widely consumed cereal globally, making it the primary source of toxin ingestion. While contamination of maize has been reported worldwide, the main risk comes from production areas with warmer climates since they are favorable for fungal growth and AFs production. In sub-Saharan Africa, at the end of the value chain, up to 100% of maize samples contain AFs, with the highest worldwide levels reported in this region. In Latin America, approximately 18–38% of maize samples are contaminated with AFs, while in Asia, the incidence rates range from 37% to 92%. These differences may be due to varying climatic conditions across regions. While the occurrence of AFs in the United States and

Europe has traditionally been low, there has been an increasing trend in contamination levels in recent years, likely due to climate change [32]. AFs contamination is frequently observed in other crops such as rice (*Oryza sativa* L.) from China, Egypt, India, Iran, Malaysia, Nepal, Pakistan, Philippines, United Kingdom, and United States [33]. It must be highlighted that 41.1%, 38.5%, and 20.9% of feed raw materials (e.g., maize, wheat, and soybean) from South Asia, Sub-Saharan Africa, and Southeast Asia, respectively, exceeded the maximum level for AFB_1 (20 µg/kg) established in the EU for feed [34]. In the RASFF portal, it was found that 66% of the notifications identified from 2020 to 2022 were related to the presence of AFs in rice from Pakistan, while an additional 10% were associated with AFs in rice from India, Sri Lanka, and the Netherlands. Sporadic notifications regarding the contamination of maize, millet, wheat, and couscous from various origins were also detected [27].

AFM_1 can be present in milk and processed milk products. Several nations including Pakistan, India, and several sub-Saharan African nations usually reported toxin levels above the European regulatory limits [35, 36]. However, in Europe, there have only been isolate incidents where this toxin was detected. As an example, from 2007 to 2012, no alerts were triggered in the EU. However, in 2012, five notifications regarding AFM_1 in milk were reported. These notifications were associated with an increased prevalence of AFs in maize from the southeast of Europe. This region experienced a severe drought during the maize growing season, which led to the higher occurrence of these toxins in maize [15]. Similarly, in 2016 AFM_1 was detected above the legal maximum six times on milk products from Italian origin [19]. In addition, residues of AFs can be detected in other animal products such as meat and eggs; however, no reports were detected [37].

10.2.2 Deoxynivalenol

DON is a type B trichothecene mycotoxin produced by various *Fusarium* species. This toxin is known to induce gastrointestinal tract toxicity in both humans and animals. Its effects are particularly notable in cases of acute intoxication, where symptoms such as anorexia and emesis can be observed. Due to its ability to induce vomiting, DON is also commonly known as vomitoxin. This toxin can be found as a contaminant in starchy staple foods, including wheat, barley, oats, rye, corn, and potatoes, making it a global concern. However, significant outbreaks of acute human disease related to DON have been predominantly reported in Asia [38].

An evaluation conducted by the EFSA revealed that approximately half of the samples analyzed in monitoring programs contained DON. The presence of this mycotoxin was observed in unprocessed grains of undefined end-use (44.6% of samples), food (43.5% of samples), and feed (75.2% of samples). Maize, wheat, and oat grains, as well as food and feed products derived from them, were frequently found to contain higher levels of DON compared to other cereal varieties. In general, the levels of DON exceeded the maximum tolerable limits in 0.8% of the food samples tested, and guidance values were surpassed in 1.7% of the feed samples. Notably, wheat bran exhibited substantially higher levels of DON compared to other wheat milling products. Finally, processed cereals such as bread, fine bakery wares, breakfast cereals, and pasta generally had much lower levels of DON compared to raw grains and grain milling products. This suggests that processing methods employed in their manufacture may help reduce the concentration of DON. In this sense, DON can be detected in 75% of bread products, 36% of cereal flakes, 58% of biscuits, and 50% of raw pasta, generally in concentrations lower than 80 µg/kg [39]. However, a low number of notification regarding DON were identified, probably because the legislation establishes 1,250 µg/kg as the maximum levels for this contaminant in unprocessed cereal grains and concentrations are usually lower. In this sense, in the period 2020–2022, seven notifications were identified, related with the presence of DON in corn grain, popcorn, wheat, quick cooking noodles, and breadcrumbs [27].

10.2.3 Zearalenone

ZEN, previously known as F-2 toxin, is a naturally occurring nonsteroidal mycotoxin with estrogenic properties. It is primarily produced by *Fusarium* species such as *F. graminearum* and *F. culmorum*, which are commonly found in soil and are prevalent in temperate and warm regions worldwide. ZEN can contaminate a variety of cereal crops, including maize, wheat, barley, oats, and rice.

In the EU, ZEN can be detected in approximately 17% of grains intended for human consumption. However, the occurrence of this mycotoxin varies significantly depending on the specific crop. For instance, maize exhibits the highest prevalence with up to 33% of samples testing positive for ZEN, with lower-bound and upper-bound concentrations ranging from 13 to 15 µg/kg. Wheat follows with approximately 22% of samples testing positive, containing concentrations between 2.4 and 5.4 µg/kg. Barley, oats, and rice show intermediate levels, with around 15% of positive samples, and concentrations ranging from 0.7 to 9.7 µg/kg. Rye generally demonstrates lower levels of ZEN contamination. In general, wheat bran, maize, and maize-derived products such as maize flour and cornflakes tend to have the highest levels. On the other hand, certain cereal fractions exhibit higher concentrations of ZEN. This toxin is typically redistributed among the milling fractions. By-products generated during the cleaning process of raw cereal grains, including dust, hulls, and other residues, have been found to contain 3–30-fold higher concentrations of ZEN compared to the cleaned cereal grains. Additionally, bran may contain up to twice the concentration of ZEN compared to other cereal fractions [40].

When examining final food products, the presence of ZEN has been detected in several goods (Figure 10.6). However, the highest ZEN levels were observed in wheat flakes ranging from 8.4 to 25 µg/kg. A relatively high contamination frequency of 33.2% was observed in biscuits, which can be attributed to the presence of wheat bran

Pasta
- 3.9%
- 0.6 to 5.8 µg/kg

ZEN

Bread and rolls
- 7.4%
- 0.9 to 5.2 µg/kg

Breakfast cereals
- Corn flakes: 17.3%
- Wheat flakes 14.3%
- Mixed breakfast cereals 12%

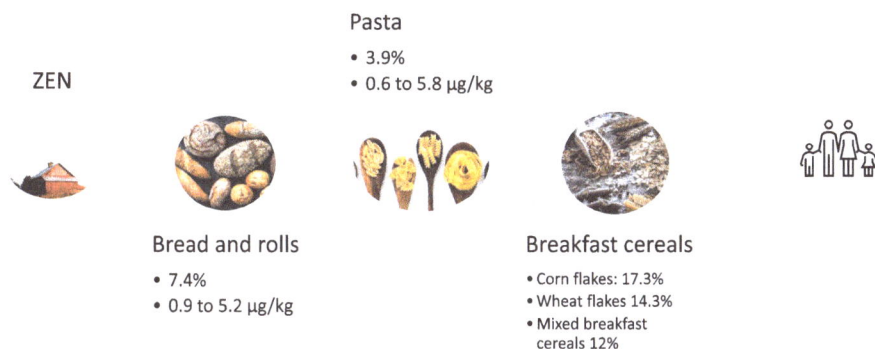

Figure 10.6: Transference of ZEN from the field to final products.

in certain types of biscuits and the use of vegetable oils containing ZEN as an ingredient. Remarkably, maize germ oil and wheat germ oil exhibited a very high contamination frequency of 86.4% [40].

The levels of ZEN contamination vary significantly depending on the country. For instance, in Southern European countries such as Spain and Portugal, contamination by this mycotoxin is relatively rare. However, in central European countries like Germany, the frequency of ZEN contamination is much higher. Similarly, in other regions of the world, such as Iran, the contamination of cereals is common [41]. However, the levels at which this toxin usually occurs are significantly lower than the legally permitted ones, leading to sporadic notifications associated with the presence of this toxin in cereals and bakery products. For instance, in 2007, there were five notifications, followed by two in 2008 and four in 2012 [10, 11, 15].

10.2.4 T-2 and HT-2 toxins

T-2 and HT-2 are type-A trichothecenes produced by *Fusarium* species, primarily *F. sporotrichioides*. This fungi, found in temperate regions, commonly grow on various grains, resulting in a global prevalence of these toxins. Among cereals, oat is particularly susceptible to contamination by T-2 and HT-2 toxins. T-2 toxin undergoes metabolic conversion to HT-2 toxin after ingestion, and both forms possess equal toxicity. Extensive research has established that T-2 toxin exhibits significant cytotoxic, immunosuppressive, and hematotoxic effects. Acute intoxication with type-A trichothecenes, particularly T-2 toxin, can lead to the development of alimentary toxic aleukia (ATA). This mycotoxicosis is characterized by symptoms such as leukopenia, agranulocytosis, bone marrow depletion, necrotic angina, and, in severe cases, death. Notably, an important outbreak of ATA occurred in Russia during World War II in the 1940s, resulting in the death of thousands of people. This historical event highlights the devastating impact of T-2 toxin poisoning on human health [42].

Recent data collected from 2011 to 2016 provide valuable insights into the presence of T-2 and HT-2 toxins in food and feed in Europe. In terms of food samples, less than 10% tested positive for T-2 toxin and less than 13% for the combined sum of T-2 and HT-2 toxins. The highest levels of T-2 and HT-2 were reported in grain-based products and breakfast cereals, particularly those containing oats. For instance, oat grains exhibited concentrations ranging from 127 to 128 µg/kg (lower bound-upper bound), while oat cereal flakes showed concentrations from 13.9 to 16.5 µg/kg. Quantifiable levels of these mycotoxins were also detected in other food categories encompassing grain-based products such as snacks and food for infants and young children. However, in non-grain-based foods like legumes and fats and oils, T-2 and HT-2 concentrations were only occasionally quantified. Regarding feed samples, approximately 12% tested positive for T-2 or HT-2 toxins. Again, oats exhibited the highest mean concentrations (lower bound mean = 401 µg/kg; upper bound mean = 405 µg/kg). Other feed categories demonstrated relatively low mean concentrations [43]. However, it is important to note that there are significant variations in T-2 and HT-2 toxin occurrences based on the specific matrix and the year of analysis. For feed samples collected between 2008 and 2017, higher occurrences were observed in grains from Eastern (48%), Central, and Northern Europe (30%) compared to Southern Europe (11.7%). South America (21.5%) and East Asia (11%) exhibited intermediate levels of contamination. In other regions of the world, less than 5% of cereals were found to be contaminated with T-2 toxin [34]. In the last years, there is only one notification in the RASFF portal related to the presence of T-2 toxin in organic corn snack with cheese and onion for babies from the Netherlands [27].

10.2.5 Fumonisins

Fusarium species, particularly *F. verticillioides* and *F. proliferatum*, produce a group of over 25 analogues known as FBs, which commonly infect maize and sorghum. However, only type B FBs are significant in terms of both toxicity and occurrence. Specifically, the FB_1, FB_2, and FB_3 constitute over 95% of the total content of FBs found in contaminated samples. Among these, FB_1 is typically present at the highest concentrations in food and feed [44].

A number of investigations have found FB_1 and FB_2 in maize and its by-products in several European nations. While these chemicals are common, their concentration levels typically vary between 0.2 and 2 mg/kg, with higher amounts observed in unprocessed materials. FB_3 is commonly detected in conjunction with FB_1 and FB_2, although its concentration usually remains approximately 90% lower than that of FB_1. Even though FBs are generally linked with maize, 16% of the wheat flour samples also contained this mycotoxin. Maize is used to make a variety of products, including tortillas, corn flakes, popcorn, flour, and oils. Contamination levels are greatly decreased during the production process. For example, the treatment used in the extrusion process reduces FBs levels in tortilla chips derived from maize flour by 59%, flour by 60%, and grits and snack products by 50%. FB_2 can also be produced by *Aspergillus niger*, a fungus known to infect grapes, wheat, and maize, thereby resulting in the contamination of these goods and their derived products. In the case of animal feed, FBs were detected in 54% of feed samples, with an average content of 1,674 µg/kg. The prevalence of FBs is significantly higher in South America, reaching 77% of positive samples. The EFSA recently study the prevalence of FBs in animal feed ingredients such as cereal grains, complex feed, forages, and roughages. These toxins were found in 23% of the samples used for monitoring programs in Europe. However, in the case of maize products, this ratio increased to 46% [44–46]. As with DON, the toxin levels at which FBs are usually found are below the maximum permitted by law. Therefore, there are only around three notifications per year [27].

10.2.6 Ochratoxin A

Ochratoxins are produced by several species of *Aspergillus* and *Penicillium*, primarily those belonging to the *Circumdati* section of *Aspergillus*. The prevalence of these fungi varies across different geographic regions. In countries with cold and temperate climates, *P. verrucosum* stands out as the major source of cereal contamination. In tropical climates, various species of *Aspergillus* infect cereals, coffee, cocoa, and edible nuts. Moreover, within the *Aspergillus* section *Nigri*, *A. carbonarius* plays a significant role in grape and wine contamination with OTA in Europe. The isolation of OTA can be traced back to 1965 when it was first discovered in cultures obtained from cereals, causing fatal outcomes in animals [47, 48].

The majority of notifications regarding the presence of OTA in food are linked to fruit and vegetables (Figure 10.7). Fruit-derived products, such as grape juice and wine, serve as significant sources of OTA intake in the EU. On average, grape juice presents an OTA concentration of 0.55 µg/kg, while wine exhibits a contamination level of 0.36 µg/kg. It is worth noting that there can be substantial variations in the prevalence of this toxin in wine from 1 year to another, although, in general, the detected levels remain below the maximum tolerable limits. In this sense, from 2020 to 2022 there was only one notification regarding the exceedance of maximum level for OTA in this matrix, more concretely in a wine from Italy. Most of the notifications (25 out of 61) in the fruit and vegetable category were due to OTA in dried figs from Turkey. The next most frequently reported product, with 13 notifications, was dried raisins from Uzbekistan, Pakistan, or Afghanistan [27].

In the case of coffee, a comprehensive meta-analysis revealed that the global pooled prevalence of OTA in coffee was 53%, with a mean concentration of 3.21 µg/kg. Notably, the lowest and highest concentrations of OTA in coffee were observed in Taiwan (0.35 µg/kg) and Turkey (79.0 µg/kg), respectively. In the European market, the estimated average OTA contamination level in ground coffee stands at 0.72 µg/kg [49–51].

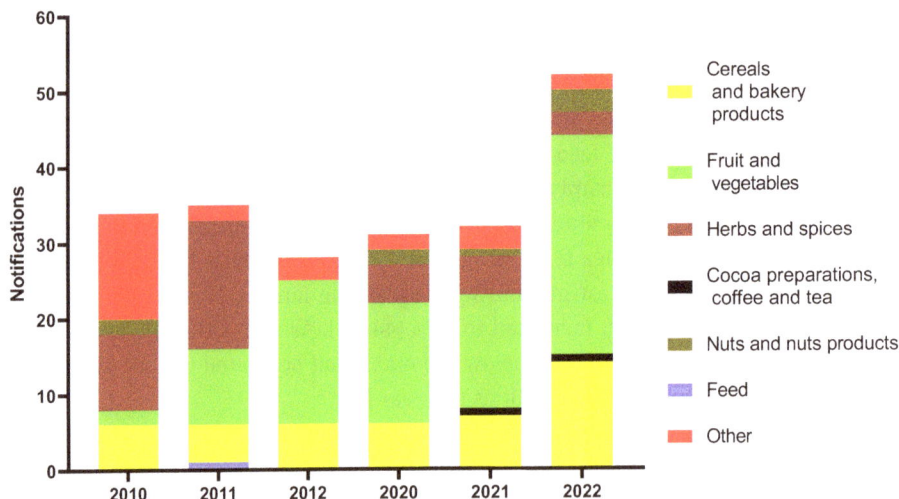

Figure 10.7: Number of AF notifications per year according to the matrix.

Regarding cocoa, a study conducted in 2002 estimated that this particular food matrix was contaminated with an average concentration of 0.24 µg/kg in the EU. However, a more recent investigation has revealed higher contamination levels, with a lower bound of 0.57 µg/kg and an upper bound of 0.83 µg/kg [52]. The identified alerts in this category include OTA in spray-dried instant coffee from Vietnam and in ground coffee from Serbia [27].

Cereals are prone to contamination by OTA, with the highest positive rates being reported in South Asia, where 60.4% of cereal samples were found to contain this toxin [50]. However, it is important to consider that there are significant variations depending on the year and the type of cereals. A comprehensive study conducted in the United States examined breakfast cereals derived from maize, rice, wheat, and oat. The results revealed that 42% of the analyzed samples were contaminated with OTA, with concentrations ranging from 0.10 to 9.30 ng/g. Notably, oat-based breakfast cereals exhibited the highest incidence of OTA contamination, affecting 70% of the samples. In contrast, only 15% of corn and rice-based breakfast cereals were found to be contaminated, while wheat-based products displayed intermediate levels of contamination at 32% [53]. In the European market, cereals and cereal products contain an average level of 0.29 µg/kg [51]. Regarding rice, a study conducted in China found that 4.9% of the samples were contaminated with OTA. However, it is important to note that in the majority of cases, the concentrations detected were within the maximum levels permitted in Europe [50]. Cereal by-products, such as beer, can serve as an additional source of OTA exposure. However, it is important to note that there is a significant reduction in the levels of this toxin during the production process [54]. Within the RASFF portal, a total of 27 recent notifications were associated with the presence of OTA in cereals and bakery products. Specifically, there were 6 notifications in 2020, 7 in 2021, and 14 in 2022. Particularly, there was a notable increase in alerts concerning basmati rice. Other products found to contain this toxin include muesli, quinoa, wheat, rye, maize, and oats from various origins [27].

The EFSA has reported that highest mean concentrations of OTA are usually recorded in categories of "Plant extract formula," "Flavourings or essences," and "Chili pepper" [52]. For example, in Pakistan, 43.9% of crushed chili samples tested positive for OTA, with a maximum concentration of 64.5 µg/kg. In Brazil, the percentage of positive dried red chili samples doubled to 85.7%, with a maximum contamination level of 97.2 µg/kg. Similarly, Malaysia showed similar findings in dried chili samples, while in Spain, 42.9% of dried chili samples were positive, with an average concentration of 11.9 µg/kg (maximum value of 73.8 µg/kg). Results from Spain are in concordance with the contamination values observed in red chili from Italy. In contrast, China displayed lower contamination rates since OTA was only detected in 20% in chili samples, with a maximum concentration of 30.7 µg/kg [55]. However, in recent years, the number of notifications has decreased significantly (Figure 10.7), ranging from three to five notifications per year. Most of these notifications are primarily linked to the presence of the toxin in chili from China and India, paprika from Spain and Portugal, and nutmeg from Indonesia.

Animal feeds often contain cereal ingredients, and as a result, OTA can be transferred and accumulated in animals through the food chain. This can lead to the detection of OTA in various animal-derived products, such as pig muscle, liver, and kidney. Consequently, by-products such as dry-cured ham and salamis may also contain OTA, albeit in relatively low levels. It is worth noting that OTA has been reported in milk products as well. However, the contribution of milk products to OTA intake in humans is considered negligible [50].

10.2.7 Patulin

PAT is generated by various fungal species, including *Penicillium*, *Aspergillus*, and *Byssochlamys*. Among these, *P. expansum*, a common spoilage fungus found in fruits, is predominantly responsible for the production of this mycotoxin. Apples and apple juice are particularly susceptible to PAT contamination. However, other mold-infested fruits such as pears, grapes, cherries, and peaches can also be affected by PAT contamination. Furthermore, PAT can be detected in figs, certain vegetables, cereals, cheeses, and even seafood [56].

Exposure to PAT poses a risk that is linked to the consumption of both visibly moldy and apparently clean products infected with the fungus. This is because the mycotoxin can accumulate not only in the visible lesions but also in other parts of the spoiled fruit. Moreover, PAT has shown resistance to the conditions applied during food processing. Consequently, it can be detected in fruit-derived products such as juices, purees, ciders, jams, marmalades, vinegars, and dried rings [57]. In recent years, there have been approximately three notifications per year in the RASFF portal, with the majority of them being associated with the presence of PAT in apple juice from Serbia and Denmark. A review of the published data has concluded that products from other European countries showed average levels within an acceptable range [58]. However, in Czech Republic, contamination levels of 328.5 µg/kg were detected in grapes in 2018. However, samples were contaminated by strains of *P. expansum* [59]. In addition, other study involving apple juices from the Romanian market showed toxin levels ranged between <0.7 µg/L and 101.9 µg/L, but only the 6% surprised the maximum limit set by the EU [60]. Iran reported a particularly high average level of 620 µg/kg in apple leather, a product created by dehydrating cooked fruit into leathery sheets. This dehydration process can lead to increased toxin concentration [61]. Similarly, studies conducted in South Africa and Turkey also identified concentrations above the recommended levels [58].

10.2.8 Citrinin

CTN is produced by various fungal species, including *Monascus*, *Penicillium*, and *Aspergillus*. This mycotoxin is primarily synthesized during storage and can be found in a range of food sources such as grains, beans, fruits, and spices. Additionally, CTN is commonly associated with red mold rice, a rice product that undergoes fermentation by *Monascus* spp., making it a significant source of this mycotoxin.

CIT has been documented to occur worldwide, including Europe, Asia, North America, and Africa. Cereals, particularly rice, are frequently contaminated with this mycotoxin. In Vietnam, 13% of rice grain samples were found to contain CIT, with average levels of 0.38 µg/kg. Similarly, in Japan and India, approximately 13% of the samples were contaminated by CIT, with levels ranging from 49 to 92 µg/kg. The highest CIT levels were detected in two samples from Canada, with concentrations of 700 and 1,130 µg/kg. However, the main risk is related with the red yeast rice, a fermented product of the *Monascus purpureus* (red yeast) grown on white rice. In this sense, some *Monascus* species, principally *M. purpureus*, produce CIT during fermentation and up to 80% of final products may contain this mycotoxin, showing, in some cases, high levels ranging between 140 and 44,240 µg/kg. In addition, CIT has also been detected in wheat and maize. However, in general, less than 12% of grains are contaminated by this mycotoxin [62, 63]. CTN can also be found in approximately 8% of fruits, fruit juices, and vegetable juices, with concentrations reaching up to 0.20 µg/L. In fruits, CTN is often found together with PAT, particularly in the presence of rotten spots areas. In the case of medicinal and aromatic herbs, this mycotoxin can be detected in as much as 61% of samples, with concentrations reaching up to 355 µg/kg [63].

10.2.9 Ergot alkaloids

EAs are produced by *Claviceps* species, primarily *C. purpurea*, a fungus that infects grasses and cereals such as rye, wheat, barley, millets, and oats. The fungus replaces the host seeds with its sclerotia, known as ergot, within which EAs are subsequently produced and accumulated. During the Middle Ages, the consumption of EAs-contaminated food led to numerous outbreaks of a human poisoning known as St. Anthony's fire. Modern agricultural practices and food processing techniques have reduced the occurrence of severe cases of this mycotoxicosis, now referred to as ergotism. However, the complete elimination of EAs from cereals remains a food safety concern. EAs are characterized by a tetracyclic ergoline ring system. Over 50 different analogues that have been identified, some of the most relevant ones include ergotamine, ergocryptine, ergocristine, as well as ergometrine [64].

A comprehensive study conducted across Europe aimed to evaluate the presence of EAs in various food and feed samples reveled that 84% of rye-based food, 67% of wheat-based food, and 48% of multigrain food samples were contaminated with EAs. Furthermore, EAs were also present in 52% of rye-based feed, 27% of wheat-based feed, and 44% of triticale-based feed samples. The total EAs levels varied considerably ranging from ≤1 (limit of quantification) to 12,340 µg/kg. Notably, the mean concentration of EAs in rye-based food was found to be 28 µg/kg, while for wheat-based food, it was 7 µg/kg. In contrast, other products analyzed in the study showed mean concentrations below 1 µg/kg [65]. Similar results were observed in Canada although important annual differences in ergot incidence are observed [66]. In the EU, eight notifications were produced though RASFF portal from 2020 to 2022, most of them due to the presence of these toxins in rye flour although EAs were also occasionally detected in enzymes, barley flour, and grain spelt spaghetti [27].

10.2.10 Emerging mycotoxins

As mentioned before, the term "emerging mycotoxins" encompasses secondary metabolites that are not currently regulated or routinely monitored. While this category includes various compounds, two of the most commonly detected ones are BEA and ENNs. These mycotoxins are structurally related compounds produced by *Fusarium* species and can be mainly found in cereal grains and their by-products. However, they have also been identified in other food matrices such as nuts, dried fruits, bananas, and medicinal herbs. In Europe, between 2000 and 2013, BEA was detected in 54% of unprocessed grain samples, while ENNs were found in 76% of those samples. However, when examining processed food samples, the contamination levels were reduced to 20% for BEA and 37% for ENNs. However, these contaminants are usually at low levels although in some occasions hundreds of milligram per kilogram can be reached [67].

10.3 Phycotoxins

Phycotoxins, also known as marine toxins, are a broad and diverse category of chemical compounds produced by phytoplankton, mostly dinoflagellates and diatoms, that pose a risk to human food safety. Phytoplankton is consumed by filtering bivalves such as mussels, oysters, clams, scallops, and cockles. When HABs occur, shellfish acquire toxins, and human intoxication results from eating of contaminated seafood. Human exposure can also occur through the consumption of contaminated fish, the inhalation of aerosols, or direct skin contact [68].

10.3.1 Lipophilic marine toxins

Lipophilic marine toxins are a class of toxic compounds that are produced by certain species of marine microorganisms. The most well-known groups of lipophilic marine toxins are summarized in Figure 10.8 [69–71].

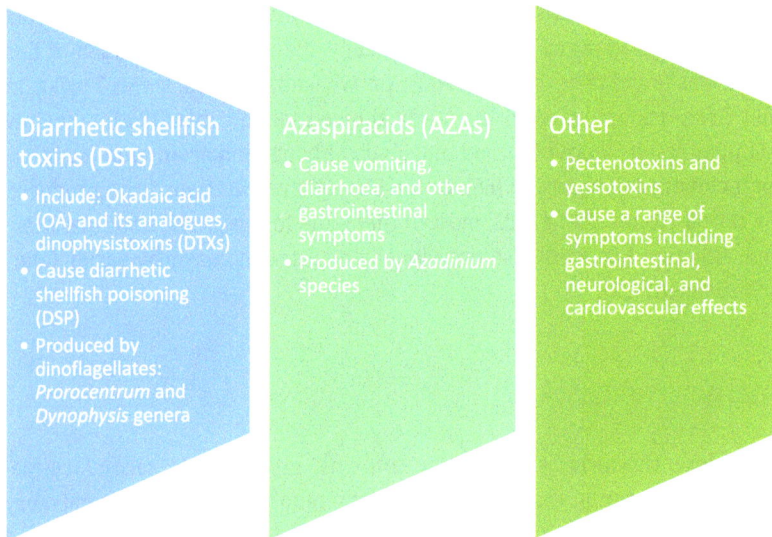

Diarrhetic shellfish toxins (DSTs)
- Include: Okadaic acid (OA) and its analogues, dinophysistoxins (DTXs)
- Cause diarrhetic shellfish poisoning (DSP)
- Produced by dinoflagellates: *Prorocentrum* and *Dynophysis* genera

Azaspiracids (AZAs)
- Cause vomiting, diarrhoea, and other gastrointestinal symptoms
- Produced by *Azadinium* species

Other
- Pectenotoxins and yessotoxins
- Cause a range of symptoms including gastrointestinal, neurological, and cardiovascular effects

Figure 10.8: Summary of lipophilic marine toxins.

The EU has established regulatory limits for lipophilic marine toxins in seafood to protect public health. Therefore, EU Member States must implement motoring programs to assess the presence of both marine biotoxins and the causal phytoplankton in shellfish production regions. However, there is always a risk that mollusks accumulate toxins above the legal limit in the period between two consecutive samplings. Galicia, located in the northwest of Spain, is Europe's largest producer of bivalve mollusks. However, this region is also highly susceptible to contamination by marine lipophilic toxins. To ensure the safety of seafood products, a comprehensive monitoring program was established in this region. Between 2014 and 2017, near to 19,000 samples of bivalve mollusks were collected and analyzed in Galicia for the marine lipophilic toxins regulated by the EU. The results of the monitoring program have shown that okadaic acid (OA) was the most frequently detected toxin, in around 10% of samples, and it was the only one that has led to harvesting closures as a single contaminant. Additionally, concentrations of lipophilic marine toxins were generally higher in raft mussels compared to other bivalves. In addition, some spatial and seasonal variations were observed. Generally, the outer areas of the estuaries are more affected by OA and DTX2 (dinophysistoxins) than the inner areas. OA levels reach their maximum in the spring, while the detection of DTX2 is almost entirely restricted to the fall-winter season and YTXs have been found to peak in August-September [72]. Similar data were observed in Portugal where the distribution DSPs shows a recurrent pattern. In the northwest coastline, both offshore and estuarine/lagunar bivalve species are highly contaminated with DSP toxins, whereas in the southwest and south coasts, high levels of DSP toxins are predominantly found in bivalves from offshore areas, with the highest levels being found in truncate donax (*Donax trunculus*) [73]. In Great Britain, toxins from the OA group were also the most frequently detected, especially in Scotland where up to the 23% of the analyzed samples contained this group of toxins. The pattern of OA group occurrence was consistent across years, peaking in summer and declining during autumn and winter, although the magnitude and abundance varied significantly. Notable interannual changes were observed in certain regions, particularly an increase in DTX2 occurrence in northwest Scotland and England was observed in recent years. Moreover, seasonal changes were identified where the dominant toxin, OA, was replaced by higher proportions of DTX2 in late summer and autumn [74]. In the Mediterranean Sea, the presence of OA has also been detected, but with lower incidences [75].

Although monitoring programs are effectively implemented human intoxications are still recorded, generally associated with illegal shellfish harvesting of mussels and donax clams picked at low tide in beaches by locals or tourists [73].

Since 2020, there have been a total of 10 notifications issued in the RASFF system. Out of these, eight notifications were due to the presence of OA primarily in mussels, although it was also detected in cockles and truncate donax. All of these notifications were related to shellfish collected within the EU, except for one case involving frozen stuffed mussels from Turkey. Additionally, in 2022, two notifications were issued regarding the presence of azaspiracids (AZAs). The samples in question were blue mussels and oysters from Ireland [27].

10.3.2 Saxitoxin group

Saxitoxin and their analogues are responsible for a group of syndromes known as paralytic shellfish poisoning (PSPs). These tetrahydropurine compounds are molecules that can be mainly categorized in four subgroups: carbamate, decarbamoyl, *N*-sulfocarbamoyl, and hydroxylated saxitoxins. Carbamoyl PSP toxins include the saxitoxin (STX), neosaxitoxin, and gonyautoxins 1–4, and all these compounds have an analogue formed by the loose of the carbamoyl group yielding the decarbamoyl PSPs, which include the decarbamoylsaxitoxin, decarbamoylneosaxitoxin, and decarbamoylgonyautoxin 1–4. Examples of *N*-sulfocarbamoyl PSP toxins include C1 and C2 toxins and gonyautoxins 5 and 6 [76].

Although some freshwater cyanobacteria are able to produce STXs, these toxins are mainly produced by species of dinoflagellates with worldwide occurrence. PSPs do not show a homogeneous distribution in terms of time and geographical location. In Europe, Spain and Portugal are more affected by STX group toxins than other areas. Some major intoxications with PSPs happened in Europe, and the largest outbreak happened in 1976 in Galicia (Spain); as a result of this event more than 100 people from several European countries were affected. These events led to the establishment of the first European monitoring program aimed to protect human health and safeguard the shellfish industry [77]. Among the bivalve species analyzed from 1995 to 2020 in Galicia, PSPs were detected over the regulatory limit in 1.6%. Similar occurrence date was observed in Portugal (1%) while in Great Britain less than 0.3% of the analyzed samples were over the regulatory limit. However, trace levels of these toxins can be detected in up to the 80% of the samples [78]. The implementation of the monitoring program across Europe allowed that only six incidents of human illness linked to PSTs happened [79]. Since 2020, only two notifications were issued in the RASFF system, in both cases due to the presence of PSPs in scallops from Norway. In general, mussels and cockles had the highest toxicities during the HABs, while the scallops had the smallest [27]. However, during the event, the overall occurrence of STX group toxins appears to be similar and did not significantly vary with bivalve species [78]. In addition, there are reports of non-traditional vectors containing PSPs including marine gastropods, crustaceans, echinoderms, tunicates, and ascidians [80].

10.3.3 Domoic acid

Domoic acid (DA) and its isomers are water-soluble phycotoxins that induce amnesic shellfish poisoning (ASP) in humans. These toxins are primarily produced by diatoms belonging to the genus *Pseudo-nitzschia* and red algae of the genus *Chondria*. *Pseudo-nitzschia* spp. are found in marine waters across both warm and cold climates worldwide. Generally, higher sea temperatures are linked to elevated production of DA, although certain strains can produce this toxin in cooler waters [81]. In Europe, the presence of DA has been reported in shellfish on the coast of Spain, Portugal, Ireland, United Kingdom, France, Italy, and Greece [82]. Scallops are the shellfish species mostly affected by DA, but these toxins can be accumulated in other products such as mussels and razor clams [83]. Intoxications have been reported in Europe, although they are relatively rare compared to other regions such as the Pacific coast of North America. DA is included in the European Monitoring program, the obtained data revel that around 13% of the samples collected from 1995 to 2020 in Galicia (NW Spain) contained DA in concentrations lower than the regulatory limit and therefore they can be commercialized. Higher prevalence was observed in specific years in Scotland (56%) and Catalonia (NE Spain) (23.8%), while lower prevalence was observed in England and Wales (2.4%) from 1999 to 2009 [84]. This toxin is only occasionally detected after-market control, for instance, in 2010 three notifications were reported, two for scallops from the United Kingdom and one for various bivalve mollusks from France [13]. Another notification was issued in 2017 regarding the detection of ASP toxins in live mussels from Ireland [20]. Since 2020, no notifications were found [27].

10.3.4 Emerging marine toxins

The geographic distribution of phytoplankton is largely influenced by sea-surface temperatures, which are being altered by global climate change. This shift in temperatures is causing an alteration in the distribution of phytoplankton, and as a result, toxins that were once found only in tropical waters are now being detected in northern areas. These toxins can contaminate shellfish and fish that may eventually reach the human food chain. While there are currently no regulatory limits for emerging toxins, legislation forbids the sale of fish products containing ciguatoxins (CTXs) and certain species of fish known to contain tetrodotoxin (TTX). However, there is no legislation currently in place for palytoxin despite it being the most toxic natural compound [6].

TTX, a potent neurotoxin, is the causative agent responsible for pufferfish/fugu poisoning, which is frequently reported in Japan. However, in 2008, an incident of TTX intoxication occurred in Spain, which raised concerns about the presence of this toxin in European waters [85]. Although there have been no reports of human illness caused by TTX poisoning, since then, the occurrence of TTX in mollusks has significantly increased in recent years. This toxin has been detected in various mollusk species, with mussels (*Mytilus edulis*) and oysters (*Crassostrea gigas*) being the most commonly affected ones. The detection of TTX in these mollusks has been reported in multiple countries across Europe, including England, Greece, the Netherlands, Spain, Italy, and France [86, 87].

CTXs are responsible for causing Ciguatera fish poisoning, a condition characterized by gastrointestinal, neurological, and cardiovascular symptoms. These toxins are commonly found in fish from the Pacific, Caribbean, and Indian Ocean regions. While occurrence data on CTXs in Europe are limited, a confirmed case of CFP occurred after the consumption of fish caught in the Canary Islands [88]. Furthermore, CTXs have been identified in fish from Madeira [89].

The group of cyclic imines (CIs) consists of more than 40 compounds and is considered emerging due to its recent discovery. Reports about the presence of these compounds in shellfish are usual across Europe, although they are not currently regulated. The largest subgroups of CIs are the spirolides and pinnatoxins (PnTXs) which are known to occur in bivalve mollusks from various regions worldwide and exhibit high acute toxicity in mice upon intraperitoneal injection. Despite their toxicity in animal models, no adverse effects in humans have been reported to date. However, there is a growing concern about the potential risk these toxins may pose to human health, highlighting the need for continued research and monitoring of CIs in seafood [90]. In 2022, an information notification for attention was issued in the RASFF portal regarding the presence of PnTXs in fresh mussels from the Netherlands [27].

10.4 Cyanotoxins

Cyanotoxins, also known as freshwater toxins, are a diverse group of toxic metabolites produced by certain species of cyanobacteria that can contaminate freshwater reservoirs and aquatic ecosystems around the world. These toxins are classified into three categories based on their mechanisms of toxicity: hepatotoxins (such as microcystins (MCs), nodularins, and cylindrospermopsins), neurotoxins (including anatoxin-a and analogues, saxitoxins (STXs), and β-N-methylamino-L-alanine), and dermatotoxins (like aplysiatoxins and lyngbyatoxins). MCs are the most extensive group of cyanotoxins, comprising about 100 analogues, with microcystin LR (MC-LR) being the most toxic analogue [91].

Although cyanotoxins are commonly associated with contaminated drinking water, they have also been found in food. Aquatic products can accumulate cyanotoxins through contaminated feed, direct contact with contaminated water, and biomagnification through the food web. For instance, anatoxin-a has been detected in Carp (*Cyprinus carpio*) and STXs in Rainbow trout (*Salmo gairdnerii*), while MCs were found in various fish species, bivalves, and crustaceans. In some cases, the amount of toxin detected exceeded the recommended tolerable daily intake level. Interestingly, plants have been found to absorb toxins from cyanotoxin-contaminated water as well, with toxins primarily detected in roots. However, there are also reports of MCs in leaves, stems, and seedlings. However, most of the available data comes from studies in which plants were irrigated with water containing high cyanotoxin content [92, 93]. Finally, individuals can be exposed to these toxins through food supplements made from cyanobacteria and microalgae. Studies have revealed that a majority of the Spirulina products available in the market contain MCs in concentrations that may pose health risks, especially for children and infants who consume them on a daily basis [94]. One notification was found regarding the presence of MCs in klamath algae powder in 2022 [27].

10.5 Detoxification strategies

In order to minimize the impact of natural toxins on the food chain, several approaches have been developed. When dealing with phycotoxins, control strategies primarily revolve around the elimination of HABs through various methods such as mechanical removal of algae, introducing biological agents that induce algal mortality or using chemicals that facilitate their precipitation. However, as a general practice, harvesting areas are typically closed until the blooms naturally dissipate and shellfish detoxification takes place [95]. On the other hand, several measures can be applied to reduce the mycotoxins content of food and feed [96, 97]. Below is a summary of the main strategies traditionally applied to minimize the contamination of different food groups. Finally, the use of nanotechnology as an emerging detoxification strategy is disclosed.

10.5.1 Cereals, grains, and cereal by-products

Cereal and cereal-based products are susceptible to contamination by a diverse range of mycotoxins, which can occur due to fungal infections either in the field or during postharvest and storage stages. Several environmental factors contribute to the presence of mycotoxins, including inadequate storage conditions, insect damage, and weather conditions such as rainfall, humidity, and temperature. Additionally, the characteristics of the food itself, including pH, composition, and water activity, also play crucial roles in determining the extent of mycotoxin contamination [98]. Good agricultural practices have a significant impact on reducing mycotoxin contamination. The Codex Alimentarius Commission recommends several key measures that can effectively mitigate the risk of mycotoxin contamination, and the most important ones are summarized in Figure 10.9 [99].

When preventive measures fail and grains become contaminated with mycotoxins either in the field or during storage, there are management options available for using the contaminated grain. During the industrial processing the levels of *Fusarium* toxins in cereal products can be reduced. However, in some cases, processing may increase the levels of mycotoxins in food or feed by-products. This is because the mechanical cleaning of cereals, such as de-hulling, can result in by-products that concentrate *Fusarium* toxins to a significant degree, leading to higher concentrations of mycotoxins in these materials compared to the cereals prior to cleaning. Recent re-

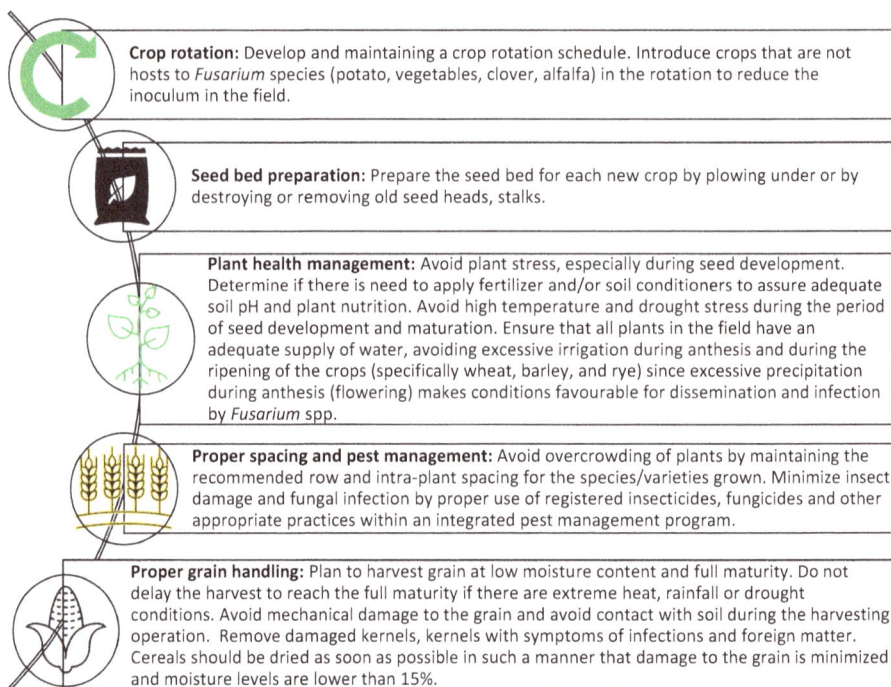

Crop rotation: Develop and maintaining a crop rotation schedule. Introduce crops that are not hosts to *Fusarium* species (potato, vegetables, clover, alfalfa) in the rotation to reduce the inoculum in the field.

Seed bed preparation: Prepare the seed bed for each new crop by plowing under or by destroying or removing old seed heads, stalks.

Plant health management: Avoid plant stress, especially during seed development. Determine if there is need to apply fertilizer and/or soil conditioners to assure adequate soil pH and plant nutrition. Avoid high temperature and drought stress during the period of seed development and maturation. Ensure that all plants in the field have an adequate supply of water, avoiding excessive irrigation during anthesis and during the ripening of the crops (specifically wheat, barley, and rye) since excessive precipitation during anthesis (flowering) makes conditions favourable for dissemination and infection by *Fusarium* spp.

Proper spacing and pest management: Avoid overcrowding of plants by maintaining the recommended row and intra-plant spacing for the species/varieties grown. Minimize insect damage and fungal infection by proper use of registered insecticides, fungicides and other appropriate practices within an integrated pest management program.

Proper grain handling: Plan to harvest grain at low moisture content and full maturity. Do not delay the harvest to reach the full maturity if there are extreme heat, rainfall or drought conditions. Avoid mechanical damage to the grain and avoid contact with soil during the harvesting operation. Remove damaged kernels, kernels with symptoms of infections and foreign matter. Cereals should be dried as soon as possible in such a manner that damage to the grain is minimized and moisture levels are lower than 15%.

Figure 10.9: Good agricultural practices that reduce mycotoxin contamination.

views have focused on physical methods employed for mitigating mycotoxins in cereals [100]:

– **Blending:** The practice of blending is continuously employed as untested lots of grain with varying levels of mycotoxins are mixed on the farm.
– **Screening and aspirating grain**: By utilizing a grain sieve and gravity separator, this method effectively eliminates numerous damaged and diseased kernels, which often contain the highest levels of mycotoxins. The process involves segregating moldy kernels, broken kernels, and fines, resulting in a significant reduction of up to 66% in FBs concentrations within a batch of grain.
– **Optical sorting machines:** These instruments detect and remove only the discolored kernels, including contaminated ones that may not show visible symptoms. This approach can achieve reductions close 80% of both AFs and FBs in maize.
– **Dry-milling**: This technique separates the grain into particulate components. In this way, the grain is segregated into more contaminated and less contaminated components. For instance, AFs are mainly found in the germ and hull fractions. FBs and other *Fusarium* toxins show the highest concentrations in the bran followed by the germ. Starch fractions tend to be low in mycotoxins. However, in damaged kernels the endosperm is much more likely to be colonized, resulting in higher mycotoxin levels in the starch fractions.

– **Washing or steeping**: This method offers the potential to effectively decrease mycotoxin levels, particularly water-soluble compounds. By elevating the pH of the solution using chemicals such as sodium carbonate, the solubility of certain toxins like ZEN can be significantly enhanced. Wet-milling, in general, is capable of removing the majority of mycotoxins from the desired fraction, which is the starch. As a result, wet-milling reduces mycotoxin concentrations to a significant extent.

Most mycotoxins are chemically and thermally stable; however, several studies have shown that thermal processing techniques involving higher temperatures, such as roasting and extrusion, can effectively reduce the levels of various mycotoxins in different food products. For example, roasting can decrease AF levels by 50–70% in peanuts and pecans, by 40–80% in maize, and the content of OTA in coffee beans by up to 97%, depending on the temperature and particle size. However, the degradation of OTA in wheat through heating or extrusion was less efficient. In the case of trichothecenes, results are not in agreement since some authors reported drastic reductions of DON under all investigated conditions, while others observed only moderate effects or no reduction. Extrusion cooking is a widely used food process to produce cereal snacks and breakfast cereals using cereals as major ingredients. This technique has been effective in reducing several mycotoxins, DON is most susceptible to extrusion cooking, followed by FB_1; while only a limited amount of AFs, ZEN, and OTA can be reduced by extrusion. However, the results between the different studies vary widely since reductions of ZEN and FBs levels in maize by 65–83% and by 34–95% have been reported. Other toxins such as CIT and EAs can also be efficiently degraded through heating [101, 102]. Other techniques including irradiation, ionizing (gamma), and nonionizing (UV, solar, microwave) can also be effective to ameliorate mycotoxin contamination by eliminating or reducing pathogenic microorganisms. For instance, it was observed that AFB_1 can be reduced up to 95% in maize when exposed to a dose of 10 kGy gamma radiation. Similarly, the photodegradation of cereals with nonionizing irradiation reduce AFs by about 40% after 3 h and by up to 75% after 30 h of direct sunlight [102].

Baking plays a significant role in the preparation of cereal-based foods like bread, cake, and cookies. The outcome of mycotoxins presents in the raw materials varies depending on factors such as the type of mycotoxin and the breadmaking process, including the flour, yeast, fermentation, and baking times as well as the temperature employed. In general, slight reduction of ZEN, AFs, and FB_1 was observed, whereas OTA was relatively stable during the breadmaking process and there are contradictory data for DON since some studies reported reductions in their levels while others suggested that DON was highly stable or even increased in this process. On the other hand, the majority of DON (50% to 95%) can be washed off during malting and brewing [102].

10.5.2 Milk and milk products

Milk and dairy products remain a source of concern when it comes to mycotoxin contamination, particularly with regards to AFM_1. It is generally believed that neither pasteurization nor sterilization processes significantly reduce the levels of AFM_1 in dairy products [103]. In the case of sterigmatocystin, there is a significant carryover to cheese due to the low solubility of this mycotoxin in aqueous media, 80% of the toxin present in milk was found in the curd and only 20% was found in the whey [104]. Biological detoxification of dairy products represents a promising tool to reduce the impact of mycotoxins. Lactic acid bacteria, including strains from genera such as *Lactobacillus*, *Streptococcus*, *Lactococcus*, and *Propionibacterium*, have been shown to have the ability to bind AFB_1. Therefore, they can be utilized as probiotics or incorporated into dairy products as starter or adjunct starter cultures leading to a reduction in AFs levels [105].

Certain varieties of cheese are frequently associated with high concentrations of mycotoxins, which can proceed from diverse sources. These sources include the milk used as a raw material as well as molds that can be intentionally added during the ripening process or act as contaminants. In this sense, certain molds are deliberately added to contribute to the desired flavor and texture of the cheese. Although controversial, optimal conditions of fermentation and cheese-making may permit the production of significant levels of mycotoxins, as implied by the high incidence of mycotoxins from *P. roqueforti* in blue cheeses [106]. Preventive measures remain the most practical and effective means to control the presence of mycotoxins in cheese. Pasteurization of cheese milk is necessary to minimize mold contamination. In the case of mold-ripened cheeses, it is important to carefully select mold strains for the ripening process based on their low or nontoxigenic nature. This selection helps to reduce the risk of mycotoxin contamination. Additionally, several preventive measures have been suggested to minimize mycotoxin presence in these cheeses. Some of these measures include the use of additives such as natamycin which presents fungicide activity, vacuum-packaging, salting, storage at low temperatures, and incorporation of natural plants or their extracts [107].

10.5.3 Nuts, seeds, and spices

Several mycotoxins can occur in nuts and seeds, but the AFs and OTA are more frequently reported, although *Fusarium* toxins such as ZEN and BEA have also been detected in these matrices [108].

The most effective preventive measure against mycotoxin contamination involves implementing proper pre- and postharvest practices. This includes thorough cleaning of the product to remove physical impurities and ensuring that it is dried to appropriate moisture levels immediately after harvesting for safe storage. Furthermore, plant-derived compounds such as eugenol, quercetin, and limonene have shown to inhibit production of mycotoxins and therefore they can be used as preventive measures against contamination [109]. The nonthermal/cold atmospheric pressure plasma has also proven to be useful to eliminate up to 70% of AFs from dehulled hazelnuts; in addition, a reduction in DON and the emerging NIV were also observed [110]. Cyclopiazonic acid levels are known to be reduced in the presence of oxygen; therefore, it is likely that diffuse oxygen into porous storage bags causes degradation of cyclopiazonic acid during the storage season [111]. Nonionizing irradiation, specifically UV light treatment, has emerged as a potential method for reducing mycotoxins in groundnuts, prunus (plums), pistachios, and almonds since light treatment can lead to near-complete degradation of AFs in these products [112].

10.5.4 Fruits and vegetables

Fruits and vegetables serve as a favorable environment for the proliferation of numerous microscopic fungi, which are able to produce mycotoxins. Each plant species exhibits a certain degree of specificity when it comes to the types of mold species that can infest them [113]. The selection of intact fruits and the removal of rotten ones is the best approach to avoid mycotoxin contamination. Alternaria species can grow and produce toxins at low temperatures. The emerging *Alternaria* toxins do not suffer any apparent losses at room temperature over 20 days or at 80 C after a 20-min period [114]. Cold plasma treatment is a promising intervention in food processing. The activated chemical species of cold plasma demonstrates remarkable efficacy in combating microorganisms, including molds. In addition, chemical breakdown of mycotoxins through various pathways happens leading to degradation products [115]. PAT is the major mycotoxin in fruit juices, but is decomposed during the production of cider [101]. In recent years, the application of novel nonthermal food processing techniques to the management of mycotoxin contamination was also studied. High pressure processing (600 MPa during 5 min) can constitute an effective tool in the removal of removal from juices, and this technique can remove ENNs from different juices (orange, strawberry, and grape) with reductions percentages of up to 75%, while the traditional thermal treatment allows reduction percentages varying from 3% to 25% [116].

10.5.5 Feed and silage

Cereals and its by-products constitute the main ingredients of finished feed. Maize is the main raw material used as an energy source; protein sources are mainly contributed by soy and soybean complex (soybeans, soybean oil, and soybean meal) [117]. Traditional decontamination methods employed for cereals have been previously disclosed. On the other hand, the use of sequestering agents has been proposed for animal feed to mitigate mycotoxin risks. These agents, which are large molecular weight compounds, have the ability to bind mycotoxins in the gastrointestinal tract, limiting their bioavailability. The main adsorbing agents are summarized in Figure 10.10 [118].

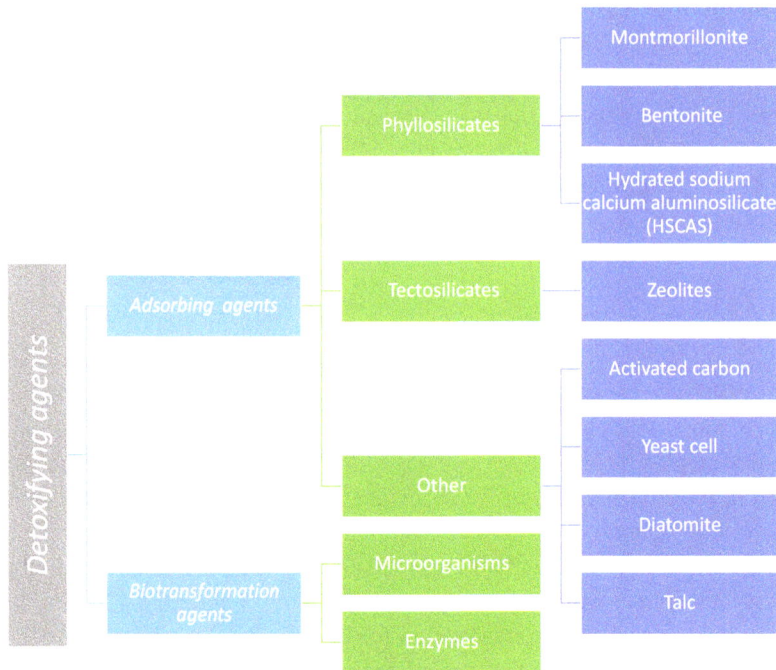

Figure 10.10: Summary of detoxifying agents.

Mineral and organic adsorbents can adsorb substances on their surface or within their interlaminar spaces. For instance, zeolites offer a large and specific binding surface where the adsorption is conditioned by the size, shape, and charge of the molecules. On the other hand, activated carbon is a highly porous nonsoluble powder formed by pyrolysis of organic materials. In aqueous solution, activated carbon adsorbs most of the mycotoxins efficiently, but essential nutrients can also be removed from treated products [119]. Among biological adsorbents, yeast cell walls have gained significant popularity. These cellular structures possess the ability to adsorb a diverse range of compounds from the surrounding environment. However, it is important to note that only yeast cell walls containing specific components such as β-D-glucans, glucomannans, and mannan-oligosaccharides are capable of effectively adsorbing mycotoxins [120]. In addition, some strains of bacteria having cell-wall peptidoglycans and polysaccharides bind mycotoxins in the small intestine [121].

An alternative approach involves the degradation of mycotoxins by certain microorganisms, including bacteria, fungi, or enzymes, known as biotransforming agents.

However, it is important to consider that in some cases, the toxicity of the metabolized compounds can be higher than that of the original mycotoxin [122].

In recent years, various "sequestrants" have been introduced in the market whose objective is to prevent the absorption of mycotoxins in the intestine of animals. These products are mixed in the feed or the silo, but their effectiveness is questionable, so it is advisable to use only products authorized by the EU. For now, there are only three authorized products: bentonite (1m558) for the reduction of AFs, a FBs esterase (1m03) for the reduction of these mycotoxins, and a strain of *Coriobacteriaceae* (1m01) for the detoxification of trichothecenes such as DON. These "sequestrants" have been the only ones that have demonstrated their effectiveness in reducing the toxic effects of mycotoxins in animals [123].

Silage, a valuable type of fodder, is composed of green forage that undergoes lactic fermentation for preservation. Nevertheless, during the feed-out process, when the silo face is exposed for the removal of silage, aerobic spoilage becomes nearly inevitable. This spoilage is primarily initiated by yeasts, which trigger a rise in pH and temperature, thereby creating a favorable environment for the proliferation of other aerobic microorganisms that have a slower growth, such as filamentous fungi. In order to mitigate the risk of mycotoxin contamination, various measures can be implemented, including the addition of pH-lowering additives to the silage. This prevents fungal growth and subsequent mycotoxin production. One effective approach involves incorporating organic acids, like propionic acid, to attain the required acidity for preservation, independent of microbial acid production. It is important to note, however, that while these measures are effective in preventing mycotoxin formation during storage, they do not eliminate toxins that may already be present in the harvested forage [124].

10.5.6 Nanotechnology

Nanotechnology is the scientific field focused on the understanding and control of matter at dimensions between 1 and 100 nm where unique phenomena enable novel applications. While the study of nanoparticles has a long history, the true emergence of nanotechnology began in the 1980s with the concept of a nanoscale "assembler," which would be able to build a copy of items of arbitrary complexity in the nanoscale, was proposed. Since then, nanotechnology has rapidly evolved into a basis for innovative industrial applications, driving exponential growth in various sectors. In this sense, nanotechnology has emerged as a promising field for effectively removing soil contaminants, addressing water pollutants, and minimizing the impact of several contaminants. The research efforts are focused on the development of both natural and synthetic adsorbents for remediation of hazardous heavy metals, pesticides, mycotoxins, antibiotics, and other emerging pollutants [125]. Various nanomaterials are being employed to eliminate contaminants through mechanisms such as adsorption, conversion, redox reactions, and stabilization [126].

Nanoscale materials have a large surface area to volume ratio which tends to increase on a higher number of adsorption sites. Besides functionalized nanoscale materials can be made by the combination of sand, clay, chitosan biochar, ferrate, metal oxides, zeolites, and other chemicals [125]. In this way, it is possible to make engineered nanomaterial which, according to the EU legislation, are defined as "any intentionally produced material that has one or more dimensions of the order of 100 nm or less or that is composed of discrete functional parts, either internally or at the surface, many of which have one or more dimensions of the order of 100 nm or less, including structures, agglomerates or aggregates, which may have a size above the order of 100 nm but retain properties that are characteristic of the nanoscale" [127]. In this sense, the term "nano-object" encompasses materials that possess one, two, or three external dimensions within the nanoscale range. If all three dimensions of the object fall within the nanoscale, it is classified as a nanoparticle. On the other hand, a nanostructure refers to a composition consisting of interconnected constituent parts, with at least one of those parts existing in the nanoscale. Last, nanostructured materials are characterized by having internal or surface nanostructures, which contribute to their unique properties and behaviors [128].

Magnetic nanoparticles (NPs) consist of magnetic elements, such as iron, nickel, cobalt, and their oxides. In most applications, particles with a size range of approximately 10–20 nm exhibit optimal performance characteristics. Each individual nanoparticle behaves as a single magnetic domain, demonstrating superparamagnetic properties, wherein they do not retain magnetization after exposure to a magnetic field. That is, they are not magnetized after the action of a magnetic file. However, NPs face certain challenges. Over extended periods, they tend to aggregate, leading to loss of magnetism and dispersibility. Additionally, bare metallic nanoparticles are susceptible to oxidation upon contact with air. To overcome these drawbacks, nanoparticles can be coated with various materials, resulting in core-shell structures. This approach involves a central core of naked nanoparticles surrounded by a protective shell formed by constituents such as surfactants, polymers, silica, or carbon (Figure 10.11). This coating enhances stability and prevents aggregation. Moreover, NPs can be dispersed within a continuous matrix, coated onto larger particles in a core-shell configuration, or form agglomerates connected through their shells. Nanostructured materials can be created by incorporating magnetite nanoparticles into biopolymer beads composed of materials like chitin, alginate, pectin, or agarose (Figure 10.11) [129].

Different nanotechnological applications were developed for the removal of toxins from different matrices. Carbon nanomaterials, such as nanodiamonds, carbon nanotubes (CNTs), and magnetic graphene oxide (MGO), have demonstrated remarkable effectiveness in the adsorption of mycotoxins. For instance, 1 mg nanodiamond aggregates (40 nm) can absorb around 10 µg of AFB_1 and 15 µg of OTA via electrostatic interactions. Similarly, single/multiwalled CNTs have shown good adsorption capacity

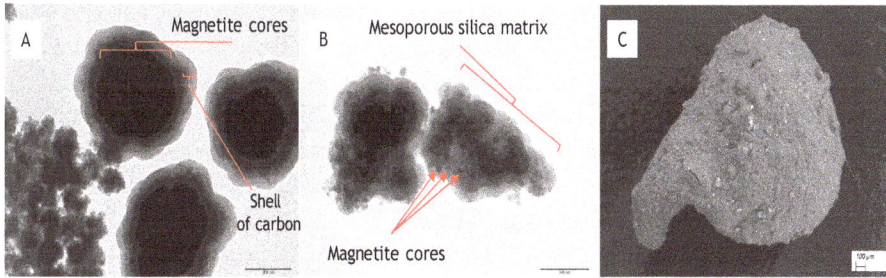

Figure 10.11: Different types of particles. Spherical shell of carbon encapsulating several magnetite (Fe_3O_4) cores (A). Mesoporous silica-loaded nanostructures with magnetite nanoparticles anchored on the surface (B). Nanostructured beads composed of magnetite nanoparticles and alginate (C).

for ZEN, trichothecenes, and AFs [130]. Surface-active maghaemite nanoparticles, a type of uncoated superparamagnetic nanoparticles, exhibit remarkable effectiveness in the removal of CTN from *Monascus*-treated food products [131].

MGO prepared by using graphene oxide and iron oxide NPs has been employed for the adsorption of various *Fusarium* mycotoxins, including DON, HT-2, T-2, and ZEN. The surface of MGO contains oxygen functional groups that enable interactions with these mycotoxins. In the case of palm kernel cake, MGO has proven to be effective in reducing the levels of DON, HT-2, T-2, and ZEN by 69.57%, 57.40%, 37.17%, and 67.28%, respectively [132]. Similarly, MGO modified with chitosan particles was developed for the simultaneous removal of AFB_1, OTA, and ZEN at 50 °C and pH 5 from water solutions [133].

Chitosan is subjected to carbonization, resulting in the production of nanoparticles. These nanoparticles are combined with rectorite to create a nanocomposite material. This nanocomposite demonstrates excellent adsorption capabilities for ZEN. The ability of the nanocomposite to adsorb the mycotoxin is attributed to two factors: the high carbon content and the significant surface area of the nanocomposite [134]. Chitosan-coated magnetite particles can remove PAT from fruit juice, with a maximum adsorption capacity of 6.67 mg/g [135]. This toxin can also be removed by the use magnetic molecularly imprinted adsorbent Fe_3O_4@SiO_2@CS-GO@MIP and absorption of 7.11 mg/g [136]. Similarly, cross-linked chitosan–glutaraldehyde polymers show high adsorption capability for AFB_1, OTA, ZEN, and FB_1, with a predicted maximum adsorption of 24.8 mg/g for OTA and 9.18 mg/g for ZEN [137].

The utilization of nanostructured particles comprising multiple absorbents has exhibited promising outcomes in the removal of various mycotoxins. Nanostructured materials incorporating activated carbon, bentonite, and aluminum hydroxide have demonstrated the capability to eliminate up to 87% of mycotoxins from aqueous solutions. These nanostructures possess a maximum adsorption capacity of 450 μg of toxin per gram of nanoparticle. Notably, the absorption capacity was maintained even in the presence of matrix such as beer [97].

Photocatalytic degradation, as an advanced oxidation technology, has also been prosed for the elimination of mycotoxins. This approach offers numerous advantages, including low cost, environmental friendliness, and the absence of secondary pollution. A wide range of nanomaterials, including graphene/ZnO hybrids, g-C_3N_4, Fe_2O_3, titanium dioxide, and UCNP@TiO_2, have been utilized for the photocatalytic degradation of mycotoxins. For instance, the WO_3/RGO/g-C_3N_4 composites exhibit enhanced photocatalytic activity in the degradation of AFB_1 when exposed to visible light irradiation [138, 139].

Nanotechnology applications in the field of phycotoxins primarily revolve around the development of detection techniques, such as nanoparticle-based immunoassays and magnetic solid-phase extraction [140]. Furthermore, the utilization of covalent organic frameworks for solid-phase adsorption of OA has been recently proposed, demonstrating the potential of these nanocomposites in water monitoring devices [141]. Remarkably, a study revealed that multicore magnetite nanoparticles coated with carbon nanomaterials exhibited impressive efficiency (>70%) in the removal of SXTs, CIs, AZAs as well as up to 38% of DSPs from water solutions and from *Gymnodinium catenatum* and *Prorocentrum lima* cultures [96]. These nanostructures have demonstrated efficacy in the removal of freshwater toxins. Notably, they were able to eliminate approximately 85% of microcystin LR, along with high percentages of other cyclic peptide toxins, even from cultures of *Microcystis aeruginosa* [96]. Chitosan-cellulose composite materials, N-doped TiO_2 nanocomposites, and magnetophoretic nanoparticles of polypyrrole were also successfully applied for the removal of various cyanotoxins from water sources [142–144]. Similarly, covalent organic frameworks have also been proposed as an alternative for the elimination of cyanotoxins [145].

Keywords: Marine toxin, mycotoxin, cyanotoxin, food, feed

Abbreviations: AFs, aflatoxins; ASP, amnesic shellfish poisoning; AZAs, azaspiracids; BEA, beauvericin; CIs, cyclic imines; CTN, citrinin; CTXs, ciguatoxins; DA, domoic acid; DON, deoxynivalenol; DSP, diarrhetic shellfish poisoning; DTXs, dinophysistoxins; EAs, ergot alkaloids; EC, European Commission; EFSA, European Food Safety Authority; ENNs, enniatins; EU, European Union; FBs, fumonisins; MCs, microcystins; MGO, magnetic graphene oxide; NPs, nanoparticles; OA, okadaic acid; OTA, ochratoxin A; PAT, patulin; PnTXs, pinnatoxins; PSPs, paralytic shellfish poisoning; RASFF, European Rapid Alert System for Food and Feed; TTX, tetrodotoxin; ZEN, zearalenone.

References

[1] Ricci A, Chemaly M, Davies R, Fernández Escámez PS, Girones R, Herman L, et al. Hazard analysis approaches for certain small retail establishments in view of the application of their food safety management systems. EFSA J. 2017;15(3):e04697.

[2] Halkier B, Holm L. Shifting responsibilities for food safety in Europe: An introduction. Appetite. 2006;47(2):127–33.

[3] Ugland T, Veggeland F. Experiments in food safety policy integration in the European Union. J Common Market Studies. 2006;44(3):607–24.

[4] King T, Cole M, Farber JM, Eisenbrand G, Zabaras D, Fox EM, et al. Food safety for food security: Relationship between global megatrends and developments in food safety. Trends Food Sci Technol. 2017;68:160–75.

[5] Medina Á, González-Jartín JM, Sainz MJ. Impact of global warming on mycotoxins. Curr Opin Food Sci. 2017;18:76–81.

[6] Botana LM. Toxicological perspective on climate change: Aquatic toxins. Chem Res Toxicol. 2016; 29(4):619–25.

[7] Commission Regulation (EC) No 1881/2006 of 19 December 2006 setting maximum levels for certain contaminants in foodstuffs. Off J Eur Union. L 364:5–24.

[8] Regulation (EC) No 853/2004 of the European Parliament and of the Council of 29 April 2004 laying down specific hygiene rules for food of animal origin. J Eur Union. L 139:55–205.

[9] Commission Regulation (EU) 2023/915 of 25 April 2023 on maximum levels for certain contaminants in food and repealing Regulation (EC) No 1881/2006. Off J Eur Union. L 119:103–57.

[10] European Commission Directorate-General for Health Consumers. The rapid alert system for food and feed (RASFF): annual report 2007: Publications Office; 2008.

[11] European Commission Directorate-General for Health Consumers. The rapid alert system for food and feed (RASFF): annual report 2008: Publications Office, 2009.

[12] European Commission Directorate-General for Health Consumers. The rapid alert system for food and feed (RASFF): annual report 2009: Publications Office, 2010.

[13] European Commission Directorate-General for Health Consumers. The rapid alert system for food and feed (RASFF): annual report 2010: Publications Office, 2011.

[14] European Commission Directorate-General for Health Consumers. RASFF, the Rapid Alert System for Food and Feed: 2011 annual report: Publications Office, 2012.

[15] European Commission Directorate-General for Health Consumers. RASFF, the Rapid Alert System for Food and Feed: 2012 annual report: Publications Office, 2013.

[16] European Commission Directorate-General for Health Consumers. RASFF, the Rapid Alert System for Food and Feed: 2013 annual report: Publications Office, 2014.

[17] European Commission Directorate-General for Health Consumers. RASFF annual report 2014: Publications Office, 2015.

[18] European Commission Directorate-General for Health Consumers. RASFF annual report 2015: Publications Office, 2016.

[19] European Commission Directorate-General for Health Consumers. RASFF annual report 2016: Publications Office, 2017.

[20] European Commission Directorate-General for Health Consumers. RASFF annual report 2017: Publications Office, 2019.

[21] European Commission Directorate-General for Health Consumers. RASFF annual report 2018: Publications Office, 2019.

[22] European Commission Directorate-General for Health Consumers. RASFF annual report 2019: Publications Office, 2020.

[23] European Commission Directorate-General for Health Consumers. RASFF annual report 2020: Publications Office, 2021.

[24] Rodríguez I, González JM, Botana AM, Sainz MJ, Vieytes MR, Alfonso A, et al. Analysis of natural toxins by liquid chromatography. In: Fanali S, Haddad PR, Poole CF, Riekkola M-L, eds. Liquid chromatography. Elsevier, Netherlands, 2017;479–514.

[25] Gruber-Dorninger C, Novak B, Nagl V, Berthiller F. Emerging mycotoxins: Beyond traditionally determined food contaminants. J Agric Food Chem. 2017;65(33):7052–70.

[26] Awuchi CG, Amagwula IO, Priya P, Kumar R, Yezdani U, Khan MG. Aflatoxins in foods and feeds: A review on health implications, detection, and control. Bull Environ Pharmacol Life Sci. 2020;9:149–55.

[27] RASFF WINDOW [Internet]. Europa.eu. [Cited 29 of May 2023]. Available at: https://webgate.ec.europa.eu/rasff-window/screen/search.

[28] Ebrahimi A, Emadi A, Arabameri M, Jayedi A, Abdolshahi A, Yancheshmeh BS, et al. The prevalence of aflatoxins in different nut samples: A global systematic review and probabilistic risk assessment. AIMS Agric Food. 2022;7:130–48.

[29] Commission Implementing Regulation (EU) 2019/1793 of 22 October 2019 on the temporary increase of official controls and emergency measures governing the entry into the Union of certain goods from certain third countries implementing Regulations (EU) 2017/625 and (EC) No 178/2002 of the European Parliament and of the Council and repealing Commission Regulations (EC) No 669/2009, (EU) No 884/2014, (EU) 2015/175, (EU) 2017/186 and (EU) 2018/1660. Off J Eur Union. L 277:89–129.

[30] Bordin K, Sawada MM, Rodrigues CEdC, da Fonseca CR, Oliveira CAF. Incidence of aflatoxins in oil seeds and possible transfer to oil: A review. Food Eng Rev. 2014;6(1):20–28.

[31] Khazaeli P, Mehrabani M, Heidari MR, Asadikaram G, Lari Najafi M. Prevalence of aflatoxin contamination in herbs and spices in different regions of Iran. Iran J Public Health. 2017; 46(11):1540–45.

[32] Jallow A, Xie H, Tang X, Qi Z, Li P. Worldwide aflatoxin contamination of agricultural products and foods: From occurrence to control. Compr Rev Food Sci Food Saf. 2021;20(3):2332–81.

[33] Mahato DK, Lee KE, Kamle M, Devi S, Dewangan KN, Kumar P, et al. Aflatoxins in food and feed: An overview on prevalence, detection and control strategies. Front Microbiol. 2019;10:2266.

[34] Gruber-Dorninger C, Jenkins T, Schatzmayr G. Global mycotoxin occurrence in feed: A ten-year survey. Toxins. 2019;11(7):375.

[35] Saha Turna N, Wu F. Aflatoxin M1 in milk: A global occurrence, intake, & exposure assessment. Trends Food Sci Technol. 2021;110:183–92.

[36] González-Jartín JM, Rodríguez-Cañás I, Alfonso A, Sainz MJ, Vieytes MR, Gomes A, et al. Multianalyte method for the determination of regulated, emerging and modified mycotoxins in milk: QuEChERS extraction followed by UHPLC-MS/MS analysis. Food Chem. 2021;356:129647.

[37] Coppock RW, Christian RG, Jacobsen BJ. Chapter 69 – Aflatoxins. In: Gupta RC, ed. Veterinary toxicology. 3rd ed. Academic Press, United Kingdom, 2018;983–94.

[38] Pinton P, Oswald IP. Effect of deoxynivalenol and other type B trichothecenes on the intestine: A review. Toxins [Internet]. 2014;6(5):1615–43.

[39] Authority EFS. Deoxynivalenol in food and feed: Occurrence and exposure. EFSA J. 2013;11(10):3379.

[40] Chain EPoCitF. Scientific opinion on the risks for public health related to the presence of zearalenone in food. EFSA J. 2011;9(6):2197.

[41] Aldana JR, Silva LJG, Pena A, Mañes VJ, Lino CM. Occurrence and risk assessment of zearalenone in flours from Portuguese and Dutch markets. Food Control. 2014;45:51–55.

[42] Kiš M, Vulić A, Kudumija N, Šarkanj B, Jaki Tkalec V, Aladić K, et al. A two-year occurrence of Fusarium T-2 and HT-2 toxin in croatian cereals relative of the regional weather. Toxins. 2021;13(1):39.

[43] Authority EFS, Arcella D, Gergelova P, Innocenti ML, Steinkellner H. Human and animal dietary exposure to T-2 and HT-2 toxin. EFSA J. 2017;15(8):e04972.

[44] Chain EPanel oCitF, Knutsen H-K, Alexander J, Barregård L, Bignami M, Brüschweiler B, et al. Risks for animal health related to the presence of fumonisins, their modified forms and hidden forms in feed. EFSA J. 2018;16(5):e05242.

[45] Schatzmayr G, Streit E. Global occurrence of mycotoxins in the food and feed chain: Facts and figures. World Mycotoxin J. 2013;6(3):213–22.

[46] Kamle M, Mahato DK, Devi S, Lee KE, Kang SG, Kumar P. Fumonisins: Impact on agriculture, food, and human health and their management strategies. Toxins. 2019;11(6):328.

[47] Abarca M, Accensi F, Bragulat M, Castella G, Cabanes F. *Aspergillus carbonarius* as the main source of ochratoxin A contamination in dried vine fruits from the Spanish market. J Food Prot. 2003;66 (3):504–06.

[48] Van der Merwe K, Steyn P, Fourie L, Scott DB, Theron J. Ochratoxin A, a toxic metabolite produced by *Aspergillus ochraceus* Wilh. Nature. 1965;205(4976):1112–13.

[49] Khaneghah AM, Fakhri Y, Abdi L, Coppa CFSC, Franco LT, de Oliveira CAF. The concentration and prevalence of ochratoxin A in coffee and coffee-based products: A global systematic review, meta-analysis and meta-regression. Fungal Biol. 2019;123(8):611–17.

[50] Li X, Ma W, Ma Z, Zhang Q, Li H. The occurrence and contamination level of ochratoxin A in plant and animal-derived food commodities. Molecules. 2021;26(22):6928.

[51] Authority EFS. Opinion of the Scientific Panel on contaminants in the food chain [CONTAM] related to ochratoxin A in food. EFSA J. 2006;4(6):365.

[52] Schrenk D, Bodin L, Chipman JK, del Mazo J, Grasl-Kraupp B, Hogstrand C, et al. Risk assessment of ochratoxin A in food. EFSA J. 2020;18(5):e06113.

[53] Lee HJ, Ryu D. Significance of ochratoxin A in breakfast cereals from the United States. J Agric Food Chem. 2015;63(43):9404–09.

[54] González-Jartín JM, Alfonso A, Rodríguez I, Sainz MJ, Vieytes MR, Botana LM. A QuEChERS based extraction procedure coupled to UPLC-MS/MS detection for mycotoxins analysis in beer. Food Chem. 2019;275:703–10.

[55] Chen J, Chen Y, Zhu Q, Wan J. Ochratoxin A contamination and related high-yield toxin strains in Guizhou dried red chilies. Food Control. 2023;145:109438.

[56] Wright SAI. Patulin in food. Curr Opin Food Sci. 2015;5:105–09.

[57] Sadok I, Stachniuk A, Staniszewska M. Developments in the monitoring of patulin in fruits using liquid chromatography: An overview. Food Anal Methods. 2019;12(1):76–93.

[58] Saleh I, Goktepe I. The characteristics, occurrence, and toxicological effects of patulin. Food Chem Toxicol: An International Journal Published for the British Industrial Biological Research Association. 2019;129:301–11.

[59] Ostry V, Malir F, Cumova M, Kyrova V, Toman J, Grosse Y, et al. Investigation of patulin and citrinin in grape must and wine from grapes naturally contaminated by strains of *Penicillium expansum*. Food Chem Toxicol: An International Journal Published for the British Industrial Biological Research Association. 2018;118:805–11.

[60] Oroian M, Amariei S, Gutt G. Patulin in apple juices from the Romanian market. Food Addit Contam: Part B. 2014;7(2):147–50.

[61] Montaseri H, Eskandari M, Yeganeh A, Karami S, Javidnia K, Dehghanzadeh G, et al. Patulin in apple leather in Iran. Food Addit Contam: Part B. 2014;7(2):106–09.

[62] Silva LJG, Pereira AMPT, Pena A, Lino CM. Citrinin in foods and supplements: A review of occurrence and analytical methodologies. Foods. 2021;10(1):14.

[63] EFSA. Scientific opinion on the risks for public and animal health related to the presence of citrinin in food and feed. EFSA J. 2012;10(3):2605.

[64]   Krska R, Crews C. Significance, chemistry and determination of ergot alkaloids: A review. Food Addit Contam: Part A. 2008;25(6):722–31.

[65]   Malysheva S, Larionova D, Diana Di Mavungu J, De Saeger S. Pattern and distribution of ergot alkaloids in cereals and cereal products from European countries. World Mycotoxin J. 2014; 7(2):217–30.

[66]   Walkowiak S, Taylor D, Fu BX, Drul D, Pleskach K, Tittlemier SA. Ergot in Canadian cereals – Relevance, occurrence, and current status. Can J Plant Pathol. 2022;44(6):793–805.

[67]   EFSA. Scientific opinion on the risks to human and animal health related to the presence of beauvericin and enniatins in food and feed. EFSA J. 2014;12(8):3802.

[68]   Rossini GP, Hess P. Phycotoxins: Chemistry, mechanisms of action and shellfish poisoning. In: Luch A, ed. Molecular, Clinical and Environmental Toxicology: Volume 2: Clinical Toxicology. Basel: Birkhäuser Basel, 2010;65–122.

[69]   Valdiglesias V, Prego-Faraldo MV, Pásaro E, Méndez J, Laffon B. Okadaic acid: More than a diarrheic toxin. Mar Drugs. 2013;11(11):4328–49.

[70]   Rodríguez I, Alfonso A, González-Jartín JM, Vieytes MR, Botana LM. A single run UPLC-MS/MS method for detection of all EU-regulated marine toxins. Talanta. 2018;189:622–28.

[71]   Gerssen A, Pol-Hofstad IE, Poelman M, Mulder PPJ, Van den Top HJ, De Boer J. Marine toxins: Chemistry, toxicity, occurrence and detection, with special reference to the Dutch situation. Toxins. 2010;2(4):878–904.

[72]   Blanco J, Arévalo F, Correa J, Moroño Á. Lipophilic toxins in Galicia (NW Spain) between 2014 and 2017: incidence on the main molluscan species and analysis of the monitoring efficiency. Toxins. 2019;11(10):612.

[73]   Vale P, Botelho MJ, Rodrigues SM, Gomes SS, Sampayo MAdM. Two decades of marine biotoxin monitoring in bivalves from Portugal (1986–2006): A review of exposure assessment. Harmful Algae. 2008;7(1):11–25.

[74]   Dhanji-Rapkova M, O'Neill A, Maskrey BH, Coates L, Teixeira Alves M, Kelly RJ, et al. Variability and profiles of lipophilic toxins in bivalves from Great Britain during five and a half years of monitoring: Okadaic acid, dinophysis toxins and pectenotoxins. Harmful Algae. 2018;77:66–80.

[75]   Bazzoni AM, Mudadu AG, Lorenzoni G, Soro B, Bardino N, Arras I, et al. Detection of dinophysis species and associated okadaic acid in farmed shellfish: A two-year study from The Western Mediterranean Area. J Vet Res. 2018;62(2):137–44.

[76]   Botana, A.M., and Rey, V. Chemistry and analysis of PSP toxins. In: L.M. Botana and A. Alfonso, eds. Phycotoxins. Chemistry and biochemistry. Oxford: John Wiley and Sons, 2015;69–84.

[77]   Campos M, Fraga S, Mariño J, Sanchez F. Red Tide Monitoring Programme in NW Spain: Report of 1977–1981. International Council for the Exploration of the Sea, CM. 1982;50:27.

[78]   Blanco J, Moroño Á, Arévalo F, Correa J, Salgado C, Pazos Y, et al. Twenty-five years of PSP toxicity in Galician (NW Spain) bivalves: Spatial, temporal, and interspecific variations. Toxins. 2022;14(12):837.

[79]   Bresnan E, Arévalo F, Belin C, Branco MAC, Cembella AD, Clarke D, et al. Diversity and regional distribution of harmful algal events along the Atlantic margin of Europe. Harmful Algae. 2021;102:101976.

[80]   Silva M, Rey V, Barreiro A, Kaufmann M, Neto AI, Hassouani M, et al. Paralytic shellfish toxins occurrence in non-traditional invertebrate vectors from North Atlantic Waters (Azores, Madeira, and Morocco). Toxins. 2018;10:9.

[81]   Jeffery B, Barlow T, Moizer K, Paul S, Boyle C. Amnesic shellfish poison. Food Chem Toxicol: An International Journal Published for the British Industrial Biological Research Association. 2004; 42(4):545–57.

[82]   Saeed AF, Awan SA, Ling S, Wang R, Wang S. Domoic acid: Attributes, exposure risks, innovative detection techniques and therapeutics. Algal Res. 2017;24:97–110.

[83]   EFSA. Marine biotoxins in shellfish – Domoic acid. EFSA J. 2009;7(7):1181.

[84]   Blanco J, Moroño Á, Arévalo F, Correa J, Salgado C, Rossignoli AE, et al. Twenty-five years of domoic acid monitoring in Galicia (NW Spain): Spatial, temporal and interspecific variations. Toxins [Internet]. 2021;13:11.

[85]   Rodriguez P, Alfonso A, Vale C, Alfonso C, Vale P, Tellez A, et al. First toxicity report of tetrodotoxin and 5,6,11-trideoxyTTX in the Trumpet Shell Charonia lampas lampas in Europe. Anal Chem. 2008; 80(14):5622–29.

[86]   Antonelli P, Salerno B, Bordin P, Peruzzo A, Orsini M, Arcangeli G, et al. Tetrodotoxin in live bivalve mollusks from Europe: Is it to be considered an emerging concern for food safety? Compr Rev Food Sci Food Saf. 2022;21(1):719–37.

[87]   Turner AD, Dean KJ, Dhanji-Rapkova M, Dall'Ara S, Pino F, McVey C, et al. Interlaboratory evaluation of multiple LC–MS/MS methods and a commercial ELISA method for determination of tetrodotoxin in oysters and mussels. J AOAC Int. 2023;106(2):356–69.

[88]   Pérez-Arellano J-L, Luzardo O, Brito AP, Cabrera MH, Zumbado M, Carranza C, et al. Ciguatera fish poisoning, Canary Islands. Emerg Infect Dis. 2005;11(12):1981.

[89]   Otero P, Pérez S, Alfonso A, Vale C, Rodríguez P, Gouveia NN, et al. First toxin profile of ciguateric fish in Madeira Arquipelago (Europe). Anal Chem. 2010;82(14):6032–39.

[90]   Noirmain F, Dano J, Hue N, Gonzalez-Jartin JM, Botana LM, Servent D, et al. NeuroTorp, a lateral flow test based on toxin-receptor affinity for in-situ early detection of cyclic imine toxins. Anal Chim Acta. 2022;1221:339941.

[91]   Svirčev Z, Lalić D, Bojadžija Savić G, Tokodi N, Drobac Backović D, Chen L, et al. Global geographical and historical overview of cyanotoxin distribution and cyanobacterial poisonings. Arch Toxicol. 2019;93(9):2429–81.

[92]   Abdallah MF, Van Hassel WHR, Andjelkovic M, Wilmotte A, Rajkovic A. Cyanotoxins and food contamination in developing countries: Review of their types, toxicity, analysis, occurrence and mitigation strategies. Toxins. 2021;13(11):786.

[93]   Lee J, Lee S, Jiang X. Cyanobacterial toxins in freshwater and food: Important sources of exposure to humans. Annu Rev Food Sci Technol. 2017;8(1):281–304.

[94]   Papadimitriou T, Kormas K, Vardaka E. Cyanotoxin contamination in commercial Spirulina food supplements. J Consum Prot Food Saf. 2021;16(3):227–35.

[95]   Reboreda A, Lago J, Chapela M-J, Vieites JM, Botana LM, Alfonso A, et al. Decrease of marine toxin content in bivalves by industrial processes. Toxicon. 2010;55(2):235–43.

[96]   González-Jartín JM, de Castro Alves L, Alfonso A, Piñeiro Y, Vilar SY, Rodríguez I, et al. Magnetic nanostructures for marine and freshwater toxins removal. Chemosphere. 2020;256:127019.

[97]   González-Jartín JM, de Castro Alves L, Alfonso A, Piñeiro Y, Vilar SY, Gomez MG, et al. Detoxification agents based on magnetic nanostructured particles as a novel strategy for mycotoxin mitigation in food. Food Chem. 2019;294:60–66.

[98]   Mousavi Khaneghah A, Fakhri Y, Gahruie HH, Niakousari M, Sant'Ana AS. Mycotoxins in cereal-based products during 24 years (1983–2017): A global systematic review. Trends Food Sci Technol. 2019;91:95–105.

[99]   Commission CA. Code of practice for the prevention and reduction of mycotoxin contamination in cereals, including annexes on ochratoxin A, zearalenone, fumonisins and tricothecenes (CAC/RCP 51–2003). Prevention and Reduction of Food and Feed Contamination. Rome, Italy: FAO, WHO, 2003;1–13.

[100] Munkvold GP, Arias S, Taschl I, Gruber-Dorninger C. Mycotoxins in corn: Occurrence, impacts, and management. In: Serna-Saldivar SO, ed. Corn. Oxford: Elsevier; 2019;235–87.

[101] Karlovsky P, Suman M, Berthiller F, De Meester J, Eisenbrand G, Perrin I, et al. Impact of food processing and detoxification treatments on mycotoxin contamination. Mycotoxin Res. 2016; 32(4):179–205.

[102] Wan J, Chen B, Rao J. Occurrence and preventive strategies to control mycotoxins in cereal-based food. Compr Rev Food Sci Food Saf. 2020;19(3):928–53.

[103] Prandini A, Tansini G, Sigolo S, Filippi L, Laporta M, Piva G. On the occurrence of aflatoxin M1 in milk and dairy products. Food Chem Toxicol: An International Journal Published for the British Industrial Biological Research Association. 2009;47(5):984–91.

[104] Metwally MM, El-Sayed AMaA, Mehriz AM, Abu Sree YH. Sterigmatocystin – Incidence, fate and production byA versicolor in ras cheese. Mycotoxin Res. 1997;13(2):61–66.

[105] Gavahian M, Mathad GN, Oliveira CAF, Mousavi Khaneghah A. Combinations of emerging technologies with fermentation: Interaction effects for detoxification of mycotoxins? Food Res Int. 2021;141:110104.

[106] Benkerroum N. Mycotoxins in dairy products: A review. Int Dairy J. 2016;62:63–75.

[107] Sengun I, Yaman D, Gonul S. Mycotoxins and mould contamination in cheese: A review. World Mycotoxin J. 2008;1(3):291–98.

[108] Sataque Ono EY, Hirozawa MT, Omori AM, de Oliveira AJ, Ono MA. Chapter 19 – Mycotoxins in nuts and seeds. In: Preedy VR, Watson RR, eds. Nuts and seeds in health and disease prevention. 2nd ed. Academic Press, London, 2020;255–70.

[109] Thanushree MP, Sailendri D, Yoha KS, Moses JA, Anandharamakrishnan C. Mycotoxin contamination in food: An exposition on spices. Trends Food Sci Technol. 2019;93:69–80.

[110] Hojnik N, Cvelbar U, Tavčar-Kalcher G, Walsh JL, Križaj I. Mycotoxin decontamination of food: Cold atmospheric pressure plasma versus "classic" decontamination. Toxins. 2017;9:5.

[111] Ezekiel CN, Ayeni KI, Akinyemi MO, Sulyok M, Oyedele OA, Babalola DA, et al. Dietary risk assessment and consumer awareness of mycotoxins among household consumers of cereals, nuts and legumes in north-central Nigeria. Toxins. 2021;13(9):635.

[112] Suman M. Last decade studies on mycotoxins' fate during food processing: An overview. Curr Opin Food Sci. 2021;41:70–80.

[113] Enikova RK, Stoynovska MR, Karcheva MD. Mycotoxins in fruits and vegetables. J IMAB. 2020; 26(2):3139–43.

[114] Scott PM, Kanhere SR. Stability of Alternaria toxins in fruit juices and wine. Mycotoxin Res. 2001; 17(1):9–14.

[115] Misra NN, Yadav B, Roopesh MS, Jo C. Cold plasma for effective fungal and mycotoxin control in foods: Mechanisms, inactivation effects, and applications. Compr Rev Food Sci Food Saf. 2019; 18(1):106–20.

[116] Pallarés N, Sebastià A, Martínez-Lucas V, Queirós R, Barba FJ, Berrada H, et al. High pressure processing impact on emerging mycotoxins (ENNA, ENNA1, ENNB, ENNB1) mitigation in different juice and juice-milk matrices. Foods [Internet]. 2022;11(2):190.

[117] Magnoli AP, Poloni VL, Cavaglieri L. Impact of mycotoxin contamination in the animal feed industry. Curr Opin Food Sci. 2019;29:99–108.

[118] Kolawole O, Meneely J, Greer B, Chevallier O, Jones DS, Connolly L, et al. Comparative in vitro assessment of a range of commercial feed additives with multiple mycotoxin binding claims. Toxins [Internet]. 2019;11(11):659.

[119] Wielogórska E, MacDonald S, Elliott C. A review of the efficacy of mycotoxin detoxifying agents used in feed in light of changing global environment and legislation. World Mycotoxin J. 2016;9(3):419–33.

[120] Pfliegler WP, Pusztahelyi T, Pócsi I. Mycotoxins – prevention and decontamination by yeasts. J Basic Microbiol. 2015;55(7):805–18.

[121] Kabak B, Dobson AD, Var Il. Strategies to prevent mycotoxin contamination of food and animal feed: A review. Crit Rev Food Sci Nutr. 2006;46(8):593–619.

[122] Loi M, Fanelli F, Liuzzi VC, Logrieco AF, Mulè G. Mycotoxin biotransformation by native and commercial enzymes: Present and future perspectives. Toxins. 2017;9(4):111.

[123] von Holst C, Robouch P, Bellorini S, González de la Huebra MJ, Ezerskis Z. A review of the work of the EU Reference Laboratory supporting the authorisation process of feed additives in the EU. [corrected]. Food Addit Contam: Part A. 2016;33(1):66–77.

[124] González-Jartín JM, Ferreiroa V, Rodríguez-Cañás I, Alfonso A, Sainz MJ, Aguín O, et al. Occurrence of mycotoxins and mycotoxigenic fungi in silage from the north of Portugal at feed-out. Int J Food Microbiol. 2022;365:109556.

[125] Hulla J, Sahu S, Hayes A. Nanotechnology: History and future. Hum Exp Toxicol. 2015;34(12):1318–21.

[126] Das S, Mukherjee A. Nanotechnology as sustainable strategy for remediation of soil contaminants, air pollutants, and mitigation of food biodeterioration. In: Singh P, Kumar V, Bakshi M, Hussain CM, Sillanpää M, eds. Environmental applications of microbial nanotechnology. Elsevier, Netherlands, 2023;3–16.

[127] Regulation (EU) 2015/2283 of the European Parliament and of the Council of 25 November 2015 on novel foods, amending Regulation (EU) No 1169/2011 of the European Parliament and of the Council and repealing Regulation (EC) No 258/97 of the European Parliament and of the Council and Commission Regulation (EC) No 1852/2001. Off. J. Eur. Union. L 327:1–22.

[128] Nanotechnologies I. vocabulary–part 1: Core terms. International Stanardisation Organisation (ISO), Technical Specification ISO/TS, 2015:80004-1.

[129] Gleiter H. Nanostructured materials: Basic concepts and microstructure. Acta Mater. 2000; 48(1):1–29.

[130] Horky P, Skalickova S, Baholet D, Skladanka J. Nanoparticles as a solution for eliminating the risk of mycotoxins. Nanomaterials. 2018;8(9):727.

[131] Magro M, Moritz DE, Bonaiuto E, Baratella D, Terzo M, Jakubec P, et al. Citrinin mycotoxin recognition and removal by naked magnetic nanoparticles. Food Chem. 2016;203:505–12.

[132] Pirouz AA, Selamat J, Iqbal SZ, Mirhosseini H, Karjiban RA, Bakar FA. The use of innovative and efficient nanocomposite (magnetic graphene oxide) for the reduction on of Fusarium mycotoxins in palm kernel cake. Sci Rep. 2017;7(1):12453.

[133] Abbasi Pirouz A, Abedi Karjiban R, Abu Bakar F, Selamat J. A novel adsorbent magnetic graphene oxide modified with chitosan for the simultaneous reduction of mycotoxins. Toxins. 2018;10(9):361.

[134] Thirugnanasambandan T, Gopinath SCB. Nanomaterials in food industry for the protection from mycotoxins: An update. 3 Biotech. 2023;13(2):64.

[135] Luo Y, Zhou Z, Yue T. Synthesis and characterization of nontoxic chitosan-coated Fe3O4 particles for patulin adsorption in a juice-pH simulation aqueous. Food Chem. 2017;221:317–23.

[136] Sun J, Guo W, Ji J, Li Z, Yuan X, Pi F, et al. Removal of patulin in apple juice based on novel magnetic molecularly imprinted adsorbent Fe3O4@SiO2@CS-GO@MIP. LWT. 2020;118:108854.

[137] Zhao Z, Liu N, Yang L, Wang J, Song S, Nie D, et al. Cross-linked chitosan polymers as generic adsorbents for simultaneous adsorption of multiple mycotoxins. Food Control. 2015;57:362–69.

[138] Mao J, Zhang Q, Li P, Zhang L, Zhang W. Geometric architecture design of ternary composites based on dispersive WO3 nanowires for enhanced visible-light-driven activity of refractory pollutant degradation. Chem Eng J. 2018;334:2568–78.

[139] Zhang X, Li G, Wu D, Liu J, Wu Y. Recent advances on emerging nanomaterials for controlling the mycotoxin contamination: From detection to elimination. Food Front. 2020;1(4):360–81.

[140] Zhang W, Yan Z, Gao J, Tong P, Liu W, Zhang L. Metal–organic framework UiO-66 modified magnetite@silica core–shell magnetic microspheres for magnetic solid-phase extraction of domoic acid from shellfish samples. J Chromatogr A. 2015;1400:10–18.

[141] Salonen LM, Pinela SR, Fernandes SPS, Louçano J, Carbó-Argibay E, Sarriá MP, et al. Adsorption of marine phycotoxin okadaic acid on a covalent organic framework. J Chromatogr A. 2017;1525:17–22.

[142] Zhang H, Zhu G, Jia X, Ding Y, Zhang M, Gao Q, et al. Removal of microcystin-LR from drinking water using a bamboo-based charcoal adsorbent modified with chitosan. J Environ Sci. 2011; 23(12):1983–88.

[143] Liu G, Han C, Pelaez M, Zhu D, Liao S, Likodimos V, et al. Enhanced visible light photocatalytic activity of CN-codoped TiO_2 films for the degradation of microcystin-LR. J Mol Catal A: Chem. 2013;372:58–65.

[144] Hena S, Rozi R, Tabassum S, Huda A. Simultaneous removal of potent cyanotoxins from water using magnetophoretic nanoparticle of polypyrrole: Adsorption kinetic and isotherm study. Environ Sci Pollut Res. 2016;23(15):14868–80.

[145] Fernandes SPS, Kovář P, Pšenička M, Silva AMS, Salonen LM, Espiña B. Selection of covalent organic framework pore functionalities for differential adsorption of microcystin toxin analogues. ACS Appl Mater Interfaces. 2021;13(13):15053–63.

Index

www.ingramcontent.com/pod-product-compliance
Lightning Source LLC
Chambersburg PA
CBHW080902220326

41598CB00034B/5448